Landfilling of Waste:
Barriers

Landfilling of Waste: Barriers

Edited by

T. H. CHRISTENSEN
Department of Environmental Engineering
Technical University of Denmark
Lyngby, Denmark

R. COSSU
Department of Land Engineering
University of Cagliari
Cagliari, Sardinia, Italy

R. STEGMANN
Institute of Waste Management
Technical University Hamburg-Harburg
Hamburg, Germany

CRC Press
Taylor & Francis Group
Boca Raton London New York

CRC Press is an imprint of the
Taylor & Francis Group, an **informa** business

A TAYLOR & FRANCIS BOOK

CRC Press
Taylor & Francis Group
6000 Broken Sound Parkway NW, Suite 300
Boca Raton, FL 33487-2742

First issued in paperback 2019

Typeset in 10/12 pt Plantin by The Universities Press (Belfast) Ltd.

ISBN-13: 978-0-419-15990-2 (hbk)
ISBN-13: 978-0-367-86366-1 (pbk)

A catalogue record for this book is available from the British Library

Library of Congress Cataloging-in-Publication data available

**Visit the Taylor & Francis Web site at
http://www.taylorandfrancis.com**

**and the CRC Press Web site at
http://www.crcpress.com**

Preface

Over the last decades the landfilling of waste has developed dramatically and today in some countries involves fully engineered facilities which are subject to extensive environmental regulations. Although considerable information and experience in landfill design and operation has been obtained in recent years, very few landfills meet the current environmental standards. In view of the increasing public awareness and new scientific understanding of waste disposal problems, the standards for landfilling of waste will improve further in years to come.

In view of the increasing demand for information regarding landfills we decided to establish a series of international reference books on landfilling of waste. There is no existing worldwide tradition for the publication of information on landfilling and this series should be seen as an attempt to establish a common platform to represent the state-of-the-art, which may entrance the implementing of improvements at actual landfill sites and help to identify new directions in landfill design and landfill research.

The first book, published in 1992, deals with leachate, the strongly contaminated wastewater which develops in landfills. The book covers aspects such as landfill hydrology, leachate characterization and composition, leachate treatment and effects of leachate in groundwater.

This book, the second in the series, deals with barriers. After an introductory chapter presenting the various issues of lining and drainage, aspects such as design and construction of lining systems, properties and quality control of clay and synthetic materials, geotextiles and geogrids, and finally drainage and collection systems are discussed in detail.

Although landfilling is not favoured due to the long-term behaviour, we have to keep in mind that also with very strict avoidance and recycling programmes, residues of unknown composition and quantity have to be landfilled in countries in Europe and North America. Most of the countries in the world have to rely entirely on landfilling, thus further research should be carried out in the field of landfilling and monitoring of the practice. Of course, landfills should not be viewed as an isolated matter but rather as part of an integrated waste management system.

This book on barriers consists of edited, selected contributions to the International Symposia on Sanitary Landfills held in Sardinia (Italy) every second year and of chapters specially written for this book. The responsibility for the technical content of the book primarily rests with the individual authors and we cannot take any credit for their work. Therefore reference should be made directly to the author and title of chapters. Our role as editors has been one of reviewing and homogenizing the chapters and of making constructive suggestions during the preparation of the final manuscripts.

We would like to thank all contributors for having allowed us to edit their manuscripts and apologize for our completely unintended mistreatment of the English language.

Finally we would like to thank Ms Anne Farmer at CISA, Cagliari, for her patient work on the many drafts of the book.

Thomas H. Christensen
Raffaello Cossu
Rainer Stegmann

Contents

List of Contributors

H. Akgun
Environtech Consultants Inc., 7301A Palmetto Park Road, West Suite 301C, Boca Raton, Florida 33433, USA

O. Artières
CEMAGREF, Division Ouvrages Hydrauliques et Voirie Parc de Tourvoie, BP 121, F 92185 Antony Cedex, France

P. W. Barker
Gundle Lining Systems, 19103 Gundle Road, Houston, Texas 77073, USA

M. Barres
Bureau de Recherches Géologiques et Minières, Department of Environment, BP 6009, 45060 Orleans Cedex, France

J. P. Benneton
IRIGM, University Joseph Fourier, BP 53X, Grenoble Cedex, France

H. C. Berkhout
Akzo Industrial Systems bv, Postbus 9300, 6800 SB Arnhem, The Netherlands

E. Biener
UMTEC—Ingenieurgesellschaft für Abfallwirtschaft und Umwelttechnik, Stresemannstrasse 52, D-2800 Bremen 1, Germany

H. Bonin
Bureau de Recherches Géologiques et Minères, Department of Environment, BP 6009, 45060 Orleans Cedex, France

M. Brune
Institut für Mikrobiologie und Leichtweiss-Institut für Wasserbau, Technische Universität Braunschweig, Postfach 3329, 3300 Braunschweig, Germany

S. T. Butchko
Tensar Environmental System Inc., 1210 Citizens Parkway, Morrow, New York, Georgia 30260, USA

M. W. Cadwallader
Gundle Lining Systems, 19103 Gundle Road, Houston, Texas 77073, USA

A. Cancelli
Department of Earth Sciences, University of Milan, Via Mangiagalli 34, 20133 Milano, Italy

D. Cazzuffi
ENEL-CRIS, Via Ornato 90/14, 20162 Milano, Italy

E. D. Chiado
Almes & Associates Inc., Box 520, Pleasant Valley Road, Trafford, Pittsburgh, Pennsylvania 15085, USA

V. E. Chouery-Curtis
Tensar Environmental System Inc., 1210 Citizens Parkway, Morrow, New York, Georgia 30260, USA

T. H. Christensen
Department of Environmental Engineering, Technical University of Denmark, Building 115, DK-2800, Lyngby, Denmark

V. Ciubotariu
Ecole Polytechnique de Montréal, CP 6079, Succursale A, Montréal, Canada H3C 3A7

H. J. Collins
Institut für Mikrobiologie und Leichtweiss-Institut für Wasserbau, Technische Universität Braunschweig, Postbach 3329, 3300 Braunschweig, Germany

R. Cossu
Department of Land Engineering, University of Cagliari, Piazza d'Armi, 09123 Cagliari, Italy

U. Dayal
Almes & Associates Inc., Box 520, Pleasant Valley Road, Trafford, Pittsburgh, Pennsylvania 15085, USA

G. J. Farquhar
Department of Civil Engineering, University of Waterloo, Waterloo, Ontario, Canada N2L 3G1

Y. H. Faure
IRIGM, University Joseph Fourier, BP 53X, Grenoble Cedex, France

F. Fernandez
University of Western Ontario, Faculty of Engineering Science, Ontario, Canada N6A 5B9

J. M. Gardner
Almes & Associates Inc., Box 520, Pleasant Valley Road, Trafford, Pittsburgh, Pennsylvania 15085, USA

J. P. Gourc
IRIGM, University Joseph Fourier, BP 53X, Grenoble Cedex, France

F. Goussé
CEMAGREF, Division Ouvrages Hydrauliques et Voirie Parc de Tourvoie, BP 121, F 92185 Antony Cedex, France

H. H. Hanert
Institut für Mikrobiologie und Leichtweiss-Institut für Wasserbau, Technische Universität Braunschweig, Postbach 3329, 3300 Braunschweig, Germany

H. E. Haxo
Matrecon Inc., 815 Atlantic Avenue, Alameda, California 94501, USA

P. D. Haxo
Matrecon Inc., 815 Atlantic Avenue, Alameda, California 94501, USA

S. E. Hoekstra
Akzo Industrial Systems bv, Postbus 9300, 6800 SB Arnhem, The Netherlands

U. Holzlöhner
Bundesanstalt für Materialforschung und -prüfung (BAM), Unter den Eichen 87, D-1000 Berlin 45, Germany

H. L. Jessberger
Institute for Soil Mechanics and Foundation Engineering, Ruhr-University Bochum, Universitätsstrasse 150, D-4630 Bochum 1, Germany

R. Kirschner
Huesker Synthetic GmbH & Co., Fabrikstrasse 13–15, Postfach 1262, D-4423 Gescher, Germany

J. Lafleur
Department of Civil Engineering, Ecole Polytechnique de Montréal, CP 6079, Succursale A, Montréal, Canada H3C 3A7

M. C. Lavagnolo
CISA, Environmental Sanitary Engineering Centre, Via Marengo 34, 09123 Cagliari, Italy

P. Lechner
Department of Waste Management, Vienna University of Engineering and Technology, Karlsplatz 13, A-1040 Vienna, Austria

B. Leclercq
IRIGM, University Joseph Fourier, BP 53X, Grenoble Cedex, France

J. B. Lewandowski
Department of Land Reclamation, Agricultural University of Poznar, ul. Wojska Polskiego 73A, 60-625 Poznar, Poland

B. M. McEnroe
School of Engineering, Department of Civil Engineering, University of Kansas, Lawrence, Kansas 66045, USA

F. Malpei
Institute of Sanitary Engineering, Politecnico di Milano, Piazza Leonardo da Vinci, 32-20133 Milano, Italy

S. Melchior
Institute for Soil Science, University of Hamburg, Allende-Platz 2, D-20146 Hamburg, Germany

G. Miehlich
Institute for Soil Science, University of Hamburg, Allende-Platz 2, D-20146 Hamburg 13, Germany

C. J. Miller
Wayne State University, 2158 Engineering Building, Detroit, Michigan 48202, USA

M. Mishra
Wayne State University, 2158 Engineering Building, Detroit, Michigan 48202, USA

J. Mlynarek
Department of Civil Engineering, Ecole Polytechnique de Montréal, CP 6079, Succursale A, Montréal, Canada H3C 3A7

A. Muntoni
CISA, Environmental Sanitary Engineering Centre, Via Marengo 34, 09123 Cagliari, Italy

A. Offredi
Department of Earth Sciences, University of Milan, Via Mangiagalli 34, 20133 Milano, Italy

A. J. Peacock
National Rivers Authority (NW Region), New Town House, Buttermarket Street, Warrington, Cheshire WA1 2QG, UK

I. D. Peggs
I-Corp International, 5920 North Ocean Boulevard, Ocean Ridge, Florida 33435, USA

H. Perrier
IRIGM, University Joseph Fourier, BP 53X, Grenoble Cedex, France

P. Pierson
IRIGM, University Joseph Fourier, BP 53X, Grenoble Cedex, France

E. Prigent
CEMAGREF, Division Ouvrages Hydrauliques et Voirie Parc de Tourvoie, BP 121, F 92185 Antony Cedex, France

R. M. Quigley
University of Western Ontario, Faculty of Engineering Science, Ontario, Canada N6A 5B9

H. G. Ramke
Institut für Mikrobiologie und Leichtweiss-Institut für Wasserbau, Technische Universität Braunschweig, Postfach 3329, 3300 Braunschweig, Germany

P. Rimoldi
Geosynthetics Division, TENAX S.p.A., Via Industria 3, 22060 Viganò (CO), Italy

A. L. Rollin
Department of Chemical Engineering, Ecole Polytechnique de Montréal, CP 6079, Succursale A, Montréal, Canada H3C 3A7

R. K. Rowe
Geotechnical Research Centre, Department of Civil Engineering, The University of Western Ontario, London, Ontario, Canada N6A 5B9

T. Sasse
UMTEC—Ingenieurgesellschaft für Abfallwirtschaft und Umwelttechnik, Stresemannstrasse 52, D-2800 Bremen 1, Germany

R. Scherbeck
Jessberger & Partner Ltd, Am Umweltpark 5, D-4630 Bochum 1, Germany

K. J. Seymour
National Rivers Authority (NW Region), New Town House, Butter-market Street, Warrington, Cheshire WA1 2QG, UK

R. Stegmann
Institute of Waste Management, Technical University Hamburg-Harburg, Harburge Schlosstrasse 37, D-2100 Hamburg 90, Germany

A. Vidovic
Ecole Polytechnique de Montréal, CP 6079, Succursale A, Montréal, Canada H3C 3A7

J.-F. Wagner
Department for Applied Geology, Karlsruhe University, Kaiserstrasse 12, D-7500 Karlsruhe 1, Germany

R. B. Wallace
Environtech Consultants Inc., 7301A Palmetto Park Road, West Suite 301C, Boca Raton, Florida 33433, USA

R. Witte
Amtliche Materialprüfanstalt (AMPA), Appelstrasse. 11A, D-3000, Hannover 1, Germany

A. Zanescu
Ecole Polytechnique de Montréal, CP 6079, Succursale A, Montréal, Canada H3C 3A7

1. INTRODUCTION

1.1 Principles of Landfill Barrier Systems

THOMAS H. CHRISTENSEN,[a] RAFFAELLO COSSU[b] & RAINER STEGMANN[c]

[a] Department of Environmental Engineering, Technical University of Denmark, Building 115, DK-2800 Lyngby, Denmark
[b] Department of Land Engineering, University of Cagliari, Piazza d'Armi, I-09123 Cagliari, Italy
[c] Institute of Waste Management, Technical University Hamburg-Harburg, Harburger Schlosstrasse 37, D-2100 Hamburg 90, Germany

EMISSION CONTROL

The main long-term emissions from landfills are leachate and gas. While the gas phase can be controlled to a high degree and environmental effects are limited, leachate control is much more difficult.

Leachate is the medium by which soluble materials inside a landfill may subsequently be transported into the environment. In order to avoid uncontrolled release into the environment, landfills are lined and the leachate is collected and treated. The main problem is that of designing and building liners which retain leachate at the landfill base and avoid leachate production through elimination of water infiltration from the outside. These aspects together with the advantages and limitations of liners will be discussed in detail in this book. However, it can already be concluded that no liner is 100% efficient, especially on a long-term basis. Thus a landfill site cannot, in the long term, be considered completely safe in spite of the installation of a barrier system. There can be no guarantee that a barrier system is leak proof and installation of a liner cannot make monitoring redundant. Moreover, several site restrictions should be respected (distance to water table, supply wells and surface water bodies) and areas with low permeability natural soils should be prefered for siting in order to minimize environmental impact in case of barrier failure. The functions of the elements of the barrier systems are as follows.

Bottom barriers should provide impermeabilization to leachate, prevent biogas from escaping into the environment, provide mechanical support for the waste mass, and avoid accumulation of leachate by means of the filtration, drainage and collection system located above the bottom barrier.

Side barriers in landfills constructed below surface level should provide impermeabilization to leachate and to external water fluxes, mechanical resistance to water pressure, drainage of leachate and, if necessary, external water, and prevent lateral migration of biogas.

Top cover should prevent biogas from escaping into the environment, avoid or reduce rainwater infiltrations through a combination of sealing and drainage functions, and provide support for aftercare options such as vegetation and erosion control.

PROPERTIES OF BARRIER SYSTEM COMPONENTS

In the construction of barrier systems different kind of materials, either natural or synthetic, can be used individually or as component of a composite manufact, according to the different functions required. In Fig. 1 the most typical materials used in landfill barrier systems are summarized.

Natural Materials

Clayey soil is the most common natural lining material. The main factors which affect the quality of clay liners are hydraulic conductivity, degree of compaction, moisture content, clay composition, field placement technique and liner thickness. Compacted clay or in-situ clay deposits must be studied and engineered in a proper way to assure good performance as a liner material in accordance with design parameters. Characteristics, performances and quality control of clay liners are discussed in Chapters 2.1, 2.2 and 2.3 of this book.

Bentonite is a general term for indicating clay minerals capable of swelling, when wet, up to 15–18 times their dry volume. Mixtures of bentonite and sandy soil can provide a low permeability liner, particularly useful in areas where natural clay soil is not available. Chapter 3.4 reports detailed information on the properties of this kind of lining material.

Sand is widely used to protect synthetic liners and to increase filter stability.

Gravel is the main material for filtering and draining. The efficiency of a granular bed depends on the porosity, grain shape and strength, the rock quality (resistance to weathering and carbonate content), stability of the filter, the drainage layer thickness and general engineering of the system. These aspects are discussed in detail in Chapter 6.1.

Synthetic Materials

In recent years synthetic materials have been used increasingly to enhance soil behaviour properties. These materials (usually called geosynthetics) are able to perform many of the functions required of a barrier system in a landfill unit, such as lining, separation of different materials, drainage, filtration and reinforcement.

The main advantages in using geosynthetics compared to natural materials are their ready availability, small volume consumption, and at times better performance. Other potential advantages such as durability, low cost and low maintenance are, on the contrary, somewhat questionable. Those categories of geosynthetics which may be applied in a landfill are: geomembranes, geotextiles, geonets, geogrids and geocomposites.

Geomembranes represent a number of different synthetic flexible membranes with very low permeability; these can be categorized into bituminous geomembranes and polymeric geomembranes. The former kind is produced by spraying asphalt or bitumen on to both woven and nonwoven textiles. Polymeric geomembranes are manufactured using several kind of polymers and elastomers (see Chapter 4.1). The most widely used material in the production of geomembranes for landfill lining is High Density Polyethylene (HDPE) mainly due to its chemical and biological resistance and its capability to be seamed. Disadvantages for this material regard mainly susceptibility to stress cracking.

Geotextiles consist of synthetic fabrics produced either by common weaving (woven geotextiles) or by arranging in a randomly structured mat (nonwoven geotextiles), as described in detail in Chapter 5.1.

Geogrids are plastics formed in a wide open, regular net-like structure, specially developed for reinforcement of soil (Chapter 5.4). *Geonets* are produced by a continuous extrusion of ribs which intersect forming a network structure, suitable for liquid drainage. The most widely used polymer in the production of geogrids and geonets is polyethylene; polyester and polypropylene are also used.

Geocomposites are products obtained by combining two or more different geosynthetics and, sometimes, natural materials. The most typical configurations of geocomposites are: geotextile + bentonite + geotextile (bentonite geocomposite) and geotextile + synthetic drainage mat + geotextile (geocomposite drain).

FUNCTIONS OF BARRIER SYSTEM COMPONENTS

The functions of barrier system components, as summarized in Fig. 1, are as follows: lining, leachate percolation and drainage (horizontal flow and vertical flow), reinforcement, mechanical protection, separation, erosion control, water filtration and drainage, and biogas migration control. In Fig. 1 these functions are evaluated according to the material adopted, as also discussed herein.

Lining is the main function in a barrier system and is based on the utilization of low permeability material including synthetic membranes (geomembranes) and natural soil liners (bentonite mixtures, clay). A combination of geotextile and bentonite (geocomposite bentonite liner) is often proposed.

Leachate percolation, i.e. the vertical movement of leachate towards the landfill bottom, should be homogeneous, in order to avoid perched water tables in the waste mass. Coarse granular material (i.e. gravel) is commonly used at the top and bottom of the landfill. Coarse gravel is proposed to cover the entire bottom of the landfill. Although geotextiles may have been proposed as suitable material, they may clog when in contact with leachate and in the opinion of the authors should not be used in this regard (see Chapter 5.1).

Leachate drainage and collection: in order to avoid accumulation leachate has to be transferred directly, as fast as possible, by a drainage (horizontal flow) and collection system. For this purpose natural granular materials in combination with drain pipes are used. The size of the granular particles and the geometry of the drainage bed should be carefully considered in order to avoid clogging of and

	FUNCTIONS									
APPLICABILITY LEVEL: ● high, ◉ medium, ○ low	Lining	Leachate percolation	Leachate drainage	Soil reinforcement	Mechanical protection	Separation	Erosion control	Water filtration	Water drainage	Biogas migration control
NATURAL MATERIALS BT. Bentonite soil mixture	●				◉					
CL. Clayey soil	●				◉					
SD. Sand		◉	○		●	○		●	○	●
GV. Gravel		●	●	○		○		○	●	●
SYNTHETIC MATERIALS GM. Geomembrane	●									
GT. Geotextile		○		◉	●	●	○	●	○	○
GT. Geonet			●	◉		◉	◉		●	○
GG. Geogrid				●			●			
GB. Geocomposite bentonite	●					◉	○			
GD. Geocomposite drain	○	◉	○	○	◉			◉	●	●

Figure 1. Level of applicability of different landfill materials with regard to the required function. The evaluation is given on the base of the material property, performance and cost.

damage to the drain pipes. Geonets and geocomposite drains have been proposed for drainage and collections but their efficiency is disputed, particularly when used as primary drainage system. Design criteria for leachate drainage and collection systems are reported in Chapter 6.1.

Soil reinforcement is important in consolidating the bottom and slope of the landfill in order to ensure a stable foundation for the lining system. Stability problems can arise along the slope of the landfill sides, from the unsuitable quality of the existing soil (for instance, deformable clay) and from existing old waste deposits. In particular, geogrids have been proposed (Chapter 5.4) as materials for mechanical support of soil, geotextiles, geonets and geomembranes.

Mechanical protection: particularly, synthetic liners such as geomembranes have to be mechanically protected from puncturing by sharp objects contained in the waste, by the pressure of granular draining material, and by the action of vehicles and mobile equipment in the landfill. Mechanical protection of synthetic liners can be achieved by layers of sand or small-sized rounded gravel. These materials can be coupled with geotextiles and geocomposites such as geonets, geotextiles, etc.

Separation: when combining granular material of different size a separation layer should be provided in order to avoid penetration of one material into the other. Nonwoven geotextiles have proven to be very efficient in this function.

Erosion control: the completed landfill surface has to be protected against water and wind erosion. Special geonets and mats offer a good solution to this problem. Often these materials are coupled with substances which assure a satisfactory support for grass and vegetation growth.

Water filtration may be necessary to divert groundwater and surface water from infiltrating into the landfill (side liners, top cover). In this case natural granular material, woven and nonwoven geotextiles and geocomposites can be successfully used. The utilization of synthetic material saves on the space required.

Water drainage and collection: the infiltrated rainwater has to be removed in order to avoid intrusion into the landfill, water pressure on the lining system and water build-up to the top cover soil. For this purpose granular material, in combination with pipes, and/or geonets can be used.

Biogas migration control: landfill biogas is normally abstracted and collected by means of specific wells drilled into the landfill. However, to control biogas migration both from the landfill surface and laterally it is often important to provide a drainage layer for biogas to be connected to the pumping system.

CONTROVERSIAL OPINIONS, LONG-TERM MONITORING AND CONSTRUCTION

In the detailed chapters of the book, the reader will encounter varying opinions on the same subject. The editors cannot straighten this out, since it reflects the state of knowledge in this field. However, it is evident that considerable research is needed to answer all open questions. In particular the long-term behaviour of synthetic materials should be investigated both in the laboratory and on a full scale. Moreover, the efficiency of each installed barrier system should be carefully investigated by monitoring under actual landfill conditions. Another weak point is the actual construction of the liner system; theory and practice vary to a high degree and although the liner might have been designed in an appropriate way, misfunctioning is often associated with unappropriate construction in the field (see also Chapter 2.12).

In many countries design criteria for different barrier systems are prescribed in detail. This may be an advantage in many cases although the matter should be open to improvement in view of the development of new barrier systems. Landfill barriers are still a new topic and therefore in rapid development.

FUTURE TRENDS

As already described in this chapter, it has to date not been possible to control emissions from a landfill; ultimately, for this and other reasons, waste management concepts have to be redeveloped.

Firstly the number of landfills should be decreased, which means that waste minimization programmes have to be implemented very strongly.

Secondly, only waste residues where no further utilization is expected should be disposed of in landfills; the residues should be pretreated in such a way that potential emissions are minimized. Steps in this direction are incineration, biological treatment and/or solidification. In order to quantify the beneficial effects these measures offer to the environment, material and emission balances have to be carried out and adequately interpreted.

Currently, as landfills still receive waste potentially generating leachate, strategies of complete hydraulic containment are being introduced in some countries. Therefore, with regard to long-term behaviour more reliable materials should be implemented and more emphasis should be placed on control and repair of barrier systems. Due to the easier access to surface liners, these will play a major role in emission control, where the bottom barrier will act as a back-up system. As a consequence, surface liners might be built as double liners with leakage detection systems. These measures will result in a 'dry landfill' that has to be operated for an unlimited period of time.

This consequence is at the root of a lot of controversial discussion, as to whether landfills can be left unsupervised in the environment on a long-term basis or not. Further research is necessary in order to quantify the long-term emissions and their effects on the subsoil and groundwater which will enable the defining of an environmentally-sound landfill strategy based on scientific background.

1.2 Engineering of Landfill Barrier Systems

RAFFAELLO COSSU

Department of Land Engineering, University of Cagliari, Piazza d'Armi, 09123 Cagliari, Italy

STRUCTURE OF THE BARRIER SYSTEM

A barrier system in a landfill may be designed adopting different safety levels, by using the materials presented in Chapter 1.1 and taking into account different situations and quality of waste.

With regard to the bottom and side barrier, in Fig. 1 the sequence of potential components of the system is schematically presented and different configurations and alternatives for the materials are reported.

Some of the most significant types of lining system adopted for landfills are illustrated in Fig. 2 and discussed herein.

(a) *Single liner of natural material (low permeability soil)*: this configuration represents the simplest lining system still most widely adopted, although it is considered acceptable only under specific and fully safe hydrogeological situations.

(b) *Single liner of synthetic material (geomembrane)*: a single geomembrane in the case of waste containment does not afford any safety guarantee and may be used only under conditions similar to those mentioned for (a).

(c) *Single composite liner (clayey soil + geomembrane)*: this structure is the most widely recommended for municipal solid waste (MSW) landfilling and is included in the guidelines of many industrialized countries. The safety level concerned is hard to

	1	2	3	4	5	6	7	8	9	10
	a	*b*	*c*	*c1*	*c2*	*d*	*e*	*f*	*g*	*h*
Separation				GT				GT	GT	
Leachate Drainage	GV	GV	GV	GN	GV	GV	GV	GN	GN	GV
Protection/Separation		GT	GT		GT	GT	SD			GT
Lining	CL	GM	GM	GM	GM	GM	CL	GM	GM	GM
Protection/Separation		GT					GT			
Leachate Drainage							GV	GN	GN	
Protection/Separation							SD			
Lining			CL	GB		CL	CL	GM	GM	CL
Protection/Separation								GT		GT
Leachate Drainage										GV
Protection/Separation										GT
Lining						GM			CL	GM
Protection/Separation										
Leachate Drainage										
Protection/Separation										
Lining					GB					CL
Protection/Separation					GN					
Water Drainage					GV					
Separation					GN					
Support					GG					

Figure 1. Different sequences of potential components of a landfill barrier system. For the materials, reference is made to Chapter 1.1, Fig. 1; a–h refer to the configurations as reported in Fig. 2; C1 = single composite liner using only geosynthetics in the case of vertical expansion over an old landfill; C2 = single composite liner for a side barrier as in Fig. 6(a).

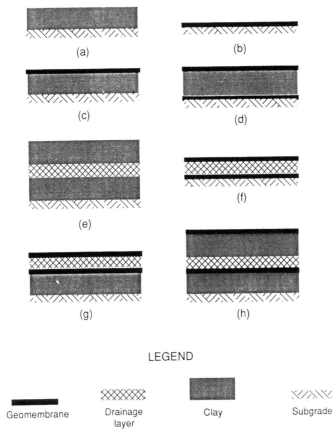

Figure 2. Types of lining systems: a = single clay liner, b = single geomembrane liner, c = single composite liner, d = single sandwich composite liner, e = double clay liner, f = double geomembrane liner, g = double semicomposite liner, double composite liner, (modified from Wallace and Akgun, 1991).

predict and is of course controversially discussed. Nowadays for MSW landfilling a minimum clay liner thickness of 1·0 m is deemed necessary with a maximum permeability of 10^{-9} m s^{-1} (Chapters 2.1 and 2.2). Mineral liners along slopes are often substituted by a geotextile–bentonite composite although many designers still prefer clay even under these conditions.

(d) *Single sandwich composite liner (geomembrane + clayey soil + geomembrane)*: this configuration increases the safety level of the single composite liner and can prevent dessication and cracking of the mineral liner. The installation of this system can be critical with regard to the emplacement technique and stability behaviour when the landfill is sited on a slope.

When compared to single liners, composite liners present significant benefits with regard to the amount of leachate which may migrate through the system even in the presence of holes in the geomembrane (Chapter 2.4).

In order to maximize the advantages of composite liners, the geomembrane should be positioned in direct contact with the top of the mineral liner, avoiding emplacement of draining material (such as geonets, pipes, gravel, etc.) between synthetic and mineral liners.

When a control drainage layer is deemed necessary the double lining system should be adopted. The reasons for doing so are: to increase the safety degree of the barrier system, to detect leachate leakage, to control the general performance of the system and to collect and drain off the leaked leachate. Thus the eventual pressure head and the leachate infiltration potential on the lower lining layer will be reduced.

(e) *Double liner of natural material*: this configuration, which is rarely used, foresees the installation of separation material (such as geotextiles) between the mineral layers and drainage layers.

(f) *Double liner of synthetic material*: the double synthetic liner has often been proposed, particularly for the side barrier systems. Synthetic materials, such as geonet, can be conveniently adopted. The efficiency of this configuration is severely influenced by the quality of the installation and long-term behaviour of the geomembranes.

(g) *Double semicomposite liner (double synthetic top liner + single natural liner)*: this structure tends to couple the advantages of the single composite liner and the double synthetic liner, thus providing a further safety level. Global efficiency still depends on the long-term behaviour of the geomembrane.

(h) *Double composite liner*: this configuration fully exploits the possibility of a control drainage for the double lining system. It is, of course, the more expensive solution and nowadays is adopted mainly for industrial waste landfill. The interest for MSW landfill applications is, however, increasing in order to minimize any adverse environmental impact.

DESIGN OF LINERS

Lining components in barrier systems should be designed specifically according to the type of material used.

Natural Materials

The characteristics of the hydraulic conductivity of mineral liners depend greatly on many factors such as construction, methodology, leachate head build-up, material quality, transport of contaminants, etc. (see Chapter 2.1). The emplaced clay liner should be carefully engineered and controlled both in laboratory tests and in the field in order to present a homogeneous performance (Chapters 2.1, 2.2 and 2.3).

During the construction phase particular care should be devoted to compaction and moisture control. Existing clay strata should be considered as a raw material and remoulded and processed in order to respect proper design parameters.

The efficiency of the base mineral liner can be improved by installing two separate mineral liners, one with a sorptive function and the other ensuring a long-term permeability (Chapter 2.5). Low-permeability natural materials other than clay have often been used but the results appeared to be controversial (Chapters 2.1 and 3.4).

Synthetic Material

The most widely used material for synthetic liners is HDPE geomembrane with a thickness of 2·0–2·5 mm. Moreover, recently LDPE (low density polyethylene) has been used due to its better flexibility. HDPE geomembranes are supplied as sheet rolls, with 4·0–10·0 m width and are seamed on site. Both material and seams have to be checked according to different tests in order to verify original properties and design performance (Chapter 4.2–4.7).

Mechanical support of synthetic liners should be adequate to avoid localized subsidence and/or differential settlement which can determine failure of the liner or significantly change the slope. A typical critical situation to this regard can be observed when a lining system is realized over an existing waste deposit. In this case foundation reinforcement is necessary (Chapter 5.4). The supporting soil for the synthetic liner should be shaped according to the design slope, well compacted and rolled. The side slope should be inferior to 85% of the frictional angle between the geomembrane and the supporting soil (Jorgensen, 1987).

Geomembrane has to be installed avoiding folding as this appears to be one of the reasons for failure (Chapter 4.7). Seaming should be carried out by welding geomembrane sheet along the direction of the leachate flux. Furthermore seaming should be parallel to the line of the maximum slope along the side. After emplacement, lining should be controlled according to a well-defined Quality Control Procedure (see Chapter 1.3).

After installation and control, geomembranes have to be mechanically protected. To this purpose sandy soil is often adopted with 20–30 cm thickness. As an alternative, particularly when a gravel drainage layer is installed above the liner, geotextile or geocomposite can be interposed between the granular and synthetic material (see Chapter 5.3).

In connection with the leachate drainage pipeline it is important to provide underneath special protective measures such as reported in Fig. 3. Special protection should also be provided underneath roads used by operating vehicles.

DRAINAGE SYSTEM DESIGN

The design criteria for an effective leachate drainage system are presented in detail in Chapter 6.1.

In many cases a drainage well for vertical drainage between different layers has been proposed (Fig. 4). It is important to have flexible structures that can adapt themselves to settling and horizontal landfill movement and can resist to potential damage by operating machines. Such structures can be realized by using a fixed cylinder (0·8–1 m diameter) made of metallic net, filled with large size gravel, or by gradually lifting a metallic cylinder. The efficiency and real need for this structure is, however, disputed. The vertical drainage well can also be combined with biogas extraction network.

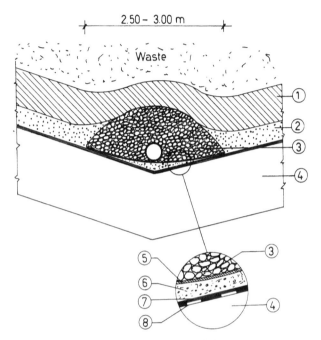

Figure 3. Cross-section of possible MSW landfill drain: 1 = raw compost layer (optional); 2 = gravel (small size); 3 = gravel (large size); 4 = clay stratum with thickness > 1 m and $k < 10^{-9}$ m s^{-1}; 5 = geocomposite bentonite; 6 = sandy-loamy soil; 7 = high density geotextile; 8 = geomembrane. Geocomposite and sandy-loamy soil are emplaced only underneath the drainage pipeline.

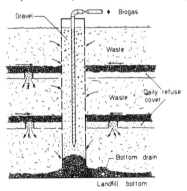

Figure 4. Example of vertical drainage well. A gravel drain (or the equivalent) has to be realized above the intermediate refuse cover.

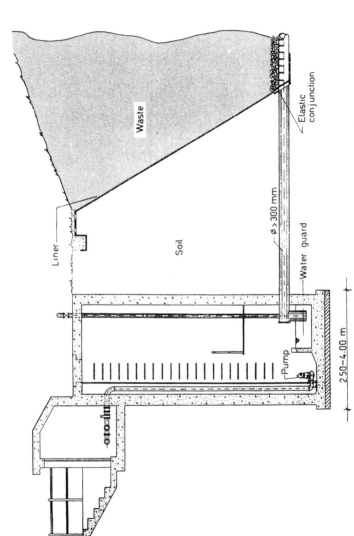

Figure 5. Example of a manhole for final leachate collection.

Leachate drained from the bottom of the landfill, has to be collected continuously in order to avoid leachate hydraulic head over the barrier system. Collection manholes must have the following characteristics:

- installation outside the waste deposit area
- dimensions allowing satisfactory maintenance and working activity
- materials should be impermeable and resistent to leachate corrosion (often pre-manufactured polyethylene and fibreglass wells are installed);
- leachate pipe must be syphoned on arrival in manholes and water guards should invariably be present in order to avoid air infiltration in the leachate drainage and consequent clogging problems.

An example of a leachate collection manhole is given in Fig. 5. According to some guidelines, in order to avoid accumulation, leachate should not be pumped but collected and conveyed only by gravity. This means that landfill should be constructed on a slope or above ground level. This approach is observed mainly in Austria and Germany, but it is slowly being introduced in other European countries also.

The engineering of side barriers often involves interception of draining water infiltrations from permeable soil layers. Typical situations are reported in Fig. 6. Pumping off of the collected water may be necessary in order to avoid water pressure on the barrier.

DESIGN OF TOP COVER SYSTEM

The engineering of top cover changes according to different local situations (climatology, morphology, etc.). Some typical configurations are summarized in Fig. 7.

Erosion control and vegetation support, coupled with a channel system for water run-off, is particularly necessary in the presence of steep slopes. It is possible to use special materials such as pre-seeded mats or sprayed in-situ mixtures of seeds and gripping solutions for this purpose.

A vegetative soil layer with a minimum thickness of 60–80 cm is

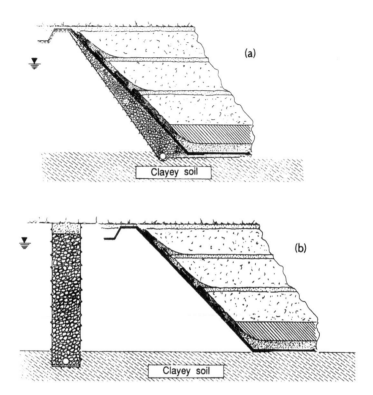

Figure 6. Interception of water infiltrations along the side barrier: (a) drainage layer integrated in the barrier system; (b) drainage trench. A sequence of potential components for case (a) is given in Fig. 1 (configuration C2).

necessary to support vegetation. However, to minimize water infiltration by storing of water from the wet season to the dry season, a thicker cover may be preferred. In some cases when trees have to be planted, the thickness should be larger, e.g. 150 cm.

A sub-surface drain may be installed to prevent ponding and allow rapid drainage of water. For this purpose either gravel, sand or geosynthetics (see Chapter 2.9) can be used. For natural draining material a minimum thickness of 30 cm is suggested.

A low permeability liner, such as clay, bentonite mat or geomembrane, allows for controlling of water infiltration into the refuse to be

		1	2	3	4	5	6	7	8	9	10
	Vegetation										
	Erosion control and vegetation support			GG	GG	GG					
VEGETATIVE SOIL		X	X	X	X	X	X	X	X	X	X
	Separation		GT		GT		GT				
	Surface Drainage	GV	GV	GD	GN	GD	GV	GV	GD		
	Separation	SD	GT				SD				
	Lining	CL	CL	GM	GM	GB	GB	CL	GM	CL	GB
	Separation	SD	GT								
	Gas Drainage	GV	SD	GD	SD	GD	SD			SD	GD
	Separation			SD		SD					SD
SOLID WASTE											

Figure 7. Different examples of potential component sequence for a landfill top cover. For the materials, reference is made to Chapter 1.1, Fig. 1. Key: 1, all natural cover; 2, natural cover with geotextile for separation; 3, all synthetic cover; 4, synthetic cover with natural biogas drain; 5, geocomposite cover; 6, natural drainage with geocomposite liner; 7, natural cover without biogas drainage; 8, synthetic cover without biogas drainage; 9, natural cover without water drainage; 10, synthetic cover without water drainage. Configurations 7–10 do not offer the best performance. (See Fig. 1 for the abbreviations used here.)

checked. An underlying drain is needed for collection and removal of biogas.

LONG-TERM PERFORMANCE AND FAILURE MECHANISMS

A barrier system with lining materials performing perfect containment and permeble materials performing perfect drainage does not exist. Quite often it has been found that liners leak and drains clog!

The individual components of a barrier system can fail according to different mechanisms.

Compacted clay liners may fail to function satisfactorily as cracking occurs due to adverse climatic conditions and deformation due to differential settlement (Chapter 2.8). Some studies have shown that certain substances can seriously affect the impermeability of clays

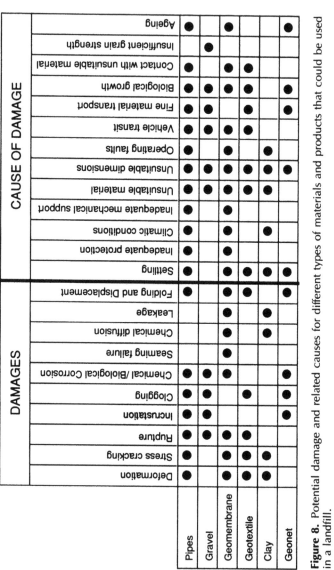

Figure 8. Potential damage and related causes for different types of materials and products that could be used in a landfill.

(Chapter 2.1), but failures with MSW leachate have not been proven. Similar problems are encountered with soil–bentonite mixtures. Strong ionic concentrations can lead to variations in the clay structure, influencing the swelling capacity and consequently the permeability (Chapter 3.5).

Ageing of geomembranes may result in accelerated stress cracking phenomena with a possibility of ruptures and leakage (Chapter 4.7). Filtering and draining material under biological activity can clog with the formation of incrustation which dramatically decreases the drainage efficiency (Chapter 6.4).

The main possible failure mechanisms, both in the short and long term, are summarized in Fig. 8 for the different materials, indicating the most common damages and the causes of failure.

When designing a barrier system it is fundamental to take into account all possibilities of failure whilst bearing in mind, as data input, a certain percentage of material damage.

REFERENCES

Jorgensen, J. B. (1987). Danish code of practice for lining. In *Proceedings Sardinia '87, First International Landfill Symposium*. CIPA, Milan, pp. 1–15.

Wallace, R. B. & Akgun, H. (1991). Leakage in double lined systems: the containment objective. In *Proceedings Sardinia '91, Third International Symposium*. CISA, Cagliari, pp. 731–740.

1.3 Quality Control Assurance for Barrier Systems

RAFFAELLO COSSU[a] & ALDO MUNTONI[b]

[a] *Department of Land Engineering, University of Cagliari, Piazza d'Armi, 09123 Cagliari, Italy*
[b] *CISA, Environmental Sanitary Engineering Centre, Via Marengo 34, 09123 Cagliari, Italy*

INTRODUCTION

Quality control assurance (QCA) is fundamental in order to obtain satisfactory performance and effectiveness of landfill barrier systems. Quality can be defined as the level of performance of the lining and leachate collection system and its components with respect to specific requirements. Quality should be controlled at different phases of the barrier system's life span. Accordingly the QCA plan should include:

- design quality control (DQC)
- construction quality control (CQC)
- operating quality control (OQC)

These are discussed in the following sections.

DESIGN QUALITY CONTROL (DQC)

Design quality control should be considered a preliminary tool for verifying the quality of a barrier system design. The aim of DQC is to verify compliance of the barrier design with technology standards and guidance.

DQC in most countries is an official procedure. A control authority verifies whether all technical design specifications meet the minimum technology guidance set according to state regulations. A decision is then made whether to issue or deny the permit according to results obtained with this analysis.

Third-party consultants are often involved in this phase by the control authority or by the landfill manager in order to have a quality assurance. Moreover, a third-party consultant may also be more restrictive than existing guidelines, suggesting design modification to better protect the environment, following rather a state of technology than the state of the art. Increasingly, landfill owners and managers use this kind of consultancy to protect themselves against eventual design faults or environmental problems which could derive from merely following the official minimal guidance.

CONSTRUCTION QUALITY CONTROL (CQC)

The aim of the CQC program, once more carried out by a third-party team, is to provide quality certification, information and reports to the control authorities and/or the owner of the landfill. The documentation should verify the compliance of the construction of the barrier to the design specifications.

This procedure is also important with regard to public opinion as the landfill realization process gains in term of credibility. Still another advantage of a CQC plan is that the experienced field personnel of the CQA team can provide proactive guidance during the construction of the barrier system (Deardoff, 1991). It has been reported from USA experience that after implementing a CQC program for barrier system, a 97% decrease in geomembrane liner defects was registered (Deardoff, 1991).

Implementation of CQC Plan

A CQC plan should include the following elements (Andreottola *et al.*, 1991; Deardoff, 1991):

Responsibilities and qualifications. The organizations and personnel involved in constructing of the landfill barrier system should be explicitly identified and described within the CQC plan. The qualifications of the CQC inspectors should meet the necessary requirements (in terms of training and experience) of the permitting authority.

Inspection activities. The CQC plan should describe all observations and tests on materials that will be carried out for each component of the facility in order to satisfy all design criteria and authority specifications.

Sampling procedures. The CQC plan should describe the correct sampling procedures (in terms of sample location, size frequency, acceptance criteria) to be applied during the construction of each facility component.

Documentation. All CQC activities should be reported in detail with, for instance, inspection data sheets, daily and final reports, acceptance reports. Minimum requirements for the documentation should be included in the CQC plan.

Inspection activities

The programme of inspection activities consists of observations and tests to be performed by the QCA officer during the construction of the barrier system. This programme, which should usually be carried out at design level and approved by the control authority, is composed of different steps.

Preliminary testing. During a preconstruction phase the design drawings and specifications of the barrier system to be constructed are reviewed. All expected site conditions are verified with the real field situations. If significant differences are registered, a modification of the barrier design and/or the construction procedure could be requested from the facility designer and/or construction contractor.

Conformance testing. Offsite materials (soil, synthetic liners, drainage material, pipes, etc.) to be used for a barrier system construction should be inspected in order to ensure that they are as specified in the design. Material inspection continues throughout the construction period. Inspection of on-site soil materials should be performed as the material is excavated. Initial inspections of soil materials can initially be performed using visual–manual soil classification techniques. In addition to these visual–manual observations, a sufficient number of samples should be tested in order to ensure that material properties meet the design specifications. Conformance testing of offsite materials should be carried out by an independent laboratory.

As far as offsite soil material is concerned, borrow area inspections may also be carried out to ensure that only suitable materials are transported to the site. Inspection activities for synthetic liners include testing of the raw materials, review of the manufacturer's quality control programme and testing of samples of the supplied material. Evaluation of satisfactory installment of equipment is also an important stage of the initial inspection activity. Different quality tests which have been assessed for the various barrier materials are discussed in Chapters 3.1–3.5 (mineral materials) and Chapters 4.1–4.7 (synthetic materials) of this book.

Construction inspection and testing.   An important activity on starting to construct a landfill is the inspection of the foundations in order to assure that these are suitable to support the barrier system and waste deposit. Particularly the slope and quality of the foundation soil should be checked.

Continuous visual observation of all construction phases is one of the major means of ensuring that a component is constructed to meet or exceed the design specifications. An adequate observation activity will identify problems resulting from incorrect handling of equipment or from an inadequate construction procedure resulting in the failure of the barrier component. Observations should be integrated by testing of the used barrier materials and by measurement of the facility dimensions in order to detect any defect.

All improper practices or inadequate materials encountered during the construction of the barrier will be reported by the QCA officer, who will apply corrective measures. Selection of appropriate test methods should be based on site-specific conditions, and on material types. Construction control and site material verification can be carried out using field methods. However, these methods are usually not precise enough to be used for acceptance testing, which will require standardized laboratory tests. The plan for conducting field tests, including methods for determining sampling frequency and location, should be described in detail in the QCA plan.

Placement of the mineral liner must be checked carefully. Materials must possess homogeneous and measured characteristics, a correct moisture content and be compacted in thin layers (>30 cm) using specific compacting equipment. In-situ verification of the permeability on a close sampling grid should be carried out.

When dealing with existing clay soil it is important to verify the homogeneity of the materials and eventually consider processing (moisture control, milling, compaction, etc.) the surface layers in order to achieve a high-quality efficiency level. Protection measures should be applied to avoid dessication or frosting of clay liners.

Gravel should be checked in order to verify rock quality, grain size distribution and absence of fine clogging materials.

The emplacement of geomembranes is critical as these materials, unlike natural liners, have no self-sealing or attenuation capacity. The quality of leak-proofing depends to a great extent on emplacement techniques (smooth soil contact, adequate seaming, positive climatic conditions, satisfactory mechanical protection). Particularly, seaming is a key element in the successful installation of a geomembrane (Chapter 4.2). Various destructive and non-destructive tests have been proposed to assess the continuity and quality of the seam (Chapter 4.4). Adverse atmospheric conditions can influence the quality of artificial liners dramatically (Chapter 2.12).

Other specific tests have been developed to assess quality of geotextiles, geonets and geocomposites. Geotextiles should undergo careful testing when used as mechanical protection for geomembranes (Chapter 5.3). A summary of the main factors to be inspected in a barrier system and the most widely used inspection methods are reported in Table 1.

Completion inspection. Once completed, each block of work is submitted to the completion inspection of the QCA personnel. QCA inspection personnel should first carry out a visual check of the completed facility component to ensure that it meets the design specification. The visual check can be integrated by tests in order to detect any defect, which could lead to the failure of the facility component. Selection of appropriate test methods should be based on site-specific conditions and on material types.

Foundation completion tests include testing and proof-rolling to ensure uniform foundation soil consistency, and surveying to check elevations, slopes and foundation boundaries.

The soil liner should be inspected for cracks, holes, defects, or any other features which may increase its permeability. All defective areas

TABLE 1. Quality Control for Barrier Systems: Main Factors to be Inspected and More Widely Used Inspection Methods (Andreottola *et al.*, 1991)

Barrier component	*Factors to be inspected*	*Inspection methods*
Low-permeability soil liner	Coverage	Observation
	Thickness	Surveying; measurement
	Clod size	Observation
	Tying together of lifts	Observation
	Slope	Surveying
	Installation of protective cover	Observation
	Soil type (index properties)	Visual–manual procedure
		Particle size analysis
		Attenberg limits
		Soil classification
	Moisture content	Oven-dry method
		Nuclear method
		Calcium carbide (speedy)
		Frying pan (alcohol or gas burner)
	In-place density	Nuclear methods
		Sand cone
		Rubber balloon
		Drive cylinder
	Moisture–density relations	Standard proctor
		Modified proctor
		Soil–cement M-D test
	Strength (laboratory)	Unconfined compressive strength
		Triaxial compression
		Unconfined compressive strength for soil–cement
	Cohesive soil consistency (field)	Penetration tests
		Field vane shear test
		Hard penetrometer
		Hand-held torvane
		Field expedient unconfined compression
	Permeability (laboratory)	Fixed wall
		Flexible wall
	Permeability (field)	Large diameter single-ring infiltrometer
		Sai-Anderson infiltrometer
	Susceptibility to frost damage	Susceptibility classification
		Soil–cement freeze–thaw test

TABLE 1. (Continued)

Barrier component	Factors to be inspected	Inspection methods
Low-permeability soil liner	Volume change	Consolidometer (undisturbed or remolded sample) Soil–cement wet-dry test Soil–cement freeze–thaw test
Flexible membrane liner	Thickness	Thickness of unreinforced plastic sheeting, deadweight method—specifications for nonrigid vinyl chloride plastic sheeting Thickness of reinforced plastic sheeting (testing coated fabrics)
	Tensile properties	Tensile properties of rigid thick plastic sheeting (standard method test for tensile properties of plastics) Tensile properties of reinforced plastic sheeting (Grab method A—testing coated fabrics) Tensile properties of thin plastic sheeting
	Tear strength	Tear strength of reinforced plastic sheeting (modified tongue tear method B—testing coated fabrics) Tear strength of plastic sheeting (Die C—test method for initial tear resistance of plastic film and sheeting)
	Bonding materials	Manufacturer's certification
	Bonding equipment	Manufacturer's certification
	Handling and storage	Observation
	Seaming	Ply adhesion of reinforced synthetic membranes, bonded seam strength in peel (machine method, Type A test methods for rubber properties, adhesion to flexible substrate) Bonded seam strength in shear of reinforced plastic sheeting (modified grab method A—testing coated fabrics)

TABLE 1. (Continued)

Barrier component	Factors to be inspected	Inspection methods
Flexible membrane liner		Bonded seam strength in shear of unreinforced plastic sheeting (modified)
	Sealing around penetrations	Observation
	Anchoring	Observation
	Coverage	Observation
	Installation of upper bedding layer	Observation
Leachate collection system		
Granular drainage and filter layers	Thickness	Surveying: measurement
	Coverage	Observation
	Soil type	Visual–manual procedure
		Particle size analysis
		Soil classification
	Density	Nuclear methods
		Sand cone
		Rubber balloon
	Permeability (laboratory)	Constant head
Synthetic drainage and filter layers	Material type	Manufacturer's certification
	Handling and storage	Observation
	Coverage	Observation
	Overlap	Observation
	Temporary anchoring	Observation
	Folds and wrinkles	Observation
	Geotextile properties	Tensile strength
		Puncture or burst resistance
		Tear resistance
		Flexibility
		Outdoor weatherability
		Short-term chemical resistance
		Fabric permeability
		Per cent open area
Pipes	Material type	Manufacturer's certification
	Handling and storage	Observation
	Location	Surveying
	Layout	Surveying
	Orientation of perforations	Observation
	Jointing solid pressure pipe	Hydrostatic pressure test
	Perforated pipe	Observation

should be removed and replaced. The installed membrane liners must be checked for leaks, using the most appropriate test technique (tracer dyes, electrical resistivity, water filling). All tested components which present defects on completion inspection will be submitted, if possible, to corrective measures otherwise the entire block of work will be rejected and newly undertaken. Every observation and/or test should be documented by a test data sheet, and eventually photographic records. Every completed block of work submitted to inspection and acceptance survey should be documented by a daily inspection report and by a block evaluation report. Once all blocks of work are completed, the QCA officer should prepare a final report, containing a summary of the QCA inspection activity and a checklist of all documentation produced.

OPERATION QUALITY CONTROL (OQC)

A correct landfill operation is as fundamental as good design and construction. In fact, incorrect operation is often one of the main causes of landfill barrier failure. There are several aspects that need to be controlled, inspected and monitored to assure a long life and efficiency to a barrier system, according to the design specifications.

During operation of a barrier system attention should be paid to the following areas:

- *Physical movements*: settling and horizontal movements are particularly important on slope-sited landfills.
- *Cracks and erosion phenomena in the earthworks*: this aspect concerns particularly the top cover system.
- *Mechanical and chemical stress in geomembranes*: as reported in Fig. 8 of Chapter 1.2, these are the most common causes of damage for synthetic liners.
- *Leachate hydraulic head above the liner*: this aspect may be connected to inadequate filtering and draining performance of the drain systems. Visual inspection using tele-controlled video-cameras should be carried out regularly along the leachate collection pipelines. A weak point of general landfill design is connected to the fact that to date no efficient systems to repair liner failures or eliminate drainage clogging have been created. In the latter case the best solution is to adopt drainage with the

highest possible permeability in order to transfer clogging problems from the drainage layer (where maintenance is impossible)
to the pipes where several kinds of cleansing techniques can be
applied.
* *Leakage detection through synthetic and neutral liners*: this can be
  performed after installation of electrical instrumentation when
  constructing the landfill either through leakage control drains or
  by leachate production monitoring both in terms of quality and
  quantity. However, the latter should be carried out on a routine
  basis.
* *Monitoring of the barrier components*: regular testing of the performance of these components including the protection systems,
  should be carried out.
* *Keeping of a logbook of all inspection repair and maintenance
  activities*: these data are enclosed in the documentation which
  should be available during the OQC inspections.

All operation activities in a landfill should be carried out according
to an operation manual which should be part of the design documentation. In the manual, forms for regular inspection and maintenance of
barrier components should be included. The operating manual should
be the primary source of information for nearly all aspects of site
operations and should be available to all personnel for convenient
reference. It should be revised and updated on a regular basis as new
procedures are developed to cope with changes in the market or other
conditions.

REFERENCES

Andreottola, G., Cannas, P. & Muntoni, A. (1991). Quality control assurance
    plan for sanitary landfill. In *Proceedings Sardinia '91, Third International
    Landfill Symposium*. CISA, Cagliari, pp. 1039–1069.
Deardoff, G. B. (1991). Construction inspection of Municipal Landfill lining
    systems: a USA perspective. In *Proceedings Sardinia '91, Third International
    Landfill Symposium*. CISA, Cagliari, pp. 741–751.

2. LINING SYSTEMS: DESIGN AND CONSTRUCTION

2.1 Experiences with Liners Using Natural Materials

GRAHAME J. FARQUHAR

Department of Civil Engineering, University of Waterloo, Waterloo, Ontario, Canada N2L 3G1

INTRODUCTION

The primary function of a clay liner is to restrict leachate seepage from the landfill by virtue of its low hydraulic conductivity. Most design reports restrict maximum values of hydraulic conductivity to 10^{-7} cm s^{-1}, although the specific value required will depend on a complex combination of parameters which may in fact dictate much lower values. Detailed analysis to determine the correct value of hydraulic conductivity for a liner is a difficult and uncertain process because it depends on information about:

- the impact of construction methodology on hydraulic conductivity and its distribution throughout the liner, contaminant type and contaminant loading to the liner throughout the active life of the landfill;
- leachate head build-up on the liner as a function of the drainage and collection system efficiency;
- contaminant transport through the liner and the hydrogeologic units beneath the landfill, and
- groundwater quality standards at the point of compliance.

The role of clay liners in some designs has been extended to include contaminant retardation within the liner. While this is known to occur for certain leachate contaminants to varying degrees, confident predictions of field values for retardation are difficult to make at this time

(Farquhar & Parker, 1989; Gray, 1989). Inclusion of contaminant retardation in clay liner design is therefore difficult to justify.

Concern about the suitability of clay liners for municipal solid waste (MSW) landfills has increased in recent times because of certain liner failures, because of difficulties in estimating hydraulic conductivity in the field and also because more stringent groundwater standards demand hydraulic conductivities which exceed practical limits for certain types of liners (US Federal Register, 1988; Quigley *et al.*, 1988). However, information presented subsequently argues that use of well designed and constructed clay liners is warranted in many cases.

The major clay barrier configurations include in-situ clay deposits, sand-swelling clay (usually bentonite) mixtures, and remoulded and compacted clay. In-situ clay soil deposits are often satisfactory as barriers provided that the large scale permeability is not adversely influenced by weathering, root penetration or continuous inclusions of coarser materials (Williams, 1987; Quigley *et al.*, 1988).

TYPES OF CLAY LINERS AND EXPECTATIONS FOR THEIR PERFORMANCE

Mixtures of sand plus 5–15% w/w bentonite compacted in laboratory permeameters have exhibited hydraulic conductivities less than 5×10^{-8} cm s^{-1} with water or 0.01 N $CaSO_4$ as the permeant (Hoeks *et al.*, 1987). Grantham and Robinson (1988) reported on the construction of a sand–Wyoming bentonite liner installed at a landfill in Britain. Bentonite was applied at a rate of 14 kg per m^3 of sand to a depth of 10 cm to produce field measurements of hydraulic conductivity on the order of 10^{-7} cm s^{-1}. The liner was underlain by three high density polyethylene (HDPE) lysimeters to measure leachate seepage. After 18 months of observation 'small' amounts of weak to moderately strong leachate were collected in the lysimeters.

The major difficulty with sand–bentonite liners appears to be incompatibility with MSW leachate (Farquhar & Parker, 1989; Madsen & Mitchell, 1989). Depending on leachate composition, the tendency exists for Ca to replace Na on the montmorillonite structure causing shrinkage and the development of cracks. As many people have found, the result is greatly increased permeability. Hoeks *et al.* (1987) used a rigid wall permeameter to measure the hydraulic conductivity of sand–Wyoming bentonite mixtures ranging from 2·5 to

7·5 w/w. Both water and leachate, the latter having concentrations of Ca and Na of 144 and 2828 mg litre^{-1} respectively, were used as permeants. At comparative experiment times, hydraulic conductivity with water was measured to be approximately 2×10^{-7} cm s^{-1}, while that with leachate was about 2×10^{-5} cm s^{-1} two orders of magnitude higher! European bentonite is more highly substituted with Ca and Mg as opposed to Na and is therefore less likely to shrink in contact with leachate (Hoeks *et al.*, 1987; Madsen & Mitchell, 1989). Performance data are, however, sparse. It would appear, based on current information, that sand–bentonite mixtures may not perform well in the longer term as landfill liners.

Compacted clay liners consisting mainly of non-swelling clays have been used rather extensively for MSW landfills with good success in many situations. There have been cases reported where compacted clay liners have failed to exhibit field values of hydraulic conductivity less than or equal to 10^{-7} cm s^{-1} (Hermann & Elsbury, 1987; Williams, 1987). However, these situations have usually resulted from inadequate design and installation procedures. In the State of Wisconsin, by contrast, the use of strict design, installation, and quality assurance guidelines has produced nearly 30 MSW landfill liners with hydraulic conductivities less than 10^{-7} cm s^{-1}. Similar control at the Keele Valley landfill in Toronto resulted in a liner with field hydraulic conductivity measurements consistently less than 10^{-8} cm s^{-1} (Lahti *et al.*, 1987).

Incompatibility between MSW landfill leachate and compacted clay liners has not been a problem when the swelling clay mineral content has been kept to a minimum (Gordon, 1987; Farquhar & Parker, 1989). In fact, liner hydraulic conductivity often decreases in the field because of sealing due to precipitate formation, solids accumulation, and biomass growth along the upper face of the liner and into cracks and fissures, if they exist (Quigley & Rowe, 1986; Daniel, 1987; Farquhar & Parker, 1989).

In summary, among the three major types of clay liners that have been reported upon in the literature, the least suitable appears to be the sand-swelling clay variety. In-situ clay deposits are generally suitable for MSW disposal but are limited to specific geographic locations. Compacted clay liners appear to be the most versatile of the three and the most likely to provide acceptable performance provided that the requirement for hydraulic conductivity is not less than 10^{-8} cm s^{-1}. As a result, the balance of the present chapter centres on compacted clay liners with some reference, where appropriate, to landfills in natural clay deposits.

DESIGN AND CONSTRUCTION OF COMPACTED CLAY LINERS

The design and construction of compacted clay liners to achieve a specified hydraulic conductivity is a science, in many respects, still in its infancy. There is no universal code to specify that unique set of conditions which will guarantee the target hydraulic conductivity. Thus, the placement of satisfactory compacted clay liners is a complex and uncertain process. There is, however, general agreement among scientists and engineers on the factors which influence hydraulic conductivity. There are also definite trends in the parameter values selected for liner designs. These factors and trends are discussed in more detail below.

Relationship of Hydraulic Conductivity to Compaction and Moisture Content

The dependence of clay liner hydraulic conductivity on degree of compaction and moisture content has been well described by Mundell and Bailey (1985), Daniel (1987) and Madsen and Mitchell (1989) among others. The nature of this dependence is shown in Fig. 1. It is first of all specific to the clay mixture being tested and cannot be applied to other soils. As a consequence, an equivalent data set must be generated for each liner material being considered. Figure 1(b) shows the traditional increased density with compaction effort and also the nonlinear dependence of density on moisture content resulting in an optimum moisture content to produce maximum density. Other compaction efforts produce other optimum conditions. The Standard Proctor and the Modified AASHTO (American Association of State Highway and Transportation Officials) are two compaction efforts frequently used in design specifications.

Measurements of hydraulic conductivity made on specific compacted samples are shown in Fig. 1(a). For individual compaction efforts, hydraulic conductivity decreases substantially at moisture contents greater than the optimum value. As a result, most designs for compacted clay liners recommend moisture contents from 0 to 3%

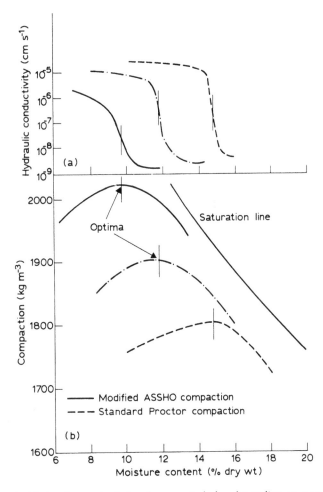

Figure 1. Hydraulic conductivity of compacted clay depending on compaction and moisture content.

'wet of optimum' and at compactions equivalent to 95% Standard Proctor or 90% Modified AASHTO (Mundell & Bailey, 1985; Gordon, 1987; Daniel, 1987; Lahti *et al.*, 1987; Williams, 1987; Gray, 1989). Strict control of moisture content is necessary, however, since major increases in moisture content reduce both the density of the clay and the ability of machinery to work the clay surface.

Clay Composition

High concentrations of swelling clays should be avoided because of their potential for shrinkage in response to cation substitution (Gray, 1989; Madsen & Mitchell, 1989). Gordon (1987) recommended that the soil contain at least 50% silt and clay size particles but did not comment on swelling clay content. The plastic index (PI) of the soil has also been used as a discriminating parameter for clay soil selection. Daniel (1987) indicated only a slight tendency toward increased hydraulic conductivity with an increased PI for PI values below 40. Gordon (1987) and Williams (1987) have recommended that the PI be at least 15. However, Lahti *et al.* (1987) stated that the clay soils used for the Keele Valley liner had PI values between 7 and 15. Field tests of the Keele Valley liner showed hydraulic conductivities to be less 10^{-8} cm s^{-1}.

Field Placement Techniques

There are frequent reports stating that hydraulic conductivity measurements on liners in place exceed laboratory measurements by one to two orders of magnitude and more. This appears to be due in large part to variability in the compacted soil fabric as a result of clay texture and compaction method. Daniel (1987), Herrmann and Elsbury (1987), Gray (1989) and Quigley *et al.*, (1988) all refer to the presence of clay clods as being detrimental to the compaction process. If these are not broken up during placement, localized zones of increased hydraulic conductivity around the clods are created and can lead to excess seepage through the liner. Improperly worked clay and failure to bond a layer to the layer below during placement will also produce higher, secondary hydraulic conductivities due to cracks and other volume defects. Thus, a proper placement technique is essential. Placement should occur in thin layers compacted by several passes of a heavy, footed roller. The roller feet should be long enough to work the layer being placed into the one below.

The liner at the Keele Valley Landfill was placed by dozer in 15-cm layers and worked in four to six passes with a 30-tonne footed roller. As noted previously, field hydraulic conductivities at Keele Valley were less than 10^{-8} cm s^{-1}.

TABLE 1. Examples of Liner Thickness

Reference	Thickness (m)	Comment
Gordon (1987)	1·2–1·5	Over 30 MSW landfills in Wisconsin
Oakley (1987)	0·92	US EPA recommended minimum
Labti *et al.* (1987)	1·2	Keele Valley Landfill Toronto
Quigley *et al.* (1988)	1·0	Minimum recommended

Liner Thickness

There is some question about whether or not the Darcy equation can be applied to soils of very low permeability (Hoeks *et al.*, 1987). Whether applicable or not, there is no doubt that seepage rates through a clay liner are increased by an increased hydraulic gradient. Consequently, increased liner thickness resulting in decreased hydraulic gradient will reduce seepage rates through the liner. In addition, the effects of construction defects and liner heterogeneities on hydraulic conductivity are reduced as the thickness increases. On the other hand, escalating costs and landfill space consumption provide upper limits for liner thickness. The relevant question then is how thick should the liner be? The tendency appears to be toward a minimum liner thickness of 1·0 m with few designs exceeding 2·0 m. Some examples are shown in Table 1.

Quality Assurance, Quality Control

Since liner permeability is so dependent on moisture content, compaction density, kneading to homogenize soil fabric, and thickness, constant monitoring of these parameters is required during placement. There is no consistent pattern to the types and frequency of field measurements reported in the literature except that most designers agree that they are essential.

In-situ measurements of density and moisture content are recommended (Mundell & Bailey, 1985; Daniel, 1987; Gordon, 1987). Permeability measurements are also required but selection of a suitable field method is difficult. A more detailed discussion of permeability assessment of compacted clay liners is presented in the following section.

MEASURING THE HYDRAULIC CONDUCTIVITY OF CLAY LINERS

The most important property of a compacted clay liner is its hydraulic conductivity. Designers make use of liner hydraulic conductivity to predict contaminant flux from a landfill in order to assess compliance at a specified boundary. Approvals for operation are based on whether or not compliance can be achieved. However, in spite of its importance, the hydraulic conductivity of compacted clay and methods to measure it are generally poorly understood.

It is not uncommon for the hydraulic conductivity measurements of a specific soil to differ from one to two orders of magnitude depending on the experimental procedure used. Rigid wall (RW) permeameters often produce values higher than confined flexible wall (FW) units especially with swelling clays and permeants other than water or $0 \cdot 01$ N $CaSO_4$. If the structure of the clay changes during permeation, shrinkage, cracking and/or piping can occur with increased flow in a RW permeameter producing high values. In contrast, the confining pressure provided in a FW permeameter can cause the clay to 'heal' in spite of structural changes such that little change in hydraulic conductivity is observed (Madsen & Mitchell, 1989).

The permeant used can also influence the test results obtained. Ideally, the experiments should be run with landfill leachate but usually the landfill is new at the time of testing and no leachate exists. Leachate can vary appreciably from one site to another and thus the use of a surrogate leachate may also introduce errors. Fortunately, the difference in permeability between leachate and $0 \cdot 01$ N $CaSO_4$ based tests is expected to be small for non-swelling clays. In these soils, the permeant selection may not be as critical.

Laboratory hydraulic conductivity tests are often run with hydraulic gradients of several hundred whereas gradients in the field are usually designed to be less than $1 \cdot 0$. It is felt that high gradients generate unnatural flow conditions which adversely affect hydraulic conductivity and therefore lead to errors (Quigley *et al.*, 1988). Unfortunately, the use of lower gradients requires dramatic increases in experimentation time.

In their experiments using FW permeameters, Korfiatis *et al.* (1987) observed a dependence of measured hydraulic conductivity on sample size and overburden stress. They found that values increased with test sample thickness and varied from one liner sample location to another. They also found that it reduced with an increase in overburden stress for a constant sample size. Measured values ranged from 10^{-6} to 10^{-8} cm s^{-1}. Korfiatis *et al.* stressed the need for statistical analyses when interpreting and reporting liner hydraulic conductivity data.

It has often been observed that field hydraulic conductivities are one to two orders of magnitude greater than values obtained in parallel laboratory testing (Daniel, 1987; Williams, 1987). This is due in large part to differences in the clay fabric between the laboratory and the field conditions. Differences in compaction techniques and the presence of field heterogeneities such as clods and compaction-induced cracks not present in the smaller laboratory tests have been cited as causes (Quigley *et al.*, 1988).

In general, it can be said of laboratory hydraulic conductivity measurements that:

- Few samples are large enough to account for field heterogeneities such as clods and cracks which produce increased hydraulic conductivities.
- Most experiments are conducted with hydraulic gradients several hundred times greater than those experienced in the field; the impact of these high gradients on hydraulic conductivity is not known.
- Most tests are not of sufficient duration to account for longer term interactions between the liner materials and the leachate and for the build-up of biomass and other solids along the liner surface.
- No acceptable procedure exists for extrapolating laboratory hydraulic conductivities to field conditions.

The only correct hydraulic conductivity is that which the liner exhibits in place and the only way to determine this is through seepage measurements in the field. All other values are approximations. Large scale infiltrometers are likely to produce realistic values because they do not require sample removal or elevated applied pressure and they can include the effects of field heterogeneities as their size increases. However, use of such large scale infiltrometers, a double ring unit as an example, is expensive and requires one to two months to complete.

They are therefore not suitable for frequent field determinations for quality control as the liner evolves. Substitute measurements such as sample removal in thin-walled Shelby tubes and subsequent laboratory testing can be calibrated against less frequent infiltrometer measurements and then used for quality control (Williams, 1987).

Shelby tube samples were collected and laboratory tested at a rate of one per every $8500 \, m^3$ of liner placed at the Keele Valley Landfill (Lahti *et al.*, 1987). A trial liner $30 \times 30 \, m$ was also built at the landfill using the same methods, equipment and testing procedures as for the actual liner. The concept of a well-tested trial liner and the use of a Shelby tube/laboratory test protocol calibrated against large scale field infiltrometer measurements would appear to be worthy of consideration at other landfills designed with compacted clay liners.

FIELD PERFORMANCE OF CLAY LINERS (EXPERIENCES)

Much has been written during the past few years about the ability of clay soils to contain leachate at waste disposal sites. In some cases, containment has been successful; in others it has not. At sites where containment has not been acceptable, the reasons can usually be determined: insufficient liner thickness (much less than $0.5 \, m$), incompatible leachates, poor placement and compaction methods, inadequate clay composition and a variety of other reasons such as placing monitoring wells and other devices through the liner. It is not instructive to dwell on these cases since the impact of the factors listed above are now fairly well understood. It is of more value to consider sites where reasonable leachate containment has been achieved.

It should be borne in the mind that successful performance is defined relative to a design expectation. At many sites success has meant achieving hydraulic conductivity values of $10^{-7} \, cm \, s^{-1}$ in the field. As demands on landfill performance requirements become more strict, hydraulic conductivities of $10^{-7} \, cm \, s^{-1}$ may not provide sufficient performance.

Although hydraulic conductivity is clearly the most critical of all clay liner properties and in spite of the extreme difficulty in predicting field values from laboratory measurements, studies to determine actual values in the field are surprisingly few in number.

Wisconsin Landfills

Gordon (1987) reported summary information on over 30 clay lined landfills in Wisconsin stating that, for most of these landfill liners, hydraulic conductivities less than 10^{-7} cm s^{-1} had been achieved. An earlier document (Gordon *et al.*, 1984) presented detailed liner performance assessments for four of the landfills. The Eau Claire County Landfill was instrumented with suction lysimeters for leachate collection within the liner and at a depth of 1·5 m below the liner. Four years after opening the landfill, contaminant concentrations showed marked increases in both stes of lysimeters as shown in Table 2.

The data indicate that some leachate migration through the liner had occurred; pH reduction and increases in hardness, chloride and conductivity are consistent with leachate contamination. The relatively small increase in COD is likely the result of biodegradation occurring at the linear surface. The seepage rate through the liner was estimated at 0·3 m yr^{-1} and thus breakthrough of the 1·2-m liner could be expected after 4 years. The data show much earlier contaminant arrival times probably caused by contaminant diffusion.

A second landfill in Marathon County had a 1·2–1·5 m compacted clay liner with hydraulic conductivity estimated to be less than 10^{-7} cm s^{-1}. Three lysimeters located immediately below the liner showed very little seepage through the liner and virtually no contaminant discharge.

TABLE 2. Lysimeter Contaminant Concentrations at the Eau Claire County Landfill (Gordon *et al.*, 1984)

| Contaminant | Concentrations on sample dates: | | | | | |
| | 26 Nov. 1979 | | 19 Nov. 1980 | | 16 Sept. 1983 | |
	Within liner	Below liner	Within liner	Below liner	Within liner	Below liner
Conductivity (μmhos, cm^{-1})	600	200	1 080	350	2 000	520
Hardness (mg litre^{-1} as CaCO$_3$)	190	79	528	262	870	290
Chloride (mg litre^{-1})	10	4	12	10	130	25
COD (mg litre^{-1})	28	16	114	15	140	—
pH	7·9	7·6	—	6·2	6·6	7·0

Boone County Test Landfills, Kentucky

Emcon Associates (1983) reported on a study conducted at the Boone County, Kentucky, Landfill in which a clay liner was exposed to MSW landfill leachate for a period of 9 years. The liner was constructed with clay soil having a PI of 20% to a depth of 0·61 m. In-situ hydraulic conductivity measurements were approximately 4×10^{-7} cm s^{-1} at the time of placement as compared to 2×10^{-8} cm s^{-1} in the laboratory. No deterioration in hydraulic conductivity was measured after the 9-year exposure period.

Keele Valley Landfill, Toronto, Ontario

The Keele Valley Landfill is located in a sensitive groundwater area and because of strict contaminant discharge constraints, requires a liner hydraulic conductivity less than 10^{-8} cm s^{-1} (Lahti *et al.*, 1987). Details of the liner construction are summarized in Table 3.

Frequent in-situ measurements were made for quality control during placement. Thin-walled Shelby tube samples of the liner were taken for every 8500 m^3 of clay placed and tested for permeability in the laboratory. Values of hydraulic conductivity generally exceeded 10^{-9} cm s^{-1}. Field values were judged to be consistently less than 10^{-8} cm s^{-1} and this appears to be the only site at which such low values have been achieved for compacted clay liners.

A hydraulic conductivity of 10^{-8} cm s^{-1} may in fact be the limit for compacted clay liners. It appears to be attainable only with a great deal of quality control during placement and with compaction methods similar to those used at Keele Valley.

TABLE 3. Details of Liner Construction at the Keele Valley Landfill

Property	*Description*
Thickness	1·2 m
Clay	PI between 7 and 15%
	20% partical size $<2 \times 10^{-6}$ m as minimum
Compaction	2–3% H$_2$O above optimum
	95% Standard Proctor
	Lifts:— 0·15 m thick
	— 4 to 6 passes with 30-tonne compactor

TABLE 4. Top 3 m Clay Quality at the Confederation Road Landfill

Property	Description
Mineralogy	34% carbonate, 25% illite, 24% chlorite, 15% quartz and feldspar and 2% smectite
Particle size	40–45% < 0·002 mm
CEC	15 meq. 100 g^{-1}
Moisture	28%, PL = 11%, LL = 32%
Fissures and root holes	None visible

Confederation Road Landfill, Sarnia, Ontario

Quigley and Rowe (1986) reported on the performance of an in-situ clay deposit as a barrier to leachate transport at a landfill in Sarnia, Ontario (see also Chapter 3.2). The deposit consisted of massive grey clay to a depth of approximately 30 m. The top 3 m of the clay were characterized by the information presented in Table 4. The hydraulic conductivity of the soil was not reported but the authors stated that, in response to a hydraulic gradient of 0·25, the average seepage front velocity was estimated to be 0·24 cm yr^{-1}.

After 15 years of exposure to MSW leachate, core samples were taken for pore water and solid phase analyses to assess the extent of contaminant transport through the clay. The results of the analyses are shown in Fig. 2. The theoretical seepage front (based on the Darcy equation) had reached a depth of 3·6 cm below the clay surface after 15 years. Notwithstanding the uncertainty of seepage front velocity calculations, it is clear that contaminant transport well beyond the seepage front had occurred. The conservative Cl ion had penetrated to a depth of 150 cm; organic matter to 110 cm; heavy metals Fe, Cu, Pb and Zn to 20 cm. Adsorbed phase Fe as high as 40 000 μg g^{-1} was measured indicating active ion exchange and/or precipitation within the clay. The black, oily layer with its Eh of −300 mV indicates the presence of active anaerobic biodegradation.

The study shows several significant trends:

- The permeability of the clay did not appear to be adversely affected by the MSW leachate.
- Diffusion of contaminants well beyond the seepage front is an important contaminant transport mechanism in clay liners.

MSW

CLAY SURFACE

Maximum penetration depth (cm)

—3·6	Theoretical seepage front
—15	Cu, Fe, Pb, Zn
—20	Black, oily deposits
—110	Organic contaminants
—150	Na, Cl

Maximum concentrations

Adsorbed phase   Fe: 40 000 μg g^{-1}

Aqueous phase   Cl: 2890 mg litre^{-1}
                    Na: 1200 mg litre^{-1}
                    Eh: -300 mV (minimum)
                    Organic matter: 40 moles m^{-1}

Figure 2. Transport of leachate contaminants through clay, Sarnia, Ontario. Adapted from Quigley & Rowe (1986).

- Attenuation by means of biodegradation and adsorption/precipitation actively retarded contaminants in the liner.

CONCLUSIONS

The use of compacted clay and in-situ clay deposits as barriers to leachate discharge from landfills will continue to be active in regions where clay soils are available or can be imported at reasonable cost. It is appropriate to expect that hydraulic conductivities slightly less than 10^{-7} cm s^{-1} are attainable if attention is paid to clay composition, moisture content, particle size, and compaction procedures. Hydraulic conductivities less than 10^{-8} cm s^{-1} appear to be possible only under strict quality control conditions.

Reliance solely on laboratory-based hydraulic conductivity measurements is likely to create overestimates of field performance. In spite of the difficulty in achieving them, field measurements of hydraulic conductivity are essential. The construction and evaluation of large scale test liners in the field using the equipment and procedures for the full scale liner is a valuable step toward liner verification.

The tendency toward reduced hydraulic conductivity due to bio-mass and other solid build-up at the surface of the liner has been observed by many investigators. This is reassuring to know but is not something that can be quantified and therefore cannot be relied upon as a part of liner design.

A similar situation exists for contaminant attenuation within the liner. It is reasonable to expect that the transport of biodegradable organic matter and certain heavy metals will be retarded by the clay. However, reliance on retardation as a design parameter cannot be justified because the processes cannot be adequately quantified. Partitioning coefficients are unique for the specific clay and leachate being considered. Testing to determine these coefficients is often not practical because the leachate is not usually available at the time of design and because its composition changes appreciably with site age.

Quigley *et al.* (1987, 1988) have shown clearly that, in low-permeability clay soils, transport through a liner by diffusion releases contaminants at rates much greater than permeability-dependent advective flow. Designs which focus only on liner permeability neglect this important fact. It is essential that contaminant diffusion through clay liners be considered as a part of groundwater impact assessment.

In August 1988, the United States Environmental Protection Agency (USEPA) published revised draft regulations for Subtitle D landfills. These landfills receive mainly MSW and nonhazardous industrial wastes, but will also accept household hazardous wastes and small quantities of hazardous waste from low volume generators (US Federal Register, 1988). The design criteria state that all 'new MSW landfill units must be designed with liners'. The type of liner or its properties are not specified in the regulations except where leachate or landfill gas condensate is being returned to the landfill. For these cases, the regulations state that the liner shall be a composite system consisting of a minimum 0.95 m compacted clay layer having a hydraulic conductivity of no more than 1×10^{-7} cm s^{-1} and overlain by a flexible membrane liner. The type of flexible membrane has not been specified.

The draft regulations also require that the 'design shall, at a minimum, achieve a groundwater carcinogenic risk level with an excess lifetime cancer risk level (due to continuous lifetime exposure) within the 1×10^{-4} to 1×10^{-7} range'. This risk-based constraint will be very difficult to contend with given the sparseness of toxicological data available and the uncertainties involved in predicting contaminant flux through landfill liners. It may be that engineers will be forced into using more conservative designs involving composite and/or double liner systems of which clay materials will be only one component.

REFERENCES

Daniel, D. E. (1987). Earthen liners for land disposal facilities. In *Geotechnical Practice for Waste Disposal '87*, ed. R. D. Woods. Geotechnical Special Publication No. 13, ASCE, pp. 21–39.

Emcon Associates (1983). Field verification of liners from sanitary landfills. *US EPA Report No. EPA*-600/2-83-046, Cincinnati.

Farquhar, G. & Parker, W. (1989). Interactions of leachates with natural and synthetic envelopes. In *Lecture Notes in Earth Sciences Volume 20: The Landfill, Reactor and Final Storage*, ed. P. Baccini. Springer-Verlag, Berlin, pp. 174–200.

Gordon, M. E. (1987). Design and performance Monitoring of clay-lined landfills. In *Geotechnical Practice for Waste Disposal '87*, ed. R. D. Woods. Geotechnical Special Publication No. 13, ASCE, pp. 500–14.

Gordon, M. E., Heubner, P. M. & Kinet, P. (1984). An evaluation of the performance of four clay-lined landfills in Wisconsin. In *Proceedings of the Seventh Madison Waste Conference*, ed. P. R. O'Leary. Professional Development Dept., University of Wisconsin, Madison, Wisconsin.

Grantham, G. & Robinson, M. (1988). Instrumentation and monitoring of a bentonite landfill liner. *Waste Management and Research*, **6**, 125–39.

Gray, D. H. (1989). Geotechnical engineering of land disposal systems. In *Lecture Notes in Earth Sciences, Volume 20: The Landfill, Reactor and Final Storage*, ed. P. Baccini. Springer-Verlag, Berlin, pp. 201–51.

Herrmann, J. G. & Elsbury, B. R. (1987). Influential factors in soil liner construction for waste disposal sites. In *Geotechnical Practice For Waste Disposal '87*, ed. R. D. Woods. Geotechnical Special Publication No. 13, ASCE, pp. 522–36.

Hoeks, J., Glas, H., Hofkamp, J. & Ryhiner, A. H. (1987). Bentonite liners for isolation of waste disposal sites. *Waste Management and Research*, **5**, 93–105.

Korfiatis, G. P., Rabah, N. & Lekmine, D. (1987). Permeability of compacted clay liners in laboratory scale models. In *Geotechnical Practice for Waste Disposal '87*, ed. R. D. Woods. Geotechnical Special Publication No. 13, ASCE, pp. 611–624.

Lahti, L. R., King, K. S., Reades, D. W. & Bacopoulos, A. (1987). Quality assurance monitoring of a large clay liner. *Geotechnical Practice for Waste Disposal '87*, ed. R. D. Woods. Geotechnical Special Publication No. 13, ASCE, pp. 640–54.

Madsen, F. T. & Mitchell, J. K. (1989). Chemical effects on clay fabric and hydraulic conductivity. In *Lecture Notes in Earth Sciences, Volume 20: The*

Landfill, Reactor and Final Storage, ed. P. Baccini. Springer-Verlag, Berlin, pp. 201–51.

Mundell, J. A. & Bailey, B. (1985). The design and testing of a compacted clay layer to limit percolation through a landfill cover. In *Hydraulic Barriers in Soil and Rock,* ASTM STP874, ed. A. I. Johnson, R. K. Frobel, N. J. Cavalli & C. B. Pettersson. ASTM, Philadelphia, pp. 246–62.

Oakley, R. E. (1987). Design and performance of earth-lined containment systems. In *Geotechnical Practice for Waste Disposal Design '87,* ed. R. D. Woods. Geotechnical Special Publications no. 13, ASCE, pp. 117–36.

Quigley, R. M. & Rowe, R. W. (1986). Leachate migration through clay below a domestic waste landfill, Sarnia, Ontario, Canada: Chemical interpretation and modelling philosophies. In *Hazardous and Industrial Solid Waste Testing and Disposal: Sixth Volume ASTM STP933,* ed. D. Lorenzen, R. A. Conway, L. P. Jackson, A. Hamza, C. L. Perket & W. J. Lacy. ASTM, Philadelphia, pp. 93–103.

Quigley, R. M., Yanful, E. K. & Fernandez, F. (1987). Ion transfer by diffusion through clay barriers. In *Geotechnical Practice for Waste Disposal '87,* ed. R. D. Woods. Geotechnical Special Publication No. 13, ASCE, pp. 137–58.

Quigley, R. M., Fernandez, F. & Row, R. K. (1988). Clayey barrier assessment for impoundment of domestic waste leachate (Southern Ontario) including clay–leachate compatibility by hydraulic conductivity testing. *Can. Geotech. J.,* **25,** 574–81.

US Federal Register (1988). *Part III Environmental Protection Agency, Solid Waste Disposal Facility Criteria: Proposed Role,* Vol. 53, No. 168.

Williams, C. E. (1987). Containment applications for earthen liners. In *Proceedings of the 1987 Specialty Conference on Environmental Engineering,* ed. J. D. Dietz. ASCE, pp. 122–8.

2.2 Geotechnical Design and Quality Control of Mineral Liner Systems

HANS LUDWIG JESSBERGER

Ruhr-University Bochum, Institute of Soil Mechanics and Foundation Engineering, Universitätsstr. 150, 4630 Bochum 1, Germany

INTRODUCTION

There is an increasing requirement for guidelines or recommendations concerned with the geotechnical aspects of waste disposal sites. Quality control must be directed, on the one hand, to the design itself, in a way similar to common construction engineering practice, and on the other hand to the performance of the work on-site, with the check-proving of soil samples etc.

The following geotechnical aspects are believed to be decisive and are therefore discussed in this chapter:

• general design principles including the multi-barrier concept
• component sealing systems
• suitability tests
• deformability of mineral seals
• resistance against erosion and suffosion
• quality assurance

The main sources for this chapter are the Geotechnical Recommendations 'GLC', Geotechnics of Landfills and Contaminated Land, prepared by the European Technical Committee ETC 8 of the International Society of Soil Mechanics and Foundation Engineering (Anon., 1990a). The ETC 8 members are (1990) Hans L. Jessberger (Chairman), Mario Manassero, Ton Puthaar, Christian Schlüchter,

Bertrand Soyez and Andrew Street. These recommendations were prepared as a supplement to the Technical Regulations on Toxic and Hazardous Waste of the German Federal Administration (Anon., 1991), and will be published in an enlarged and revised edition (Anon., 1993).

SOME PRINCIPLES OF THE GEOTECHNICAL DESIGN

The preparation of the geotechnical design must generally examine the following design aspects of safety components:

- subsoil as a site barrier
- basal lining system
- waste body (including any inserts)
- capping system
- disposal site environment
- geotechnical aspects of site operation
- geotechnical concerns regarding the removal of gas and leachate liquid from the landfill
- geotechnical aspects of recultivation
- site closure and subsequent aftercare measures
- supervision and long-term monitoring

Figure 1 gives details of the possible components of landfill. These components are incorporated into the design with respect to the so called multi-barrier concept.

The safety elements must be set out both in their independent functions and in their mutual influence and overall safety effect. The effect of failure of one of the safety elements must also be discussed from the point of the view of long-term behaviour. Necessary measures must be described and the relevant working conditions must be indicated.

COMPONENTS OF LINER SYSTEMS

It is the German approach to combine a mineral sealing layer with a synthetic flexible membrane (geomembrane) placed directly on top of it. The drainage blanket that is positioned over the sealing layers,

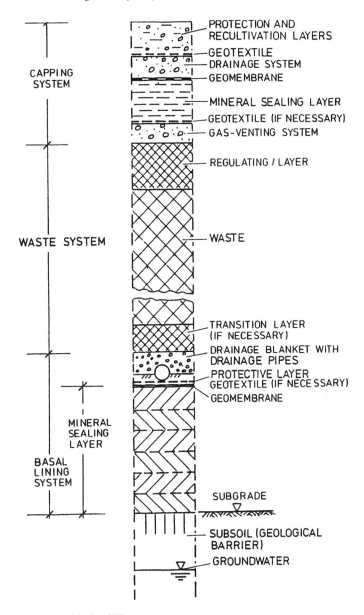

Figure 1. Possible landfill components.

which may be covered with a protective layer, is integrated in to the sealing system.

In principle, the effects of the sealing layers as a geotechnical barrier and of the drainage system are as follows:

Mineral sealing layers:

- minimisation of seepage and diffusion, related to the choice of material, compaction and thickness of the layer;
- resistance to erosion and water penetration;
- resistance to leachate, related to swelling clay mineral content;
- heavy metal adsorption capacity related to the clay mineral or organic content;
- non-susceptibility to settlement and self-healing ability related to the material's plastic characteristics, which are determined by clay content and particle size distribution;
- effects of swelling and shrinkage related to the hydrogeological conditions.

Geomembrane:

- non-susceptibility to settlement related to stress–strain behaviour;
- prevention of leakage;
- long-term chemical resistance depending on the material used and the thickness of the membrane in combination with the mineral sealing layer.

Interface between the sealing layers:

- at perforation points in the geomembrane prevention of the lateral spread of leachate at the surface;
- prevention of significant water pressure behind the geomembrane;
- sealing effect in the interface related to:
  - the smooth, fine-grained character of the surface of the mineral sealing layer,
  - the load-dependent deformation behaviour of the geomembrane and the mineral substratum,
  - effect of change in gradient of the subgrade related to the load-dependent deformation of the geomembrane.

Protective layer:

- permanent distribution of concentrated stresses on the geomembrane by the granularity of the previous blanket and the protective effect of the geotextile, if any, plus chemical resistance to leachate liquid and resistance to slipping, if appropriate.

Drainage system:

- the drainage system permits the collection and removal of leachate from the waste, thereby preventing the build-up a larger pressure head above the sealing layers.

Material specification may in certain circumstances conflict with the construction methods chosen. For instance, the need for high plasticity in the mineral sealing layer and need to prevent surface rutting. These requirements must be considered and reconciled in the construction design.

The scope of validation testing necessary for a composite lining system depends on the specific design and stage of construction. This should be laid down in the design, and the corresponding suitability check undertaken by a qualified geotechnical expert. This should include a recommendation for in-house testing during the construction phase. The regulatory authority will normally define the minimum scope of in-house and external testing required.

Although there are some arguments against the component sealing system, especially above inclined slope surfaces, today it is believed in Germany that this system should be used especially for landfills of toxic and hazardous wastes (Anon., 1991). According to these technical regulations, the component sealing systems must be chosen for basal lining and for capping systems with the following measures.

(a) Basal lining system:

- mineral sealing (low permeability soil layer) $d > 1.5$ m; clay mineral content $> 10\%$, $k < 5 \times 10^{-10}$ m/s, inclination $> 3\%$
- geomembrane $d > 2.5$ mm, specified material

(b) Capping system:

- mineral sealing $d > 0.5$ m; clay mineral content $> 10\%$, $k < 5 \times 10^{-10}$ m/s, inclination $> 5\%$
- geomembrane $d > 2.5$ mm, specified material

The drainage blankets with drainage pipes are incorporated into the sealing systems with the following measures: drainage blanket grain size 16/32 mm, $d > 0.3$ m, $k > 1 \times 10^{-3}$ m/s, drainage pipes DN 300.

More or less the same specifications, with reduced linear thickness, will be applied to municipal solid waste (MSW) landfill (TA-Siedlungsabfall) and will be incorporated in the 'Technical Instructions for MSW Landfills'.

SUITABILITY TESTS

Laboratory Tests

The suitability of all materials used in the construction and rehabilitation of landfills and contaminated sites must be proved. The suitability of minerals to be used to seal landfills generally should be proved separately for each specific application. The number of the random samples necessary for this is laid down by the qualified geotechnical expert. The following investigations should be carried out:

- determination of the nature and composition of the sealing material;
- establishment of the placement criteria;
- determination of the characteristics of the sealing materials when in place.

Soil classification and stress–strain behaviour of sealing materials:

- gran size distribution
- consistency limits
- organic constituent content
- grain density
- calcium carbonate content
- water intake
- moisture content
- density
- compression tests
- swelling
- triaxial compression test or direct shear test
- uniaxial compression test

Specifications for permeability testing:

- permeability test in triaxial equipment; back pressure
- percolation from bottom to top
- water saturation
- evaluation of permeability coefficient k at hydraulic gradient $i = 30$
- test duration until $k = $ constant
- is possible, use of a synthetic leachate or special test liquid in addition to water

In order to describe the mineralogical composition of the clay mineral fraction of sealing materials and also to ascertain the potential influences on permeability of chemical interaction with leachate, clay mineralogical and geochemical tests are necessary in addition to the suitability tests.

The tests may be conducted either on fresh samples or on pre-contaminated sample material.

The procedures described are a modified form of the routine methods used in sediment petrography:

- grain size distribution
- clay mineralogical testing
- determination of the ion exchange capacity
- X-ray defraction analysis
- water intake
- carbonate ratio

Field Trials for Mineral Basal Seals and Caps

Preparation of a field trial for the construction of a mineral seal for a landfill should be regarded as a large-scale suitability test in which the external and in-house examina should be involved. The field trial demonstrates the following points using sealing material tested in the laboratory for suitability:

- suitability of the material under site conditions;
- suitability of the methods of extraction, treatment and preparation;
- suitability of the methods of placement and compaction;
- adherence to the requirements for permeability, water content and density of the mineral sealing material in a large-scale basis;

- establishment of reference parameters for quality assurance.

For each field trial the following should be recorded:

- origin, type and condition of the mineral sealing material;
- methods used for extraction, transport, treatment and placement of the material;
- type, operating principle, weight and major dimensions of the equipment used;
- diameter and length of the roller; operating weight with and without ballast; operating speed of the compaction equipment; frequency and energy of vibrating rollers; length, cross-sectional area and arrangements of the studs used to achieve a kneading effect in the case of sheep's foot rollers;
- number of passes with the roller, indicated separately for each type where various types of roller are used;
- type, dimensions and characteristic values of any soil conditioning equipment used;
- number of rotavations and operating speed of the machine;
- methods used for breaking up clods of soil, maximum permissible clod size and degree of breaking up achieved;
- methods of checking and, if necessary, correcting the moisture content of the soil to be placed, origin of the added water, period elapsing between distribution of the water and the commencement of compaction;
- thickness of the lift before and after compaction;
- if necessary, addition of bentonite, powdered clay or other additives; quantities used, method of batching; number of rotavations or duration of mixing in the pressure mixer.

DEFORMABILITY OF MINERAL SEALS

In the design procedure for waste landfills it has to be proved that the mineral sealing systems are able to follow without increasing permeability uniform or non-uniform settlements of the subsoil or of the waste. This problem was investigated recently at Ruhr-University Bochum using the large geotechnical centrifuge. Results from these tests performed with and without overburden are presented in Chapter 3.6.

RESISTANCE TO EROSION AND SUFFOSION

The erosion and suffosion resistance of mineral sealing material is an important factor in ensuring their long-term sealing properties and therefore may need to be taken into account in the landfill design.

Mineral sealing materials are not always stable as filters (Terzaghi's filter rule) in relation to the adjoining strata. Specific tests may be necessary, particularly in the following cases:

- low content of swelling in clay minerals
- high hydraulic gradient
- gap-graded materials.

Definitions

Erosion means the removal and transport of the particles in a soil (fine-grained and coarse-grained components) as a result of liquid flow. A distinction is made between (see Fig. 2):

- external erosion: the force of the liquid flow along the surface of the stratum removes the soil particles;
- internal erosion (piping): regressive removal of soil particles along internal pore channels;

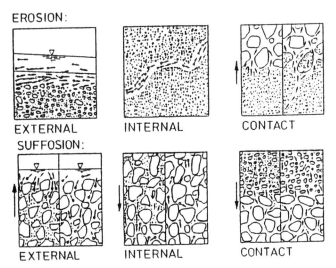

Figure 2. Definition of erosion and suffosion.

- contact erosion: movement of particles at the contact between fine- and coarse-grained strata.

In suffosion only the fine-grained components of a soil are eroded to leave a coarse-grained granular skeleton. Just as for erosion, a distinction is made between (see Fig. 2):

- external suffosion
- internal suffosion
- contact suffosion

Clogging (internal) or blinding (surface) are the terms used to describe the blocking up of soil pores by particles of the soil or precipitation products released elsewhere as a result of erosion or suffosion.

Evaluation

Resistance to erosion.   The risk of external erosion affects any element of the liner and capping system exposed on steep slopes. It can be prevented by structural measures, so no special evaluation is necessary for this purpose.

For clay materials normally used in sealing landfill sites there have been no documented failures caused by internal erosion. However, verification can be achieved by the pin-hole test (Busch & Luckner, 1974) used in earth dam construction.

Since Terzaghi's filter rule applies only to the stability at the contact zone between fine- and coarse-grained strata, it is recommended that the cohesive forces between the clay particles be taken into account for assessing resistance to contact erosion. The cohesive forces may be adversely affected by chemcial action. Resistance to contact erosion may be estimated (Davidenkoff, 1979) with the following equation, assuming horizontal mineral layers:

$$\eta = \frac{15c_0}{D_{50}(\gamma' + \gamma_w i)}$$

in which

$$\eta = \text{safety factor}$$
$$c_0 \ (\text{kN/m}^2) = \text{tensile strength of the (cohesive) soil}$$

D_{50} (mm) = average grain size of the coarse-grained stratum

i = hydraulic gradient of leachate

γ' (MN/m³) = submerged density

γ_w (MN/m³) = density of water

Resistance to contact suffosion. To assess the resistance to contact suffosion, Davidenkoff's equation can be employed, in which the average grain size of the coarse-grained skeleton of the mineral sealing layer is used instead of the average size of the adjoining coarse-grained stratum; the tensile strength relates to the fine-grained content of the stratum deposited or carried away by suffosion. Where the grain structure is well graded this demonstration is not required.

For checking resistance to contact suffosion, a permeability test may be done, in which the filter stone on the outlet side of the sample has maximum pore diameter of 40–100 μm (corresponding to P100 according to ISO 4793). This permeability test is conducted over a 10-day period with $i = 30$. To avoid clogging in the filter stone, the latter must be cleaned or exchanged at least once during the test.

QUALITY ASSURANCE

To ensure the quality of the overall structure of a landfill, the individual components must meet the quality standards. Quality assurance must relate to both the quality of the material used and to the quality of the workmanship in accordance with the existing state of technology.

Quality assurance should comprise:

- in-house testing by the contractor
- external testing by an independent party

If appropriate, the regulatory authority may request tests on random samples. All tests must be supervised by a geotechnically qualified expert with the extensive knowledge in the field of waste disposal techniques.

These tests comprise:

- initial testing of the construction materials to be processed;
- tests on the processing of the materials;

- supervision of all work, material characteristics and functions that determine quality.

The method for testing adopted, presentation of test data and in-house/external review should be adopted to meet the requirements of the particular construction process.

Acceptance should be based on the results of external testing. Full-time site supervision by the contractor's inspector and by an external inspector is required. All testing, both in-house and external, should be fully documented.

The holder of the planning permission for a specific waste disposal site should apply to the regulatory authority for final acceptance. This should be supported by full documentation of results relating to the site, and include tests relating to:

- construction of elements of the work and the completed structure;
- adherence to the requirements of the quality plan.

After acceptance of the subgrade and successful completion of the suitability test together with investigations on the field trial, work can commence on construction of the mineral sealing layer. To ensure quality is achieved, field and laboratory tests should be conducted.

The following range of tests will normally be required. In certain circumstances other tests, if proven, may be acceptable:

- characteristics of the materials to be used, including grain-size distribution, consistency limits, water intake and water content (every 1000 m^2);
- moisture content on placement, homogeneity of the material placed, number of passes with the roller, quantity of water added, if any (every 1000 m^2);
- minimum clod size, cutting depth and quantity of additives or dosage in the case of multiple component mixtures (every 1000 m^2);
- thickness of the individual lifts, evenness of the lift surfaces and adherence to proposed levels and dimensions (every 500 m^2);
- degree of compaction and homogeneity achieved in the sealing layer for each lift by determination of density, moisture content, grain-size distribution and plasticity, if appropriate, and by survey (every 1000 m^2);
- determination of the permeability of the sealing layer for each lift (every 2000 m^2).

In order to achieve tighter control over the sealing layer, it may be necessary—particularly in the case of non-uniform construction materials—to reduce the size of the test grid. More extensive investigations may be specified in particular cases, depending on the type of seal.

CONCLUSIONS

The geotechnical landfill design is directed towards pollutant emission control for all phases of construction and operation for a lining system. The main steps are:

- formulating the requirements,
- establishing appropriate material properties,
- analysing the functions of the sealing elements.

The state of technology is described in the Recommendations of the Committee on Geotechnics of Landfill Design and Remedial Works (Anon., 1990*a*). This chapter has highlighted some of these recommendations. A revised and enlarged edition of these recommendations is available in 1993 (Anon., 1993).

ACKNOWLEDGEMENT

The fruitful cooperation of the members of the European Technical Committee ETC 8 of International Society of Soil Mechanics and Foundation Engineering (ISSMFE) on Geotechnics of Landfill Design and Remedial Works is gratefully acknowledged.

REFERENCES

Anon. (1990*a*). Geotechnics of Landfills and Contaminated Land; Technical Recommendations 'GLC', ed. German Geotechnical Society for the International Society of Soil Mechanics and Foundation Engineering. Verlag Ernst & Sohn, Berlin.
Anon. (1991). Technical instructions for the storage, chemical/physical and

biological treatment, incineration and disposal of hazardous waste. Bundes-anzeiger, Köln.

Anon. (1993). Geotechnics of Landfill Design and Remedial Work; Technical Recommendations 'GLR', ed. German Geotechnical Society for the International Society of Soil Mechanics and Foundation Engineering. Verlag Ernst & Sohn, Berlin.

Busch, K. F. & Luckner, L. (1974). *Geohydraulik*. Verlag Enke, Stuttgart.

Davidenkoff, R. N. (1979). *Anwendung von Filtern im Wasserbau*. Verlag Ernst & Sohn, Berlin.

2.3 Quality Control of Clay Liners

KEITH J. SEYMOUR & ANTHONY J. PEACOCK

*National Rivers Authority (NW Region), New Town House,
Buttermarket Street, Warrington, Cheshire WA1 2QG, UK*

INTRODUCTION

In recent years there has been an increasing requirement for some
form of earthworks to be employed on landfill sites. These requirements have usually been incorporated into site licence conditions to
protect the water environment. From experience in the drafting of
such conditions, it has become apparent that a set of guidelines to
cover the testing and placing of natural materials for basal liners,
bunds and final caps is required. The guidelines described below were
originally prepared by the North West Water Authority, prior to the
formation of the National Rivers Authority (NRA), and have been
adopted successfully in the north-west of England. They are seen as
complementing Sections 4.52 and 4.53 of the UK Department of the
Environment Waste Management Paper No. 26 (Department of the
Environment, 1986) and are based on well-established civil engineering practices employed during the construction of highways and earth
dams. Since the containment of leachate within a landfill site is equally
as important as preventing a reservoir from leaking, the same
standards of design and workmanship should apply. However, unlike
the case of dam construction, there is no statutory requirement in the
UK for such works to be supervised or certified by a 'competent
engineer' acting on behalf of a landfill operator. The guidelines
presented here are intended to assist those operators who lack the
necessary in-house expertise to arrange appropriate soils testing and to
ensure an adequate standard of emplacement. They describe geotech-

nical considerations which need to be taken into account at the design, testing and construction stages.

THEORY

The permeability of any soil depends on a number of variables including:

- soil density (degree of compaction)
- moisture content
- degree of saturation
- nature of the permeating fluid

The relationship between permeability, dry density and moisture content is shown in Fig. 1. For a given compactive effort, the minimum permeability occurs a little above the Optimum Moisture Content.

The effect of varying the compactive effort on the dry density and moisture content of a given soil is shown in Fig. 2 and is described in more detail by Head (1980). The two British Standard compaction tests used to determine these relationships are BS 1377: Tests 12 and 13 (BSI, 1975), using 2·5 kg and 4·5 kg hammers respectively. The lighter hammer was originally intended to reproduce in the laboratory the compactive efforts that could be achieved in the field when the test was first introduced by Proctor in 1933. However, with the availability of modern, larger compaction plant and equipment, the Modified Proctor (AASHO) test, using the heavier hammer, has been adopted in UK civil engineering earthworks practice.

MINIMUM REQUIREMENTS

In most situations the minimum requirements of the North West Water Authority for natural clay liners on sites containing leachate forming wastes are:

- a maximum permeability of 10^{-9} m/s, as recommended in Waste Management Paper No. 26 (Department of the Environment, 1986) and in accordance with standards used in the United States;
- a minimum thickness of basal liner and final cap of 1 m. This represents the thinnest clay layer which could be emplaced practically to provide reliable continuity of seal. Furthermore, if the liner thickness falls below 1 m, calculations using Darcy's Law

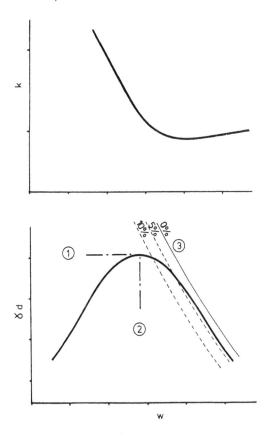

Figure 1. Density/moisture content/permeability relationship. k = Permeability; γd = dry density; w = moisture content; 1, γd maximum; 2, optimum w; 3, per cent air voids. After Lambe (1958) and Harrop-Williams (1985).

indicate significant liquid throughputs for clays with the above permeability. In sensitive situations greater thicknesses of liner or a composite system incorporating a synthetic membrane may be considered necessary to protect water interests (North West Waste Disposal Officers Landfill Liner Sub-Group, 1988);

- leachate-retaining bunds should be designed and constructed to remain stable, both during and after the operational life of the landfill and should have a minimum crest width of 2 m to enable proper tracking by compaction plant.

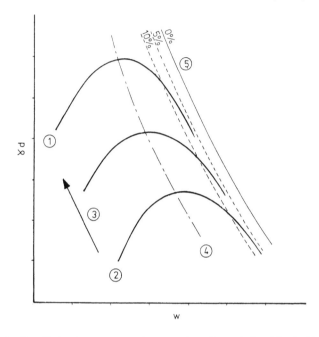

Figure 2. Density/moisture content/compactive effort relationship. 1, Modified Proctor (4·5 kg hammer); 2, Standard Proctor (2·5 kg hammer); 3, increasing compactive effort; 4, line of optimum w; 5, per cent air voids; γd = dry density; w = moisture content. After Harrop-Williams (1985).

APPROACH

It is generally impractical to carry out a large number of permeability measurements on the completed structure because of the high cost and long delays that would be involved. Perhaps of more importance, from a regulatory point of view, is the difficulty in obtaining values representative of the mass of the emplaced clay (Daniel, 1984; Day & Daniel, 1985).

It is for these reasons that the following method is proposed. This is a summary of the full set of guidelines (NRA North West Unit, 1989). They use the relationship between permeability and more rapidly and easily measured 'standard' soil index parameters, whilst adopting Table 6/4 of the Department of Transport Specification for Highway Works, Part 2 (Department of Transport, 1986), which gives certain maximum thicknesses of layers and minimum numbers of passes for selected compaction equipment (Table 1).

PROCEDURE

Stage 1: Laboratory and Field Testing

Representative samples of the material from each source to be used in any liner, final cap or specified bund (referred to as the 'seal') should be tested in an approved soils laboratory to satisfy the regulatory authorities that it is capable of being compacted in the field to an extent that will achieve the target permeability.

Is is recommended that the method of soils testing should be in accordance with the following schedule:

Compaction. The density/moisture content relationship of the material is determined in accordance with BS 1377, Tests 12 and 13. The appropriate size of hammer (2·5 kg or 4·5 kg) should be selected to reflect the actual compaction equipment to be used on site. *Where there is any uncertainty, the lighter hammer should be used.*

Permeability. The permeability (k) of the recompacted sample should be measured for each moisture content increment of the compaction test. This will involve at least five permeability values being obtained for each sample. This establishes the range of permeabilities that may be achieved in the field at varying moisture contents for either Standard or Modified Proctor densities (see Fig. 1).

When testing recompacted clays, the permeability should be measured directly by the falling head method, preferably using the BS compaction mould (Head, 1982).

Alternatively, tests may be carried out using:

• falling head cell (Head, 1982)
• oedometer cell (modified to allow direct measurement of k)
• triaxial cell under falling or constant head (Tavenas *et al.*, 1983)

Indirect evaluation of the permeability from consolidation tests should not be accepted as this method can lead to an underestimate of the k value (Tavenas *et al.*, 1983).

Classification. The natural moisture content, liquid limit and plastic limit of each sample should be measured in accordance with BS 1377, Tests 1, 2 and 3 respectively. These results are plotted on a classification chart (Fig. 3) and used to 'fingerprint' the material. The

TABLE 1. Method Compaction for Earthworks Materials Plant and Methods: (DoT 'Specification for Highway Works, Part 2' Table 6/4 (abstract)).[c]

Type of compaction plant	*Ref. No.*	*Category*	*Method 1* D^a	N^b
Smooth wheeled roller		Mass per metre width of roll:		
(or vibratory roller	1	over 2 100 kg up to 2 700 kg	125	8
operating without	2	over 2 700 kg up to 5 400 kg	125	6
vibration)	3	over 5 400 kg	150	4
Grid-roller		Mass per metre width of roll:		
	1	over 2 700 kg up to 5 400 kg	150	10
	2	over 5 400 kg up to 8 000 kg	150	8
	3	over 8 000 kg	150	4
Tamping roller		Mass per metre width of roll:	225	4
	1	over 4 000 kg		
Pneumatic-tyred		Mass per wheel:		
roller	1	over 1 000 kg up to 1 500 kg	125	6
	2	over 1 500 kg up to 2 000 kg	150	5
	3	over 2 000 kg up to 2 500 kg	175	4
	4	over 2 500 kg up to 4 000 kg	225	4
	5	over 4 000 kg up to 6 000 kg	300	4
	6	over 6 000 kg up to 8 000 kg	350	4
	7	over 8 000 kg up to 12 000 kg	400	4
	8	over 12 000 kg	450	4
Vibratory roller		Mass per metre width of a vibrating roll:		
	1	over 270 kg up to 450 kg	unsuitable	
	2	over 450 kg up to 700 kg	unsuitable	
	3	over 700 kg up to 1 300 kg	100	12
	4	over 1 300 kg up to 1 800 kg	125	8
	5	over 1 800 kg up to 2 300 kg	150	4
	6	over 2 300 kg up to 2 900 kg	175	4
	7	over 2 900 kg up to 3 600 kg	200	4
	8	over 3 600 kg up to 4 300 kg	225	4
	9	over 4 300 kg up to 5 000 kg	250	4
	10	over 5 000 kg	275	4
Vibrating plate		Mass per m² of base plate:		
compactor	1	over 880 kg up to 1 100 kg	unsuitable	
	2	over 1 100 kg up to 1 200 kg	unsuitable	
	3	over 1 200 kg up to 1 400 kg	unsuitable	
	4	over 1 400 kg up to 1 800 kg	100	6
	5	over 1 800 kg up to 2 100 kg	150	6
	6	over 2 100 kg	200	6

TABLE 1. (Continued)

Type of compaction plant	Ref. No.	Category	Method 1 D^a	N^b
Vibro-tamper		Mass:		
	1	over 50 kg up to 65 kg	100	3
	2	over 65 kg up to 75 kg	125	3
	3	over 75 kg up to 100 kg	150	3
	4	over 100 kg	225	3
Power rammer		Mass:		
	1	100 kg up to 500 kg	150	4
	2	over 500 kg	275	8
Dropping-weight compactor		Mass of rammer over 500 kg height drop:		
	1	over 1 m up to 2 m	600	4
	2	over over 2 m	600	2

[a] D = Maximum depth of compacted layer (mm).
[b] N = Minimum number of passes.
[c] Source: DoT's *Specification for Highway Works*. Reproduced by kind permission of HMSO—Crown copyright reserved.

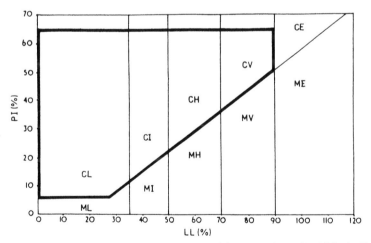

Figure 3. Plasticity chart. □, Range of acceptable material; LL, liquid limit; PL, plastic limit; PI, plasticity index (LL–PL); C, clay; H, silt. Plasticity range: L, low; I, intermediate; H, high; V, very high; E, extremely high.

particle size distribution, and in particular the clay content, should be determined by BS 1377, Test 7.

Specific gravity. The specific gravity of the soil particles is determined using BS 1377, Test 6(A) or 6(B).

Other tests. Additional tests may need to be carried out to enable the operator to design the structure in terms of slope stability and to avoid problems of unworkability and shrinkage, particularly with high plasticity clays.

Testing of in-situ material. Where it is proposed to leave soil or rock *in situ* in its natural, undisturbed state to form part of the seal, it should be tested to establish whether it has a permeability of not more than the required target. In the case of soils, permeability measurements should be carried out either in the field using variable head or constant head tests in accordance with BS 5930, Section 21.4 (BSI, 1980), or in the laboratory by the falling head method (see above).

Reporting results. Before earthworks proceed, the operator is required to submit to the regulatory authorities for approval a report containing the results of all laboratory and field tests.

Material suitability. Where material is to be recompacted to form the seal it should only be accepted as suitable when the above testing programme proves that it is capable of achieving a permeability not greater than the target. The material should have a minimum clay content (particle size less than $0.002\,\text{mm}$) of 10%. The liquid limit and plasticity index (see Fig. 3) should not exceed 90% or 65% respectively for reasons of workability.

Stage 2: Emplacement

Method selection. The operator states in advance of the site work.

- the minimum moisture content to be achieved during material emplacement,
- the type of compaction plant to be used,
- the maximum thickness of each layer,
- the minimum number of passes.

These items are selected to ensure that the target permeability and a maximum of 5% air voids are not exceeded. The first item is based on the laboratory test results. The other items should comply with or exceed the minimum requirements of Table 6/4 (Method 1) of the DoT Specification of Highway Works, Part 2, 1986 (see Table 1). A greater number of passes or thinner layers may be necessary to ensure adequate compaction and that all discontinuities are removed.

Material control. The seal is then constructed using only suitable material of the type and from the source previously tested and approved by the regulatory authorities. The following materials should be specifically excluded:

- peat, timber and other vegetable matter,
- imported waste,
- rocks, boulders or concrete having a volume of greater than $0.05\,\text{m}^3$.

Moisture control. The moisture content of the material, when placed, should be monitored and controlled to ensure it is at or above the minimum stated above, but not so high as to render the material unworkable using the method and plant selected.

Supervision. The operator is required to provide a suitably qualified and experienced geotechnical or civil engineer to plan, design and supervise the construction of the earthworks.

Seal protection. Completed seals should be protected to prevent drying out and the subsequent formation of cracks in the placed material.

Stage 3: Quality control

Testing schedule. On completion of the structure, the following quality control tests should be carried out on the placed material:

- in-situ density (BS 1377: Test 15),
- moisture content (BS 1377: Test 1),
- classification tests (BS 1377: Tests 2, 3 and 7),
- depth profiling—either by augering or excavation (and backfilling

with grout/clay/bentonite), or by levelling surveys before and after the earthworks.

The above tests should be carried out on a regular grid pattern across the structure and at depths as agreed with the regulatory authorities.

Results Acceptability

The operator is required ot take undisturbed samples for falling head permeability measurement in the laboratory, should the above tests identify areas where:

* the material differs significantly in terms of its liquid and plastic limits from those established above;
* the material has a clay content of less than 10%;
* the material contains air voids in excess of either those actually achieved during the laboratory compaction tests or 5%, whichever is greater;
* the material has a moisture content less than that stated above.

In the event of these permeability measurements exceeding the target figure or if there are any discontinuities apparent on the surface of the structure, the suspect area must be excavated and recompacted to the above standard.

CONCLUSIONS

* The guidelines represent *minimum* standards set by the North West Water Authority for the design, testing and emplacement of those earthworks on landfill sites required to avoid water pollution.
* They were drawn up to meet a practical need to increase the awareness of site operators to the fact that the approach to the preparation of landfill sites should be the same as in the construction of earth dams and highways.
* Accordingly, the guidelines are based on well-established soil mechanics theory and standard soil tests, and adopt methods used in such civil engineering earthworks contracts.
* They use the relationship between permeability, moisture content

and dry density, along with classification indices to provide simpler and quicker quality control over the site works than could be achieved by relying on a limited number of permeability tests on the completed structure.

• The guidelines recognise the fundamental importance of both the method and standard of material emplacement and the need for careful control over its moisture content. Consequently, the works should be designed and supervised by a suitably qualified and experienced geotechnical or civil engineer.

REFERENCES

British Standards Institution (1975). *BS 1377: Methods of Test for Civil Engineering Purposes.* HMSO, London.

British Standards Institution (1980). *BS 5930: Code of Practice for Site Investigations.* HMSO, London.

Daniel, D. E. (1984). Predicting hydraulic conductivity of clay liners. *Journal of Geotechnical Engineering*, 110(2), 285–300.

Day, S. R. & Daniel, D. E. (1985). Hydraulic conductivity of two prototype clay liners. *Journal of Geotechnical Engineering*, 111(8), 957–70.

Department of the Environment (1986). *Waste Management Paper No. 26. Landfilling Waste.* HMSO, London.

Department of Transport (1986). *Specification for Highway Works, Part 2.* HMSO, London.

Harrop-Williams, K. (1985). Clay liner permeability: evaluation and variation. *Journal of Geotechnical Engineering, ASCE*, III(10), 1211–25.

Head, K. H. (1980). *Manual of Soil Laboratory Testing. Vol. 1: Soil Classification and Compaction Tests.* Pentec Press, Plymouth.

Head, K. H. (1982). *Manual of Soil Laboratory Testing. Vol. 2: Permeability, Shear Strength and Compressibility Tests.* Pentec Press, Plymouth, pp. 449–457.

Lambe, T. W. (1958). The engineering behaviour of compacted clay. *Journal of Soil Mechanics and Foundations Division, ASCE*, SM2, 1655-1 to 1655-35.

NRA (1989). Earthworks on Landfill Sites. Warrington.

NWWDO Landfill Liners Sub-Group (1988). Guidelines on the use of landfill liners. Lancashire Waste Disposal Authority, Preston.

Tavenas, F., Leblond, P., Hean, P. & Leroueil, S. (1983). The permeability of natural soft clays. Part 1: Methods of laboratory measurement. *Canadian Geotechnical Journal*, 20(4), 629–44.

2.4 Leakage in Double Liner Systems: The Containment Objective

ROBERT B. WALLACE & HALUK AKGUN

Environtech Consultants Inc., 7301A Palmetto Park Road, West Suite 301C, Boca Raton, Florida 33433, USA

INTRODUCTION

One of the benefits of a double lining system is that it allows the designer and operator to quantify the performance of the system (see Chapter 1.2). Significant theoretical and empirical design procedures which allow modelling of the lining system exist. Based on certain application-derived experience, relevant assumptions can be made to provide predicted performance levels of the liners against which measured performance levels can be weighed.

A rationale has been presented (Bonaparte *et al.,* 1989; Giroud & Bonaparte, 1989*a,b*) for calculating the leakage through a hole in a geomembrane for the cases of geomembrane-only liners and geomembrane/clay composite liners. In truth, clay is a misnomer in the generic sense, because the soil component of a composite liner can take a variety of forms, including: compacted or remolded and recompacted clay; fabricated geotextile/bentonite composites; modified or treated low permeability soils; and now even bonded geomembrane/bentonite composites.

It is inappropriate herein to reproduce these design procedures in detail, but it is significant to note, in particular, the effect for a given set of parameters, the difference between the calculated leakage through the three main types of individual liners (rather than the system), namely: a clay-only liner, a geomembrane-only liner and a geomembrane/clay composite.

LEAKAGE CALCULATION

Calculation of the leakage through a clay-only liner can be carried out based on the hydraulic conductivity of the soil and the head of leachate above the liner. However, as has been thoroughly documented over the years, there are many parameters that affect the structure of a clay liner. There are limitations, especially in the case of exposed liners. Freeze/thaw and wet/dry cycles can lead to the build-up of a secondary structure within the clay. These in turn can create preferential flow paths either directly causing or at least abetting leakage, in spite of the potential for healing of cracks upon wetting.

The leakage through a clay-only liner can be calculated theoretically using Darcy's Law:

$$Q_s = k_s i A \tag{1}$$

where Q_s is the leakage rate through the clay (soil) liner, expressed in m^3/s; k_s is the hydraulic conductivity of the clay (m/s); i is the gradient (dimensionless), expressed as the ratio of the leachate head (h, expressed in m) to the liner thickness (L, expressed in m); and A is the area of liner considered (m^2).

To allow comparison of the results of leakage through a clay liner to the leakage through a hole in a geomembrane liner, a unitized leakage rate is selected, which is defined as the leakage rate through a given area, herein taken as $1000\,m^2$, and expressed as litres per thousand square metres per day equivalent to ml/m^2 day. This unitized leakage rate, q, is given in eqn (2), for the case of a clay-only liner:

$$q = Q_s \quad \text{for } A = 1000\,m^2 \tag{2}$$

The unitized leakage rate results for a $1\cdot0$-m-thick clay liner with a hydraulic conductivity of $1\cdot0 \times 10^{-9}\,m/s$ subjected to leachate heads of $10\,m$, $1\,m$, $0\cdot1\,m$, and $0\cdot01\,m$ are illustrated in Table 1. This covers the spectrum of likely possibilities above the clay liner below a pond, or in a landfill with different configurations of leachate collection layers. It is assumed that: (1) the clay component of the liner is intact, i.e. there is no pre-existing hole or other preferential path through which the leachate migrates; and (2) the flow through the layer is vertical and hence, for simplicity, each case assumes a horizontal liner.

TABLE 1. Steady-State Unitized Leakage Rates (ml/m² day)

Liner type	Liner thickness (m)	Leachate head (m)			
		10·0	*1·0*	*0·1*	*0·01*
Clay-only (eqns (1) and (2))	1·0	864	86·4	8·64	0·86
Geomembrane-only (eqns (4) and (4))	0·002	2280	720	227	72
Geomembrane/clay composite (eqns (5) and (4))	1·002	8·83	1·11	0·14	0·02

The leakage through a single hole in a geomembrane-only liner can be calculated (using eqn (3) (Bonaparte *et al.*, 1989).

$$Q_G = C_B a \sqrt{2gh} \tag{3}$$

where: Q_G is the leakage rate through a hole in the geomembrane; C_B is a dimensionless coefficient dependent on the shape of the orifice, assumed herein to be 0·6; a is the area of the hole (m²); g is the acceleration due to gravity (m/s²); and h is as defined above.

The unitized leakage rate for a hole in a geomembrane-only liner is given by eqn (4):

$$q = nQ_G \tag{4}$$

where: Q_G is as defined above; and n is the number of holes per 1000 m².

Specific assumptions have been utilized in the derivation of eqn (3) (Bonaparte *et al.*, 1989). In general, for the purposes of this comparison, a frequency of 10 holes per hectare has been assumed, each having a diameter equal to the thickness of the geomembrane, herein taken as 2 mm. The media above and below the geomembrane have been assumed to have very high permeabilities, in any case several orders of magnitude higher than that of the geomembrane. The quantities of calculated leakage using eqn (3) are presented in Table 1, after conversion to the unitized rates using eqn (4).

Lastly, for the case of a geomembrane/clay composite liner with very good geomembrane/clay contact (Bonaparte *et al.*, 1989), the leakage through a hole in the geomembrane above an intact clay liner is expressed by eqn (5). The unitized leakage rates for this case are also given by eqn (4).

$$Q_G = 0 \cdot 21 a^{0 \cdot 1} h^{0 \cdot 9} k_s^{0 \cdot 74} \tag{5}$$

where Q_G, a and h are defined as above, and k_s is the hydraulic
conductivity of the clay (m/s).

These values, illustrated in Table 1, reveal the benefits of a
composite liner with regard to the quantity of leakage that can migrate
through the liner. Leakage rates can be reduced by one to two orders
of magnitude by the use of a composite liner. However, these numbers
are not absolute, and in fact eqn (5) is inaccurate for the case when
$h > L$. The parameters have been selected to make the point that
composite liners perform in a superior manner to either of the
clay-only or geomembrane-only options. Many factors affect the
leakage of a liquid through a liner. Different geometries, configurations and cross-sections will give different results, but with the same
relative conclusions. The hydraulic conductivity of the materials above
and below these liners has been assumed to be very large, at least
relative to the hydraulic conductivities of the liner materials. It is also
assumed that no lateral gradients exist; an unlikely situation in a
landfill.

The assumptions that have been made for these calculations are, in
a landfill, simplifications of the types of conditions that could exist in a
double lined system. The consideration of the cross-sectional design of
the lining system is affected by the quantity of leakage that can be
tolerated. Clearly, the 10 m leachate head option should not apply for
a landfill, but rather would be illustrative of a containment structure
holding liquid, such as a pond. Difficulties and constraints peculiar to
that configuration exist, discussion of which is not within the scope of
this Chapter. The three cases of leachate heads of 1 m, 0·1 m and
0·01 m are all quite plausible within a landfill, depending on the
design and in particular the nature of the leachate collection layer
above the liner. The transmissivity of this layer and the lateral gradient
both affect the leachate head. If the recharge capacity of the leachate
(the source of the leachate) exceeds the ability of this layer to remove
the leachate, whether on a transient or steady-state basis, significant
heads of 1 m or greater are certainly possible. On the other hand, if a
highly transmissive medium is utilized, then this drainage layer could
be extremely efficient in removing the leachate.

In addition, the numbers represented in Table 1 are not truly
representative of a lining system, since the media above and below do

have finite hydraulic conductivities and will be resistive to migration. However, every configuration cannot be modelled herein. These calculations simply show the relative leakages through the three types of liners. The principal advantage of a double lining system is that leakage through the top liner can be intercepted by the leakage collection system between the two liners (see Chapter 1.2). The driving forces causing leakage, including leachate head, can be easily controlled within this layer. In that case, leakage through the primary liner can be tolerated because it has still been contained. Once leakage through a single liner exists containment has been lost.

The point of this comparison is to examine these relative leakage levels through isolated liners and examine how the factors causing leakage and affecting the quantity of leakage can be controlled in a given system. This discussion of quantity of leakage leads into a discussion of leakage detection.

LEAKAGE DETECTION

Detection and Measurement of Leakage Within the Barrier System

The objective of containment can be preserved if any leakage that occurs does not exit the system. This necessitates the definition of where the system ends. In some cases, it can be clinically defined as the bottom of the lining system itself. On the other hand, if materials underlying the installed lining system have similar characteristics (e.g. if the bottommost component of the system is a compacted clay, which is underlain by extensive deposits of in-situ clay), then containment may be defined as extending beyond the system itself. In such a case, however, this additional containment is not controlled, and may in fact be 'flawed' by desiccation cracks, some other secondary structure, or zones of sandy or silty materials having much higher hydraulic conductivity which could afford a preferential drainage path. In any event, containment must be assumed to have been violated if any leakage occurs from the artificial liner.

Leakage can be detected in a variety of ways in the case of a single liner, regardless of its components. The problem is that it cannot be detected and measured within the barrier system. Observation or monitoring wells can detect leakage originating with the containment system, but by the time such occurs, it is too late to preserve the containment. Measurement and quantification of the leakage is extremely desirable since it provides a performance measure of the lining system. As previously shown, it is possible to calculate the expected leakage through a liner given a reasonable set of assumptions, but it is also necessary to verify the validity of those assumptions in practice. The provision of a double lining system offers the ability to entrap the leakage through the primary liner and, depending on the configuration of the leakage detection system, measure the quantity of that leakage. This is simply not possible in a single-lined system.

Rapid Leak Detection

Regardless of the quantity of leakage through the primary liner of a double lined system, there exists a time factor which is critical to the designer's reaction to a leak. In the case of landfills, once operations have commenced, it becomes extremely impractical to attempt to remedy any leaks that are detected. Exhumation of the waste is really not worthy of consideration as a feasible alternative. On the other hand, if for any reason an intolerably large leak is detected (usually the result of a major tear or other damage during construction which, for whatever reason, escaped the scrutiny of construction quality assurance), that is of such severity that operations cannot continue, then exhumation may be required. The potential effects of such an activity can be mitigated by designing a system that will minimize the time before a leak can be detected.

Several considerations directly contribute to rapid leak detection. These include:

- *Gradient within the system.* There must be some slope to direct the leakage to a sump where it can be detected and retrieved. Geometry usually restricts this gradient to 1 or 2%, but 5% would be much more desirable. Higher gradients are only feasible when the flow path is relatively short, which is not always attainable.
- *High transmissivity and velocity.* The drainage medium within the leakage detection system must have high transmissivity and high velocity. Geonets afford both of these characteristics necessary for rapid leak detection. A layer of sand can have identical transmissivity but cannot provide the velocity required, at least not before it is saturated. Tens of thousands of litres of leakage may be required to saturate the sand layer and potentially months or years could pass before there was any indication that the leak existed.

- *Size of the cell.* Small landfill cells with separate sumps contribute to rapid leak detection since the flow path is minimized. Large cells, which may have a flow path of 250 m from one end to another would require an elevation drop of 2·5 m for only a 1% gradient within the leakage detection system. Breaking up an area into several smaller cells may have cost implications, but can also have cost benefits, since precipitation falling on empty cells need not be treated as leachate.
- *Monitoring.* Periodic monitoring, and preferably continuous monitoring of flow into the leakage detection sump can indicate the incidence of a leak. In fact, plotting of precipitation versus time on a parallel chart to leakage detection versus time can provide quantification of the time it takes for a leak to be detected, from the time the rain falls.

Rapid leak detection really refers to the transit time of the leakage within the leakage collection system. The time taken for a drop of water to get from above the lining system to the leakage collection system is another factor altogether. Clearly, the transit times through different liner types will be significantly different. A 1 m thick clay-only liner, with a hydraulic conductivity of $1·0 \times 10^{-9}$ m/s, would exhibit a theoretical transit time of approximately 30 years (assuming that the clay was initially saturated, or a steady-state condition existed). The transit time for a 2-mm-thick geomembrane with a hole would effectively be a few seconds. Given that one of the purposes of the system is to detect whether or not there is a hole or flaw present (assuming that if one was known to exist, it would have been repaired before operations), and given that it is almost certain that saturated conditions would not exist in the clay, the time to breakthrough of a leak would probably be even longer. In fact, it is conceivable in the case of a landfill that the cell could be closed before the leak would be detected.

However, rapid detection of leakage does not take precedence over the containment objective. The leakage through a composite liner has been demonstrated to be considerably less than that for a geomembrane-only liner. Similarly, the leachate collection system continually removes leachate from above the primary liner, so that the leachate head continually decreases with occasional recharge due to rain, based on the assumption that the system is properly sized. Hence, if for the transient duration of a given maximum head, the leakage under that head is minimized, then a greater proportion of the leachate will be removed through the leachate collection system and will not even penetrate the lining system, let alone exit it. Clearly, the smaller the quantity of leachate entering the lining system, the smaller the head build-up on the secondary liner, and the easier it will be to preserve containment and remove the leakage through the leakage sump.

There are trade-offs which must be customized into a particular design. It may be preferable to select a geomembrane-only primary liner and handle a greater quantity of leakage or select a composite primary liner (with its additional costs but lesser leakage), depending on: (1) the total quantities of precipitation; (2) the forms of precipitation; (3) the frequency and duration of heavy precipitation; (4) the gradients which can be built into the lining system; (5) the availability of cohesive soils; and (6) the cost factors associated with construction and monitoring. Many designers, faced with this decision, will select the composite primary liner and its consequent one or more orders of magnitude lower detected leakage, since regulators also must be satisfied, and lower leakage levels under control are superior to greater leakage levels under control.

CONCLUSIONS

The preservation of containment is the primary political issue to be considered in the design of a landfill lining system. Factors such as proper design, quality of construction, and due attention to maintenance being equal, a single lined system cannot provide the level of confidence that may be required, which can be provided by a double lined system.

One of the advantages of a double lined system is that there exists a second line of defence and leachate can penetrate the lining system without violating the containment. Single lined systems are, of necessity, somewhat more unforgiving of a leak, since once the leachate penetrates the liner, the containment will ultimately be compromised. Utilization of composite liners reduces the leakage rates by up to two orders of magnitude, for a given set of assumptions.

The variables affecting leakage through the primary liner and the variables affecting flow within the leakage detection system can be controlled.

Leachate head tends to be the main driving force causing leakage through the liner. The magnitude of that head can be controlled to a degree by the design of the leachate collection system.

REFERENCES

Bonaparte, R., Giroud, J. P. & Gross, B. A. (1989). Rates of leakage through landfill liners. In *Geosynthetics '89 Conference*. San Diego, California, February 1989, pp. 18–29.

Giroud, J. P. & Bonaparte, R. (1989*a*). Leakage through liners constructed with geomembranes—Part I. Geomembrane liners. In *Geotextiles and Geomembranes*, **8**, 27–67.

Giroud, J. P. & Bonaparte, R. (1989*b*). Leakage through liners constructed with geomembranes—Part II. Composite liners. *Geotextiles and Geomembranes*, **8**, 71–111.

Lambe, T. W. (1958). Compacted clay: Engineering behavior. *Journal of the Soil Mechanics and Foundations Division, ASCE*, Proceedings Paper 1655, **84** (SM4).

Mitchell, J. K., Hooper, D. R. & Campanella, R. G. (1965). Permeability of compacted clay. *Journal of the Soil Mechanics and Foundations Division, ASCE*, **91** (SM4) 41–65.

Wallace, R. B. (1988). Construction quality assurance of lining system installations. *Geotechnical News*, **6**(1), 16–19.

2.5 Concept of a Double Mineral Base Liner

JEAN-FRANK WAGNER

*Department for Applied Geology, Karlsruhe University,
Kaiserstrasse 12, D-7500 Karlsruhe 1, Germany*

INTRODUCTION

The general objective of all base liners is to achieve the highest possible retention of landfill leachate, and the noxious substances contained therein. This is why a sealing system has to meet the requirements of the lowest possible permeability and the highest possible sorption capacity.

The base liners known, i.e. conventional sealing materials such as plastic foil and mineral liner, both show a very low permeability. For a mineral liner of a swelling clay mineral mixture (e.g. bentonite) there is, in addition to the very low permeability, also a very high sorption capacity. For the above-mentioned materials, however, the stability required over historical periods so as to ensure safe operation of the landfill in the case of a chemical attack cannot be definitely proven (Finsterwalder, 1989).

For a mineral base liner of bentonite or bentonite-like clay mixtures, permeability slowly increases as a result of the sorption processes. This has been proven in practice through clay mineral changes appearing after several years (Echle *et al.*, 1988), and can also be predicted by theoretical consideration. Based on crystalchemical and thermodynamic considerations, Lippmann (1979), Lahann and Roberson (1980), May *et al.* (1986) and Grauer (1988) came to realize that illite and smectite represent completely unstable phases. The increase in permeability of bentonite sealing materials that had been treated with organic substances was proved by the experiments of several authors (e.g. Anderson, 1982; Fernandez & Quigley, 1985; Alther, 1987; Hasenpatt, 1988; Hasenpatt *et al.*, 1988; further references in Madsen & Mitchell, 1988 and Wienberg, 1990) (see also Chapters 3.1 and 3.5). The interaction between inorganic pollutants and smectitic clays can lead to cracking (Wagner, 1988) and thus influence the permeability of the clays. The installation of several separate heavy-metal ions inside the intermediate layer of the smectites or the simple increase of the electrolyte concentration can lead to the loss of the swelling ability and the plasticity as well as to significant volume changes.

Finally, long-term stability of the smectite, and thus the mineral bentonite liner, do not seem to be achieved because of the ever-proceeding interaction with the leachate components. This is why mineral sealing materials of more stable clay mineral mixtures (e.g. kaolinitic clays) are used increasingly. These, however, only show minimal sorption capacities, and so only extract a minimum of pollutants from the leachate.

This led to the consideration of developing a sealing system which on the one hand guarantees the highest possible removal of noxious substances from leachate of domestic and hazardous landfills. On the other hand, it has to ensure the long-term stability of a minimum permeability towards leachate. In the following sections, a new concept of a 'double mineral base liner' (DMBL) is presented, where two separate mineral sealing layers with two specifically different functional mechanisms are integrated. The installation of the DMBL allows a very high filter efficiency for pollutants as well as a low and stable permeability towards water.

FUNCTION OF THE DOUBLE MINERAL BASE LINER (DMBL)

An upper, highly reactive, 'active' layer, which is oriented towards the waste should extract as much pollutant as possible from the leachate. These noxious substances will then be fixed either on the surfaces inside the crystal or as new mineral phases. The reactions within the active layer will, however, lead to a complete reorganization of this mineral sealing layer, combined with a change of the soil-mechanical characteristics, i.e. most of all an increase in permeability. A constantly low permeability of the sealing system will be ensured with the installation of a second layer, the so-called 'inactive' layer, which is

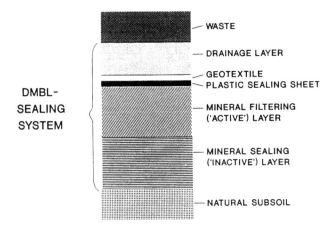

Figure 1. Schematic of a vertical section through a double mineral base liner (DMBL).

not reacting with the pollutants. The installation of an inactive layer underneath the sorption layer allows the leachate to be held back as long as possible and thus maximizes the reaction time between active layer and pollutants. The lower sealing layer produces a constant, very low percolation rate of the pollutants through the sealing system, whereas the upper filtering layer operates like some kind of 'sewage treatment plant' for landfill leachate (Fig. 1).

The solute transport in the inactive layer can be kept very small, given an adequate granular grading, an appropriate compaction and a low hydraulic gradient. Wagner and Böhler (1989) have shown that for very fine-porous media such as mineral base liners, the advective solute transport—transport by means of the carrier material water—at a low hydraulic gradient with a hydraulic conductivity (k) between 10^{-10} and 10^{-11} m/s, is almost the same as the diffusive solute transport—transport by means of a concentration difference. Figure 2 shows when, for a pollutant (in this case lead), a purely advective, purely diffusive, and convective–dispersive transport, depending on the hydraulic conductivity k, can be expected. Supposing a realistic hydraulic gradient $i = 1·5$ for a mineral base liner with k smaller than 5×10^{-11} m/s and a thickness of 1 m, the dominant transport mechanism is diffusion.

The diffusion in a porous medium is given through an effective

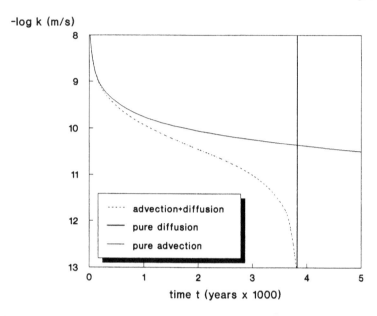

Figure 2. Migration time (for $C/C_0 = 0.5$) as a function of hydraulic conductivity k for the advective and diffuse transport of lead in a well-sorbing, mineral base liner (hydraulic gradient $i = 1.5$; effective diffusion coefficient $D = 9 \times 10^{-8}$ cm^2/s; porosity $n = 0.5$; thickness $l = 1.0$ m).

diffusion coefficient D, which is related to the diffusion coefficient in free solution D_0 as follows:

$$D = \tau D_0 \tag{1}$$

where the tortuosity factor τ represents a geometric characteristic of the porous media and is independent of the examined substance. τ describes the diffusion obstacles by the various forms and sizes of the diffusion paths within the pore system. An appropriate compaction allows a further reduction of τ and thus a reduction of the effective diffusion coefficient D and of the diffusive transport of pollutants.

When installing the lower sealing layer, the k values only have to be lowered until the advective solute transport has become slightly smaller or equal to the diffusive solute transport. Expensive complementary measures producing extremely low k values are superfluous, as the migration of pollutants, which is then exclusively controlled by diffusion, will not be affected thereby.

It is possible that the advective solute transport can even be negligibly small at higher k values, if for low hydraulic gradients a prelinear range and a stagnation gradient exist. As for the question of whether Darcy's Law is still valid for fine porous sediments, several authors have doubts (see references in Wagner & Egloffstein, 1990). A group of authors postulate a non-linear flow below a certain hydraulic gradient; another group even assumes a stagnation gradient area. Until today, the flow behaviour in fine porous sediments has not been definitely clarified. Theoretically, given the 'diffuse double layer' theory, a stagnation gradient is thinkable. However, the measurement of a range without hydraulic flow can also be faked through experimental uncertainties. The appearance of a stagnation zone would be an advantage as regards DMBL, although the diffusive transport of pollutants cannot be excluded thereby.

COMPOSITION OF THE DMBL

The DMBL is composed of two separate mineral layers; one assuming the sorption function, the other one ensuring a long-term low permeability (Fig. 1).

The upper, active layer is composed of minerals having a very large specific surface and thus a high sorption capacity and/or very reactive mineral phases, which rapidly form stable compounds with the noxious substances or act as reaction-accelerating catalysers. Suitable are materials like natural clays, enriched and upgraded with the aforementioned minerals. A primary high content of smectite (bentonite) or carbonate can be an advantage.

The lower, sealing layer consists of 'inactive', chemically very stable sealing materials. It is composed of a very fine-porous material which shows a very low permeability. Suitable materials are in the first place natural kaolinitic clays or clays that have been upgraded with kaoline, if possible without a smectite or carbonate component. If in the inactive layer the use of smectitic material cannot be avoided, one has to take care that calcium-smectite is present, as this is not as 'reactive', i.e. it will not be exchanged as easily as the sodium alternative. An upgrading with quartz sand or water glass may even improve the clays. Attention must be paid to the fact that these materials of the lower layer in any case represent chemically very stable mineral phases of low solubility.

The thickness of the active layer should be designed so that most of the pollutants can react with the existing sorption material and thus can be extracted from the leachate. This means that the reaction potential of the filtering layer should be higher than or equal to the elution potential of the landfill. The thickness of the inactive layer has to be chosen so that the percolation rate is slightly lower than the diffusive transport rate.

The installation of a draining layer (see Fig. 1) is recommended, as it drains off a great quantity of leachate from the landfill body. If drainage exists, most of the noxious substances will be disposed of in this way. The main function of the DMBL is in this case to assure a complementary security if the drainage should fail or be stopped after a certain operation period, i.e. when the landfill will be kept to itself.

The installation of a plastic liner as shown in Fig. 1 is not imperative but corresponds to the 'TA Abfall' (German Technical Guideline Waste; Bundesminister für Umwelt, Naturschutz und Reaktorsicherheit, 1991), (see also Chapter 2.2).

If the mineral sealing layer is installed as DMBL, the hydraulic conductivity $k \leq 10^{-10}$ m/s has to be observed for the whole DMBL system, i.e. as the permeability is controlled by the inactive sealing layer alone, its respective k value has to be calculated correspondingly.

Moreover, according to the 'TA Abfall', the composition and nature of the clay minerals of the mineral sealing material have to be coordinated to the necessary adsorptive capacity of the particular case. As concerns the DMBL this would mean that the adsorptive capacity necessary for the active layer should be the highest possible and for the inactive layer the lowest possible.

The thickness of the DMBL has to be determined according to the aforementioned criteria; in practice, however, it should not exceed the minimum thickness required by the 'TA Abfall'. The thickness of the filtering layer depends first of all on the landfill's volume of pollutants, but after the first rough estimates it will only be some 10 cm to a maximum of 1 m thick, so that the DMBL will be built with almost equal thicknesses of filtering and sealing layers.

The complementary installation of further mineral sealing layers or the complementary installation of foils underneath or above the DMBL layers can further improve the function of the sealing systems; economically, however, it hardly seems to be justifiable.

SUMMARY AND CONCLUSIONS

The installation of the DMBL, i.e. two mineral layers with different function principles, ensures a very high filtering efficiency for pollutants and a low and stable permeability towards the leachate.

The philosophy of the DMBL concept is derived from the observations of natural circumstances, notably the embedding of an ore deposit in a rock. In this case the ore body reacts with the surrounding rock under a gradual decrease in concentration and will thus be preserved over millions of years. A similar principle should be applied to the deposit of noxious substances. An enclosure in a foil can actually be impermeable for years or decades; a sudden fail, however, may be compared to a time bomb. This is why the landfill body should be installed so that it can react with the surrounding rock or with the installed mineral liner in such a way that allows the landfill body to cut itself hermetically off from the hydrosphere and the biosphere.

So far, the installation of a single mineral sealing layer corresponds to the state of the art. Previously bentonites were mostly used, as they process a high sorption capacity as well as a low water permeability. The permeability, however, can increase as a result of sorption processes (see above). Later on, to counteract this, more stable clays with a lower sorption capacity have been installed as monolayers. However, the essential functions of a base liner, i.e. highest possible elimination of pollutants and long-term stability, cannot be ensured by one single layer, as one excludes the other.

Other works (e.g. Hasenpatt, 1988) propose a sealing layer composed of mixtures of stable illitic and kaolinitic clays and unstable bentonite. However, a sealing layer that has been mixed in this manner will, given the chemical reactions, fail in the same way as a pure bentonite layer (i.e. permeability will increase). A multi-layer arrangement of these mixed layers will also lead to a permeability increase, because cracking, as a result of the sorption processes, will not be compensated by the stable clay minerals added. This is why, in any case, attention must be paid to the fact that a separate installation of the differently-acting mineral sealing layers should occur in the sequence described above.

The inversed installation, i.e. sealing layer above and filtering layer below, also leads to a completely different function principle. In this case the pollutants will pass only very slowly through the sealing layer, but they will not be fixed (simply a certain concentration decrease occurs as a result of hydrodynamic dispersion). In the beginning, the sorption layer situated below also shows a filter efficiency. But as a result of the interaction of the pollutants with the sorption layer (e.g. cracking), the latter rapidly becomes permeable and thus loses its retention capacity for the following pollutants. This means that the pollutants leaving the inactive layer arrive unfiltered at the soil. As regards the DMBL concept (filtering layer above), a cracking in the active layer does not have any negative effects on the sealing system, as the inactive sealing layer lying below makes the solution of pollutants remain within the cracks for a very long time. The pollutants will react with the material of the crack surface and diffuse from the cracks into the active layer and be fixed here.

Both functions (sorption capacity and stability) will only be achieved by an appropriate combination (sorption layer above, chemically-stable sealing layer below) of two mineral layers with a different function principle (see DMBL).

REFERENCES

Alther, G. R. (1987). The qualification of bentonite as a soil sealant. *Engineering Geology*, **23**, 177–91.

Anderson, D. (1982). Does landfill leachate make clay liners more permeable? *Civil Engineering, ASCE*, **9**, 66–9.

Bundesminister für Umwelt, Naturschutz and Reaktorsicherheit (1991). Storage, chemical/physical and biological treatment and incineration of wastes requiring special supervision. *Technical Guideline Waste Part 1*, Bonn, pp. 114 (in German).

Echle, W., Cevrim, M. & Düllman, H. (1988). Clay mineralogical, chemical and physical changes in a basal clay liner test area of the Geldern-Pont waste deposit. *Clays in Environmental Technology*, ed. K. A. Czurda & J.-F. Wagner. *Schriftenreihe Angewandte Geologie Karlsruhe*, **4**, Karlsruhe, pp. 99–121 (in German).

Fernandez, F. & Quigley, R. M. (1985). Hydraulic conductivity of natural clays permeated with simple liquid hydrocarbons. *Can. Geotech. J.*, **22**, 205–14.

Finsterwalder, K. (1989). Processes of solute transport in encapsulations of contaminated land—Influence on the design of liners. *Reconnaissance and Remediation of Contaminated Land*, ed. H. L. Jenberger. Balkema, Rotterdam, pp. 99–106 (in German).

Grauer, R. (1988). *Chemical Behaviour of montmorillonite in repository backfill material. NAGRA Technischer Bericht* 88-24, Baden (in German).

Hasenpatt, R. (1988). Soil mechanical changes of pure clays by the adsorption of chemical compounds. *Mitt. Institut f. Grundbau und Bodenmechanik ETH Zürich*, **134** (in German).

Hasenpatt, R., Degen, W. & Kahr, G. (1988). Flow and diffusion in clays. *Mitt. Institut f. Grundbau und Bodenmechanik ETH Zürich*, **133**, 65–76 (in German).

Lahann, R. W. & Roberson, H. E. (1980). Dissolution of silica from montmorillonite: effect of solution chemistry. *Geochim. Cosmochim. Acta*, **44**, 1937–43.

Lippmann, F. (1979). Stability relations of clay minerals. *N. Jb. Miner. Abh.*, **136**, 287–309 (in German).

Madsen, F. & Mitchell, J. K. (1988). Chemical effects on clay fabric and hydraulic conductivity. In *The Landfill, Reactor and Final Storage*, ed. P. Baccini. Springer, Berlin, pp. 201–51.

May, H. M., Kinniburgh, D. G., Helmke, P. A. & Jackson, M. L. (1986). Aqueous dissolution, solubilities and thermodynamic stabilities of common aluminosilicate clay minerals: Kaolinite and smectites. *Geochim. Cosmochim. Acta*, **50**, 1667–77.

Wagner, J.-F. (1988). Migration of lead and zinc in different clay rocks. Int. Symp. Hydrogeology and safety of radioactive and industrial hazardous waste disposal, IAH, Orléans, Doc. B.R.G.M. No. 160: Orléans, 617–28.

Wagner, J.-F. & Böhler, U. (1989). Suitability of tertiary clays from the border of the Rhine Valley (Wiesloch and Eisenberg basin) as barrier rock for a landfill. *Ber. 7 Nat. Tag. Ing. Geol.*, 257–64 (in German).

Wagner, J.-F. & Egloffstein, Th. (1990). Advective and/or diffusive transport of heavy metals in clay liners. In Proc. 6th Int. Congr. Int. Assoc. Eng. Geol. 6–10 August 1990, Vol. 2; ed. D. G. Price. Balkema, Rotterdam, pp. 1483–90.

Wienberg, R. (1990). On the influence of organic contaminants on landfill clays. *Abfallwirtschafts Journal*, **2**, 222–30.

2.6 Design Options for Hydraulic Control of Leachate Diffusion

R. KERRY ROWE

Geotechnical Research Centre, Department of Civil Engineering, The University of Western Ontario, London, Ontario, Canada N6A 5B9

INTRODUCTION

It is becoming increasingly more common to install a layer of highly permeable granular material beneath the primary liner in a landfill. However, this layer may be used in two quite different ways.

Most commonly, the layer is intended to allow detection and collection of leachate which escapes through the primary liner (see Fig. 1). In this application, there will be an outward hydraulic gradient across the primary liner and the design involves minimizing the hydraulic head on the secondary (either natural or manmade) 'liner' by collecting as much leachate as possible from the secondary leachate collection system.

A second approach to using the granular layer is to maintain a hydraulic gradient across the primary liner and *into* the landfill; thereby creating an engineered 'hydraulic trap' in which the egress of leachate is inhibited by the inward flux of water from the hydraulic control layer, as shown in Fig. 2.

The objective of this chapter is to discuss some of the technical advantages and disadvantages of these two approaches.

ILLUSTRATIVE CASE

For the purposes of quantitatively illustrating a number of points, consideration will be given to the design of a hypothetical landfill with

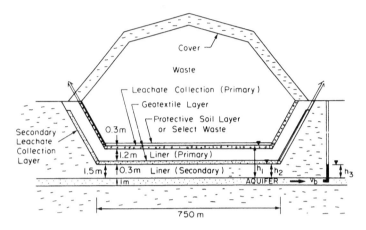

Figure 1. Schematic showing a primary liner underlain by a leak detection secondary leachate collection system. Advective flow is downward through the primary liner.

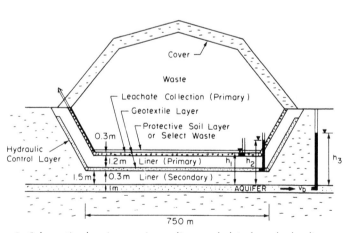

Figure 2. Schematic showing a primary liner underlain by a hydraulic control layer. The landfill is designed as a hydraulic trap with advective flow into the landfill.

an average waste thickness of 15 m which is constructed above a natural sand aquifer as illustrated in Figs 1 and 2. For simplicity of illustration, it is assumed that the design consists of (from the waste down) a 0·3-m-thick primary leachate collection system, a 1·2-m-thick compacted clay liner, a 0·3-m-thick granular layer (for secondary leachate collection or hydraulic control) and a 1·5-m-thick secondary (natural) clay liner which is underlain by a 1-m-thick granular aquifer. The landfill is assumed to be 750 m long in the direction of groundwater flow and it is assumed that the primary piping and slope on the leachate collection system is out of the plane being considered (i.e. the cross-section being examined is the critical cross-section). It is noted that some slope from the left to right of the cross-section is assumed; however, this detail is not shown on the schematics.

Consideration will be given to the migration of chloride assuming an initial source value of 1500 mg/litre and that the mass of chloride represents 0·2% of the total dry mass of the waste. The infiltration through the landfill cover is assumed to be 0·15 m/annum. The diffusion coefficient and effective porosity for chloride through the compacted clay liner are assumed to be 0·019 m^2/annum and 0·35 respectively. The corresponding values for the secondary clay liner are 0·015 and 0·25 m^2/annum respectively.

Consideration is also given to the migration of dichloromethane assuming an initial source concentration of 1500 μg/litre and a sorption parameter in the primary liner of $\rho K_d \approx 2$. As a first approximation, the diffusion coefficient for dichloromethane is taken to be the same as that of chloride and the mass of contaminant is assumed to be proportional to the initial concentration.

The hydraulic conductivity of the compacted liner is assumed to be 3×10^{-8} cm/s. Two values of hydraulic conductivity of the natural liner are considered, namely 10^{-7} cm/s and 10^{-8} cm/s.

Unless otherwise noted, the Darcy velocity in the aquifer is assumed to be 1 m/annum (i.e. a gradient of 0·003 and hydraulic conductivity of 10^{-3} cm/s in the aquifer).

The leachate mound in the hydraulic control layer is assumed to be 0·3 m above the top of the compacted clay liner (i.e. $h_1 = 3\cdot3$ m, measuring head relative to the top of the aquifer). The head in the secondary leachate collection/hydraulic control layer (h_2) and in the aquifer (h_3) will vary depending on the hydrogeologic conditions being considered.

All the analyses reported herein were performed using a finite layer contaminant transport model (Rowe & Booker, 1985, 1987, 1990a) as implemented in computer program POLLUTE v5 (Rowe & Booker, 1990b).

Potentiometric Surface in Aquifer Below Secondary Collection System

For situations where the water table and potentiometric surface in the underlying aquifer is well below the base and base of the landfill (e.g. see Fig. 1), the construction of a permeable drainage system, which is located beneath the compacted clay liner, serves two purposes. Firstly, the drainage layer functions as a secondary leachate collection system which can remove a portion of the leachate that escapes through the liner (and some escape is to be expected through any liner system where there are downward gradients). Secondly, this layer serves to reduce the hydraulic gradient through the underlying soil.

A key question in the design of these liner systems is 'what will be the impact of the contaminant which will migrate through the natural soil?' The answer to this question will depend on the properties of the natural soil, the drainage system and the liner, the geometry of the landfill, the design of the primary and secondary leachate collection system, and the properties of any underlying aquifer. Each case must be considered as a unique situation; however, it is important to recognize that some contaminant migration through the natural soil must be anticipated for designs such as that shown in Fig. 1. Even in the limiting case where no leachate mounding occurs in the secondary leachate collection system and all the leachate that escapes through the liner is collected, there will still be diffusion of contaminants into the natural soil from the secondary leachate collection layer. In most cases, there will also be downward advective contaminant transport since not all the leachate escaping through the liner is likely to be collected. Indeed, situations can readily occur where the majority of the leachate migrating through the primary liner also migrates down through the natural soil.

Ideally, the separation between the aquifer and the landfill would be as large as possible for a design involving a secondary leachate collection system such as the one shown in Fig. 1. However, in many practical situations the actual thickness may be quite thin. Under these circumstances, it cannot be assumed that even a perfectly operating liner, and primary and secondary leachate collection system will necessarily prevent contamination of groundwater in an underlying aquifer. To illustrate this, consider case 1 where it is assumed that the top of the secondary liner (which may be a natural clay) has been chosen to correspond to the potentiometric surface beneath the aquifer which, for this case, is assumed to correspond to the head of $h_3 = 1 \cdot 5$ m at the downgradient toe of the landfill. This ensures an adequate factor of safety against 'blowout' of the secondary liner and, if there is no mounding of leachate in the secondary collection layer (i.e. $h_2 = h_3 = 1 \cdot 5$ m), creates a situation where there is no inward or outward flow through the secondary liner. Thus for this case there will be downward advective transport through the primary liner ($K = 3 \times 10^{-8}$ cm/s, $i = 1 \cdot 25$) corresponding to a Darcy velocity of $0 \cdot 012$ m/annum. For this scenario, all leachate should be collected and contaminant transport through the secondary liner is by the process of molecular diffusion.

The results of analyses performed for case 1 are summarized in Table 1. Although this case represents perfect secondary leachate collection (i.e. no advective escape through the secondary liner), it is evident that the process of molecular diffusion through the secondary liner results in significant impact in the aquifer.

The impacts evident for case 1 would be unacceptable. One means of reducing impact would be to place a geomembrane above the compacted clay liner to create a composite primary liner. The geomembrane may be expected to reduce the advective flow through the liner. Based on Giroud and Bonaparte (1989), for the situation examined here (i.e. operating a primary leachate collection system), the escape of leachate by advective transport is likely to be of the order of $0 \cdot 1$ mm/annum (or less) for a well-constructed liner system (for so long as the liner system remains intact). Under these conditions, the primary transport mechanism is likely to be diffusion. Very few data have been published concerning the diffusion of contaminant through geomembranes. Diffusion coefficients (which would appear to represent the diffusive flux nD for a unit concentration gradient) of $2-4 \times 10^{-8}$ cm^2/s have been reported by Lord *et al.* (1988) and Hughes and Monteleone (1987). Recent unpublished research at the University of Western Ontario suggests that the diffusion coefficient range of $2-4 \times 10^{-8}$ cm^2/s is reasonable for dichloromethane but may be quite conservative (i.e. high) for chloride. Additional studies to better define this parameter are currently in progress. Based on a value of nD of 1×10^{-4} m^2/annum (i.e. about 3×10^{-8} cm^2/s) and an

TABLE 1. Summary of Cases Examined

Case no.	Hydraulic conductivity of secondary liner (cm/s)	Heads			Darcy velocity[c]			Peak impact in aquifer			
								Chloride		Dichloromethane	
		h_1 (m)	h_2 (m)	h_3 (m)	v_{a_1} (m/annum)	v_{a_2} (m/annum)	v_b (m/annum)	Concentration (mg/litre)	Time (annum)	Concentration (µg/litre)	Time (annum)
1	10^{-7} or 10^{-8}	3·3	1·5	1·5	0·012	0	1	>240	160	>70	550
2[a]	10^{-7} or 10^{-8}	3·3	1·5	1·5	0·000 1[d]	0	1	125	230	30	880
3[b]	10^{-7} or 10^{-8}	3·3	1·5	1·5	0·000 1[d]	0	1	10	310	<2	1 150
4	10^{-8}	3·3	1·5	3·3	0·012	-0·004	1	70	140	20	500
5	10^{-7}	3·3	3·95	4·2	-0·005	-0·005	1	15	160	<5	580
6	10^{-8}	3·3	3·49	4·2	-0·0015	-0·0015	1	85	190	20	670
7	10^{-8}	3·3	4·2	4·2	-0·007	0	1	40	170	10	630
8	10^{-8}	3·3	5·2	4·2	-0·015	0·002	1	10	170	3	580

Note: All concentrations greater than five have been rounded to the nearest multiple of five and all times rounded to the nearest decade.

[a] nD in secondary leachate collection system 10^{-3} m²/annum.
[b] nD in secondary leachate collection system 10^{-5} m²/annum.
[c] v_{a_1}, Vertical Darcy 'velocity' (flux) through the primary liner (positive down).
v_{a_2}, Vertical Darcy 'velocity' (flux) through the secondary liner (positive down);
v_b, Horizontal Darcy velocity (flux) in the aquifer.
[d] Composite primary liner involving a 2-mm-thick HDPE geomembrane and 1·2-m clay.

advective flow of 0·0001 m/annum in the primary composite liner, the impact was calculated for two cases (cases 2 and 3).

Case 2 assumes that the granular layer between the primary and secondary liner is a sand, and that a significant portion of the sand contains water held by capillarity. For this case, even though the advective transport is very small, there is significant potential for contaminant transport through the secondary leachate collection systems by diffusion. Assuming an effective value of nD of 0·001 m²/annum in the partially saturated sand, the contaminant impact on the aquifer is reduced by a little less than 50% for chloride and a little over 50% for dichloromethane. The concentrations of dichloromethane are still quite large and would be unacceptable based on Ontario's Reasonable Use Policy (MoE, 1986).

Relatively little research has been conducted concerning diffusion through a humid, moist granular layer such as the secondary leachate collection system. It may be anticipated that if an open granular stone was used (rather than sand) then the potential diffusion through water trapped in soil pores would be reduced. At present, the extent to which diffusion would be reduced has not been clearly established and more research is required. To illustrate the potential effect this could have, case 3 assumes that the diffusion through the secondary leachate collection system is reduced by two orders of magnitude from that in case 2. As is apparent from Table 1, this results in a substantial reduction in impact on the aquifer. It is apparent that the major role for the unsaturated secondary leachate collection system in this case is to act as a barrier to diffusion and that any significant failure of the geomembrane (which is restricting advective flow through the liner) would result in significant impact on the aquifer even if all the leachate was collected by the secondary leachate collection system.

From the foregoing, it is evident that diffusion is a major consideration in the design of systems such as that shown in Fig. 1; this is particularly true if a geomembrane liner is used to minimize advective transport through the liner system. Considerable research has been conducted into the diffusion of contaminant in clayey barriers (e.g. see Rowe *et al.*, 1988; Barone *et al.*, 1989). However, much more research is required into the diffusion of contaminants through geomembranes and through unsaturated granular (or geosynthetic) secondary leachate collection systems since this is likely to control impact for systems such as that shown in Fig. 1. Such research is now in progress at the University of Western Ontario.

Other Considerations

The examples considered in the previous section assumed that the potentiometric surface in the aquifer coincided with the top of the secondary liner and hence there was no advective component of flow in the secondary liner (i.e. $v_{a_2} = 0$ for cases 1–3 in Table 1). In many cases it will not be practical to select the base contours of the landfill such that this condition is satisfied.

In some cases, the potentiometric surface in the aquifer will be below the base of the secondary leachate collection system (i.e. $h_2 > h_3$) and there will be some downward advective transport of contaminant which enters the secondary leachate collection system. Advective–diffusive transport through the soil beneath the secondary leachate collection system is an important consideration in the design of these facilities. If the hydraulic conductivity of the secondary liner is of the order of 10^{-7} cm/s, then a substantial proportion of the leachate may escape through the secondary liner rather than be collected by the secondary leachate collection system. This should be considered in the design of, and assessment of impact for, these facilities.

In some cases the potentiometric surface in the aquifer will be above the base of the secondary leachate collection system. Under these circumstances, it may be necessary to reduce heads in the aquifer during construction to ensure an adequate factor of safety against blowout. By pumping, one could ensure that the potentiometric surface is maintained at, or below, the base of the secondary leachate collection system. Potentially, this would involve pumping for hundreds of years. Alternatively, the potentiometric surface could be allowed to recover after construction. This would give rise to inward gradient into the secondary leachate collection system. The disadvantage of this is that the volume of fluid collected by the secondary collection system would not provide a good indication of the volume escaping through the primary liner since it would be difficult to distinguish the different components of flow to the layer. The advantage of inward flow to the secondary leachate collection system is that it would resist outward diffusion of contaminant. To illustrate the potential effect, case 4 is essentially the same as case 1 (i.e. no geomembrane) except that the potentiometric surface in the aquifer is assumed to correspond to the design level of leachate mounding in the landfill (i.e. $h_3 = h_1 = 3.3$ m, $h_2 = 1.5$ m). Assuming all other parameters are the same, the calculated impacts given in Table 1 are reduced by more than a factor of three compared to case 1, even for a relatively small inflow of 4 mm/annum (i.e. assuming a hydraulic conductivity of the secondary liner of 10^{-8} cm/s). Ironically, a higher hydraulic conductivity of the secondary liner would result in larger inflows and hence even smaller contaminant impact. However, it is important to recognize that the inflow is controlled by both the hydraulic conductivity of the secondary liner and the hydraulic capacity of the hydrogeologic system to provide water. For example, if the hydraulic conductivity of the secondary liner in the system examined here were 10^{-8} cm/s or higher, it may not be possible for the head h_3 in the aquifer to recover to the original value of $h_3 = 3\cdot3$ m unless the aquifer is highly permeable or there is an adequate supply of water from a second deeper aquifer.

HYDRAULIC CONTROL

An alternative to using the granular layer beneath the primary liner as a secondary leachate collection system is to use it as a hydraulic control layer. For example, suppose that the potentiometric surface in the aquifer shown in Fig. 2 is such that $h_3 > h_2 > h_1$. In this case, there is both a natural hydraulic trap (i.e. water flows from the natural soil into the hydraulic control layer) and an engineered hydraulic trap (i.e. water flows from the hydraulic control layer into the landfill). Where practical, this design has the following advantages. Firstly, since there is inward flow to the hydraulic control layer and a relatively impermeable clay liner, it may be possible to design the system such that the engineered hydraulic trap is entirely passive. That is, all water required to maintain an inward gradient is provided by the natural hydrogeologic system and no injection of water to the hydraulic control layer is required. Secondly, because of the two-level hydraulic trap, there will be substantially greater attenuation of any contaminants that do migrate through the primary liner. Thirdly, since fluid can be injected and withdrawn from the hydraulic control layer, it is possible to control the concentration of contaminant in the layer (and hence the impact at the boundary) in the event of a major failure of either the liner or primary leachate collection system.

There are three factors that must be considered in the design of this engineered hydraulic trap. Firstly, the head in the hydraulic control layer must be controlled such that 'blowout' of either liner does not occur during or after construction. Secondly, the volumes of water collected by the 'hydraulic trap' must be manageable and the hydrogeologic system must have the capacity to provide the water required to maintain the hydraulic trap (if this is not the case, then the head in the aquifer will drop and the effectiveness of the trap may deteriorate with time). Thirdly, although there is a hydraulic trap, some outward diffusion of contaminants is to be expected in most cases. Contaminant migration analyses are required to assess what (if any) impact may occur under these conditions. If the impact at the site boundary is not acceptable then it can be reduced by pumping water through the hydraulic control layer (i.e. injecting fresh water at one end and extracting contaminated water at the other end). The volume of fluid to be pumped can be assessed by appropriate modelling. Models are available (e.g. Rowe & Booker, 1988, 1990b) which readily allow the designer to estimate potential impact as a function of the flow in the hydraulic control layer.

To illustrate the effect of a hydraulic control layer, cases 5, 6 and 7 each examine the case where the total head, h_1, on the landfill liner is 3·3 m (i.e. 0·3 m of leachate mounding on the liner; see Fig. 2) and the total head in the aquifer is 4·2 m. This induces an inward gradient across the liner system. In each case, the primary liner is assumed to have a hydraulic conductivity of 3×10^{-8} cm/s. In cases 5 and 6, the secondary liner is assumed to have a hydraulic conductivity of 10^{-7} and 10^{-8} cm/s respectively and the hydraulic control layer is assumed to be operating as a natural hydraulic trap (i.e. no human introduction or removal of water from the hydraulic control layer). Under these circumstances the head, h_2, in the hydraulic control layer is established by the hydraulic system and depends on the relative hydraulic conductivity of the primary and secondary liner.

As might be expected, the flows in the system with the higher permeability secondary liner (case 5) are larger than for the lower permeability secondary liner (case 6) and hence the resistance to outward flow is also greater. The greater the inward flow, the greater the resistance to outward diffusion and, consequently, the impact for case 5 is less than for case 6. It is interesting to note that the impact for case 5 with a clay primary liner system and hydraulic control is similar to that for the system with a very efficient secondary leachate collection system and a composite (geomembrane/clay) primary liner.

Cases 7 and 8 examine the behaviour of an engineered hydraulic trap where water is introduced to increase the head in the hydraulic control layer to 4·2 m and 5·2 m respectively. In case 7, there is the maximum gradient across the primary liner without creating an outward gradient across the secondary liner. This reduces impact compared to the corresponding passive case (case 6) with a 10^{-8} cm/s secondary liner. Case 8 relies more heavily on the induced pressure in the hydraulic control layer resisting outward movement of contaminant through the primary liner at the cost of causing an outward advective gradient through the secondary liner.

DISCUSSION AND CONCLUSIONS

The role of advection for allowing the migration of contaminants from waste disposal facilities is well recognized. Both the conceptual design shown in Figs 1 and 2 minimize outward advective flow through the secondary liner and of the cases examined, only one (case 8) involves any outward advective movement through the secondary liner. However, there is still significant calculated contaminant impact on the aquifer for most of the cases examined. Based on regulations in the Province of Ontario (MoE, 1986), only three of the scenarios considered are even close to being acceptable (viz. cases 3, 5 and 8). In all cases, diffusion of contaminant is a major transport process.

The results of this preliminary study, which is based on available data, suggest that there can be significant diffusion of contaminant through a HDPE geomembrane. In fact, the major potential barrier to diffusive transport for the system shown in Fig. 1 is the secondary leachate collection system. However, very little is known about the diffusion of contaminant through an unsaturated layer in a humid environment typical of that anticipated when this layer is located between two liners beneath a landfill. More research is required to determine relevant parameters; however, it is evident that diffusive transport through geomembranes and secondary leachate collection systems is an important consideration in the design of these facilities.

Diffusion is a slow process. Inspection of Table 1 shows that for the cases considered, the time prior to peak impact being reached in the underlying aquifer ranges from around 140 to 310 years for a conservative species such as chloride and from 500 to 1150 years for

an organic species such as dichloromethane which experience retardation by the soil. For this 15-m-thick landfill, the contaminant lifespan
(i.e. the period of time during which there could be unacceptable
impact if the engineering features did not function as designed) is in
excess of 100 years. Thus careful consideration must be given to the
amount of human intervention and the length of time the engineered
systems are likely to last. Of the cases considered, case 5 which
involves a hydraulic control layer and passive hydraulic trap requires
the least intervention. The implications of contaminating lifespan and
the design of hydraulic traps are discussed in more detail by Rowe
(1991*a,b*).

The results presented herein suggest that it may be possible to
design a landfill which, under operating conditions, will have negligible impact on groundwater quality. This design could involve the use
of a granular layer which provides a 'barrier' by being kept unsaturated
or by being maintained at a hydraulic head greater than that on the
base of the landfill. Preference for one system or another will depend
on the hydrogeologic conditions and, in particular, on the potentiometric surface in any underlying aquifer. The processes involved in
contaminant migration for hydraulic control systems are better understood than the processes of migration for systems that use the
granular layer as a secondary leachate collection system.

ACKNOWLEDGEMENTS

The preparation of this chapter was funded by the award of a Steacie
Fellowship and Grant A1007 by the Natural Sciences and Engineering
Research Council of Canada. The computations were performed on
an IBM R6000 workstation which forms part of a donation to the
Environmental Hazards Programme, at The University of Western
Ontario, by IBM Canada.

REFERENCES

Barone, F. S., Yanful, E. K., Quigley, R. M. & Rowe, R. K. (1989). Effect of
   multiple contaminant migration on diffusion and adsorption of some

domestic waste contaminants in a natural clayey soil. *Canadian Geotechnical Journal*, **26**(2), 189–98.

Giroud, J.-P. & Bonaparte, R. (1989). Leakage through liners constructed with geomembranes: Part I—Geomembrane liners; Part II—Composite liners. *Geotextiles and Geomembranes*, **8**, 27–68, 71–112.

Hughes, J. W. & Monteleone, M. J. (1987). Geomembrane/synthesized leachate compatibility testing. In *Geotechnical and Geohydrological Aspects of Waste Management*, ed. D. J. A. Van Zyl, S. R. Abt, J. D. Nelson & T. A. Shepherd. Lewis, Michigen, pp. 35–50.

Lord, A. E., Koerner, R. M. & Swan, J. R. (1988). Chemical mass transport measurement to determine flexible membrane liner lifetime. *Geotechnical Testing Journal, ASTM*, **11**(2), 83–91.

MoE (Ministry of the Environment) (1986). Ontario Policy—Incorporation of the reasonable use concept into MoE groundwater management activities. Policy 15-08-01, May 1986.

Rowe, R. K. (1991*a*). Contaminant impact assessment and the contaminating lifespan of landfills. *Canadian Journal of Civil Engineering*, **18**, 244–53.

Rowe, R. K. (1991*b*). Some considerations in the design of barrier systems. *Proceedings. First Canadian Conference on Environmental Geotechnics*, Montreal, 15–16 May; ed. R. P. Chapuis & M. Aubertin, pp. 157–64.

Rowe, R. K. Booker, J. R. (1985). 1-D pollutant migration in soils of finite depth. *Journal of Geotechnical Engineering, ASCE*, **111** (GT4), 479–99.

Rowe, R. K. & Booker, J. R. (1987). An efficient analysis of pollutant migration through soil. In *Numerical Methods for Transient and Coupled Systems*, ed. R. W. Lewis, E. Hinton, P. Bettess & B. A. Schrefler. John Wiley & Sons, New York, pp. 13–42.

Rowe, R. K. & Booker, J. R. (1988). MIGRATE—Analysis of 2D pollutant migration in a non-homogeneous soil system: Users manual, Rep. No. GEOP-1-88. Geotechnical Research Centre, University of Western Ontario, London, Canada.

Rowe, R. K. & Booker, J. R. (1990*a*). Contaminant migration through fractured till into an underlying aquifer. *Canadian Geotechnical Journal*, **27**, 484–95.

Rowe, R. K. & Booker, J. R. (1990*b*). POLLUTE v.5—1D pollutant migration through a non-homogeneous soil: User's manual. Rep. No. GEOP-1-90. Geotechnical Research Centre, University of Western Ontario, London, Canada.

Rowe, R. K., Caers, C. J. & Barone, F. (1988). Laboratory determination of diffusion and distribution coefficients of contaminants using undisturbed soil. *Canadian Geotechnical Journal*, **25**, 108–118.

2.7 Hydrological Studies on the Effectiveness of Different Multilayered Landfill Caps

STEFAN MELCHIOR & GÜNTER MIEHLICH

Institute for Soil Science, University of Hamburg, Allende-Platz 2, D-20146 Hamburg, Germany

INTRODUCTION

Landfill caps have to prevent infiltration of rainfall into the disposed waste, enable controlled gas collection, and often must be suitable for recultivation. Furthermore they must be long-term effective and resistant against loads and stresses such as subsidence, erosion, biopenetration, frost, desiccation and clogging of the drainage system. Therefore landfill caps are usually designed as multilayered systems with a combination of topsoil, drainage layers and one or more barrier layers.

Three types of barrier layers are commonly used and can be combined: flexible membrane liners (usually high-density-polyethylene, HDPE), compacted soil liners and/or capillary barriers. Most of the information available on barrier systems has been derived from laboratory experiments that deal with the percolation of liquids through compacted soil liners (Grube *et al.*, 1987) and the permeation of substances through synthetic liners. Differences between laboratory data and much higher hydraulic conductivities of soil liners observed in the field have been reported (Daniel, 1984).

But even though a number of caps have been planned and constructed during recent years, only a few studies deal with the hydrological performance of different caps in the field (Andersen *et al.*, 1988; Cartwright *et al.*, 1987; Hoeks *et al.*, 1987; Warner *et al.*, 1987; Wohnlich, 1987). Almost nothing is known on the long-term effectiveness of flexible membrane liners and cover systems in general.

The purpose of the study described here is to measure the water balances of different systems on a technical scale in the field and to collect data on the physics of water movement within barrier layers. Long-term monitoring of the different systems and the application of numerical simulation models on the water and solute transport through those systems are planned. The data shall be used to improve our understanding and to predict the efficiency of planned systems, to enable an improved control of the factors which determine their efficiency, and to optimize the design of covering systems in general.

APPROACH

The Water Balance of Landfill Covers

The water balance of a landfill cover is shown in Fig. 1. Precipitation (P) is the major input flux into the cover. Part of the precipitation evaporates directly from the above-ground parts of the vegetation

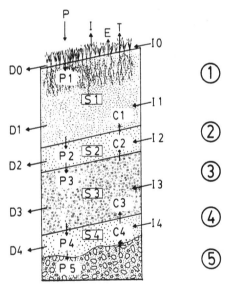

Figure 1. The water balance of a landfill cover. 1, Topsoil; 2, drainage layer; 3, compacted soil liner; 4, gas drainage layer; 5, waste; P, precipitation; I, interception; E, evaporation; T, transpiration; Pn, percolation to layer n; Cn, capillary rise into layer n; Sn, water storage in layer n; In, lateral inflow in layer n and on the surface (I0); Dn, lateral drainage of layer n and surface run-off (D0).

(interception, I) and sometimes there can be surface run-off (D0); the rest infiltrates into the topsoil (P1) and increases the water storage in that layer (S1). The water content in the topsoil decreases due to evaporation from the bare soil surface (E) as well as transpiration of the vegetation (T) and percolation to the drainage layer (P2). The temporal change of soil water content in the layers below the topsoil (dSn, n indicating the number of the individual layers) is given by the balance of percolation and capillary rise in and out of the layer (Pn – Pn + 1, respectively Cn – Cn – 1). In addition, if looking at a segment of a slope, one always has to take into account the balance of lateral inflow and outflow within the different layers and on the soil surface (Dn and In). The percolation into the disposed waste (P5) can therefore be described by the complete water balance equation (I, E and T are summarized in the term evapotranspiration, ET):

$$P5 = P - ET - \sum Dn + \sum In + C4 - \sum dSn \qquad (1)$$

The basic concept for the determination of the water balance of these systems is the use of clearly-defined test fields, so called lysimeters. Within a lysimeter, precipitation as well as soil water content and all the discharges across the lysimeter boundaries are measured, while uncontrolled inflow into the test field has to be prevented. If all the other parameters are measured, evapotranspiration, which cannot be measured directly, can be determined as the only parameter remaining in eqn (1).

The construction of a multilayered cover on the Georgswerder landfill in Hamburg (Germany), part of the remedial action on that site, gave the opportunity to integrate six test fields to determine the water balance of the Georgswerder cover and several other cover designs in full scale, on-site.

Basic Theoretical and Technical Considerations

The design of the test fields is based on the following considerations:

- *Which factors control the efficiency of a cover?* Under humid climatic conditions like in Central Europe there are several possibilities to optimize a landfill cover:

  (a) one can try to maximize evapotranspiration by establishing a water-consumptive type of vegetation and by storing a maximum of the water-surplus in winter times in a thick layer of topsoil with a large storage capacity;

(b) one can try to drain off laterally as much water as possible by controlling the slope, thickness and hydraulic conductivity of the drainage layers;

(c) one can optimize the quality of the barrier layers by optimizing the properties of the materials used, the layer design, construction technology and quality assurance.

Under the climatic conditions mentioned, however, there will always be times with a positive balance between percolation and capillary rise across the lower boundary of the root zone, especially if macropore flow occurs within the topsoil. Therefore concept (a) may be used to support the efficiency of a cover, but cannot substitute the construction of a barrier system. Under normal conditions, the quality of the barrier layers is the crucial factor, followed by the quality of the drainage system. Therefore these two factors should be tested. In case of malfunctions of cover components, the probability of detecting them (quality assurance during construction and long-term monitoring) as well as the redundancy of the system (the existence of components that can compensate for the malfunction of others) are important criteria for any kind of risk assessment, but not factors to be tested in a field study.

- *What kinds of barrier systems are available and why should they be tested?*

(a) Combined systems with flexible membrane liners on top of compacted soil layers are used very often on landfills and should be tested. They are expected to show no leakage. The long-term stability of the flexible membrane is unknown. Rather sophisticated technology and quality assurance as well as suitable weather conditions are needed to construct these systems.

(b) Compacted soil liners are often used on sites with a relatively low toxic potential. This is not the only reason why they are worth being tested. They can also be considered as a long-term state of systems like (a), if the flexible membrane liner should have lost its effect. Compacted soil liners are sometimes suspected to fail due to macropore flow through desiccation cracks or other leaks, not detected during construction. Concerning quality assurance and weather conditions, the same applies as to (a) above.

(c) Capillary barriers consist of two highly-permeable layers. The boundary between a fine-grained sand layer (the so-called capillary layer) on top of a coarse-grained gravel layer (capillary block) represents the actual barrier. Water which infiltrates into the capillary layer increases the hydraulic conductivity of that layer and is kept from percolating into the relatively drier capillary block by the power of surface tension that is working at the pore-discontinuity between the water-filled pores of the capillary layer and the relatively larger, air-filled pores of the capillary block. Under slope conditions the water then drains off laterally within the capillary layer (so-called 'wick effect') (see also Chapter 2.5). Capillary barriers perform well unless more water is infiltrating into the capillary layer than can be drained off laterally within that layer. Situations such as those after heavy rainfall or the melting of snow, for example, lead to saturation above the layer boundary and to the development of a hydraulic head that exceeds the power of surface tension and therefore results in a break-through of water across the barrier. If situations with intensive infiltration into the capillary layer can be avoided, a capillary barrier should be able to work as good as systems like (a). The stability of the layer boundary against subsidence is unknown. Only theoretical estimations are available on design criteria like suitability of materials, thickness of layers and maximum drainage rates and drainage length.

- *What kind of methodological problems are expected?*

(a) The most important water flow to be determined is the leakage through the barrier systems. This flow is of a very low intensity compared to the other flows within a cover. It must be measured directly and cannot be computed by eqn (1) because the probable error in computing evapotranspiration with meteorological data is high. Furthermore, leakage through barrier systems may show a high spacial variability if preferential flowpaths occur such as macropores in a compacted soil liner or leaks in a flexible membrane liner. Therefore, the areas tested must be large enough to detect the leakage due to random or systematically distributed inhomogeneity in a representative way.

(b) The barriers tested must be constructed with representative construction technology (compaction equipment, quality control, etc.).

(c) Any kind of material cutting through the liners to define the lateral boundaries of the lysimeters can produce unacceptable preferential flowpaths, so called 'lysimeter boundary effects'. The same may apply to the installation of instruments.

All these considerations have led to a comparative testing concept with different variants and with a measuring concept that allows the plausibility of the data to be checked. The design of the barrier systems and the factor slope are varied in the different test fields in such a way that the influence of these factors on the water balance can be determined separately. The influence of the other factors mentioned above shall be evaluated by the use of a numerical simulation model.

Design of the Test Fields

Each test field is 10 m wide and 50 m long in slope direction (Fig. 2). The lateral discharges in the different layers are collected at the lower end of the slope. Gravel-filled underdrains consisting of welded HDPE liners are installed below the barrier systems. They enable the collection and direct measurement of the barrier system leakage rates.

To avoid boundary effects, a new boundary design was developed in cooperation with the engineering office IGB (Hamburg, Germany). A combination of HDPE liners and seals of compacted soil above the barrier systems define the test field boundary and prevent lateral infiltration of water into the test field. Figure 3 shows a cross-section through the boundary between two test fields and the chronology of its construction. First, the gravel-filled underdrains of each test field are built. Then, the compacted soil liner is constructed in three lifts over the whole area. To reach the third stage, an alternating construction process is necessary. First, a lift of soil is compacted between the test fields, followed by the installation of the drainage layer on the test fields (the test field on the right is equipped with an additional flexible membrane liner). Then, alternating, a second and third lift of compacted soil is installed between the fields and, accordingly, the first and second lift of topsoil on the fields. After the third stage, parts of the now existing seal of three lifts of compacted soil between the fields are excavated and replaced by a HDPE liner which will be

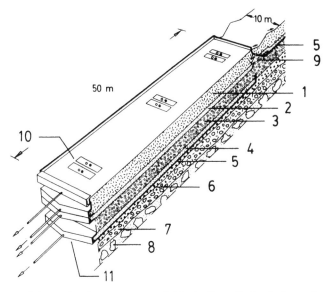

Figure 2. Schematic view on a test field. 1, Topsoil; 2, drainage layer; 3, compacted soil liner; 4, underdrain; 5, flexible membrane liner (HDPE, welded); 6, gas drainage layer; 7, former cover; 8, waste; 9, seal (compacted soil); 10, soil hydrological measuring unit; 11, gutters to collect the discharges.

covered by drainage layer and topsoil. In the end, the test field boundary is defined by the edges of HDPE liner that reach out of the topsoil. Rainfall that infiltrates outside the test fields is drained off and cannot move into the fields. Water, moving within the fields is kept there because of the HDPE liner and the remaining loam seal. In addition, the HDPE liners of the underdrains below the barrier systems overlap the test field boundaries by 40 cm to be sure to catch the total amount of water that percolates through the tested barrier system. Due to such a design, it was not necessary to cut through the barrier systems. Furthermore, it was possible to construct the compacted soil liners of the test fields continuously over all fields with the same equipment, technique and quality control like on the whole landfill cover. Preferential flow along instruments, which are installed in the liners, is prevented by flanges.

The test fields are located in two areas with different slopes (F-fields with 4% slopes, S-fields with 20% slopes, Fig. 4). All fields are designed as multilayer systems with a combination of topsoil, drainage

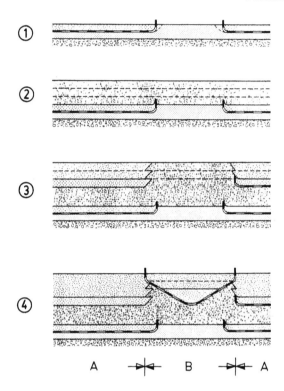

Figure 3. Cross-section through the test field boundary. 1, 2, 3, 4, Construction stages; A, area of test fields (10 m wide); B, area between two test fields (3 m wide).

and barrier layers. Due to the comparative testing concept, the topsoil and the drainage layer of all fields are designed alike (75 cm of sandy loam respectively, 25 cm of sand/gravel, separated by a geofabric to protect the drainage layer). The following barrier systems are tested on the different fields:

S1, F1:   A compacted soil liner, consisting of three lifts of glacial till, each 20 cm thick after compaction. The till has the following properties: it is composed of 17% clay, 26% silt, 52% sand and 5% gravel, carbon content 9·8% CO_3, and no organic matter; composition of the clay minerals: 50% illite, 30% smectite, 20% kaolinite and chlorite. Liquid

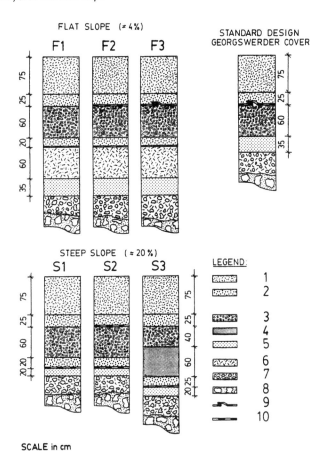

Figure 4. Layer design of the test fields. 1, Topsoil; 2, drainage layer above and underdrain below of barrier layers; 3, compacted soil liner; 4, capillary layer; 5, gas drainage layer; 6, protective layer; 7, former cover; 8, waste, 9, flexible membrane liner (HDPE, 1·5 mm, not welded, but installed overlapping in slope direction); 10, flexible membrane liner (HDPE, 1·5 mm, welded).

limit of the till is 20·4%, plasticity index 8·9, consistency index 0·8, bulk density 1·950 g/cm³ and water content 12·1% dry weight. Proctor density is 2·039 g/cm³ and optimum water content is 9·6%, the till having been compacted to >95% of the Proctor density on the wet side

of the optimum with a smooth roller under vibration. Total
pore volume of the liner is 27·0%, the degree of saturation
87%. The geometric mean of the saturated hydraulic
conductivity, measured in the laboratory during construction, was $2·4 \times 10^{-10}$ m/s. Due to its graded particle size
distribution, its low clay content and the dominance of
relatively inactive clay minerals, the potential for shrinkage
of the till is rather low compared to other clay liners.

S2, F2: A combined barrier system composed of a welded flexible
membrane liner (HDPE, 1·5 mm thick) on top of a
compacted soil liner (the soil liner has the same design as
on F1 and S1). The quality of the HDPE liners has been
thoroughly controlled during the construction. Therefore,
they are assumed to be impermeable for water, at least
during the first years of the study. During that time, these
fields serve as control plots concerning the percolation of
the barrier system. They allow the precise detection of pore
water discharge due to the consolidation of the soil liners
after the installation of the upper drainage layer and
topsoil.

F3:   Standard design of the Georgswerder cover. Unlike F2 and
S2, HDPE liners, however, are not welded but are
installed overlapping in the slope direction on top of a
compacted soil liner.

S3:   A compacted soil layer (40 cm thick), lying above a
capillary barrier consisting of the capillary layer (60 cm,
very well sorted fine-grained, pleistocene sand) on top of
the capillary block (25 cm, coarse sand/fine-grained
gravel). The particle size distribution of the capillary layer
is 1% clay and silt ($<63 \mu$m), 5% very fine sand (63–
100 μm), 85% fine sand (100–200 μm), 10% middle sand
(200–630 μm). The capillary block has 2% clay and silt
($<63 \mu$m), 1% fine sand (63–200 μm), 7% middle sand
(200–630 μm), 51% coarse sand (630–2000 μm), 35%
fine gravel (2–63 mm), and 4% middle gravel (63–
200 mm). In that extended capillary barrier design, the
upper soil liner provides a low intensive infiltration into the
capillary layer. Therefore, the capillary barrier can operate
under unsaturated conditions using the wick effect even
after heavy rainstorms or after the melting of snow.

Quality Assurance

Quality assurance begins with the selection of materials and technology, ends with the end of construction, and should be followed by some kind of long-term monitoring. To build the test fields, which is part of the long-term monitoring of the Georgswerder landfill, in a representative way, the same materials (except the capillary layer on S3), technology and quality control as for the construction of the whole landfill cover in Georgswerder have been used. The design of that cover and accordingly the design of the quality assurance program have been carried out by the engineering office IGB in Hamburg on behalf of the Hamburg Environmental Protection Agency. It cannot be presented here in detail. In general, the materials and the construction technology have been selected after detailed suitability tests. During construction, several parameters were tested in the laboratory (i.e. particle-size distribution, plasticity, saturated hydraulic conductivity, Proctor-test, carbon content). Water content and bulk density were tested in the field. The tests were performed by two parties; one on behalf of the construction company, the other on behalf of the Agency. The limits to be reached for the different parameters were derived from the suitability tests. The most important values, concerning the compacted soil liners, were a saturated hydraulic conductivity of less than 1×10^{-9} m/s, measured in the laboratory, a water content of 5% wetter than optimum, a wet bulk density measured on five points of a construction unit with four values higher than $2 \cdot 18$ g/cm^3, the fifth with at least $2 \cdot 15$ g/cm^3. The construction units were defined in advance as areas that had to be constructed continuously (average area around 5000 m^2).

In order to get a representative quality of the barrier systems to be tested on the test fields, no improvements in the quality assurance program concerning the barrier systems were made with the following exceptions:

- The quality of the HDPE liners on S2, F2 and F3 was controlled thoroughly with regard to perforations of the liners, because, even on 500 m^2, it is impossible to get a representative distribution of those leaks (i.e. only one single perforation has been found and repaired on those three fields, together covering 1500 m^2).
- Each test field group (around 2500 m^2 with the boundary areas) represents a construction unit with regard to the compacted soil liners. Nevertheless, water content and wet bulk density were

measured on five points per test field layer ($500\,m^2$), in order to be able to compare the data of the different fields.

All those parts of the test fields that serve to collect the discharges (e.g. gutters, measuring shafts, pipes between gutters and shafts, the features that protect those pipes against subsidence, gravel-filled HDPE-underdrains below the barrier systems) were controlled in excess with always at least one person on-site.

Measuring Program

The measuring program covers the following range:

- The discharges (surface run-off, lateral drainage in topsoil and drainage layer as well as leakage through the barrier system) are collected in gutters (see Fig. 2) and measured individually and with high temporal resolution within measuring shafts using a combination of water-level meters, pumps and paddle-wheel flowmeters for the high intensive flows and tipping-bucket gauges for the low intensive flows. To check the automatically recorded data, the total amount of flow is, in addition, registered manually every workday.
- In some fields, various soil hydrological measuring units with a total of 24 neutron probe tubes and 531 automatically-recording tensiometers are installed (e.g. in Fig. 1) to define moisture content and hydraulic potential in high temporal and spatial resolution.
- A weather station serves to automatically measure precipitation, air temperature, humidity, wind velocity, radiation balance and soil temperature to determine the water input by precipitation and all parameters that are necessary to compute evapotranspiration using the Penman/Monteith equation. In addition, five precipitation gauges are installed as pit gauges with splash protection and the gauge rim placed parallel to the inclined soil surface.

Furthermore, several physical, chemical and mechanical soil parameters are analyzed in the laboratory or determined by additional experiments in the field to get more information about the spacial variability of several parameters and processes (i.e. particle-size and pore-size distribution, saturated and unsaturated hydraulic conductivity, bulk density, Proctor-compaction-test, plasticity, shear strength, clay mineralogical composition, total element content, inorganic constituents of the discharges, tracer and infiltrometer experiments to determine preferential flowpaths and transit times).

RESULTS

The research and development program was started in 1986. It took one year to plan the test fields and another year to build them. The measurements to determine the water balances of the fields (precipitation, discharges and soil water contents) started at the end of 1987. The measurements will continue at least until the end of 1993, followed by a less intensive long-term monitoring program. The water balance data of the first three years of measurements (1988–1990) and the results concerning the efficiency of the different liner types are summarized here.

Water Balances of the Test Fields

In 1988, 1989 and 1990 the climatic water balance (precipitation – potential evapotranspiration calculated with the Haude equation) has been close to the long-term average in the Georgswerder area. A higher potential evapotranspiration only occurred during the summer of 1989. There were also significantly fewer days with snow and frost in all three winters.

Figure 5 shows the water balances of test fields S1 and F1. All parameters have been measured, except evapotranspiration, which was calculated solving eqn (1). The following phenomena can be described:

- Two parameters, evapotranspiration and drainage above the liner, account for more than 98% of the water input by precipitation under consideration of the change in soil water storage. The average precipitation has been 830 mm/a. Evapotranspiration takes an average of 65% on both fields in all three years, in the drainage layer an average of 33% of the water input is discharged.

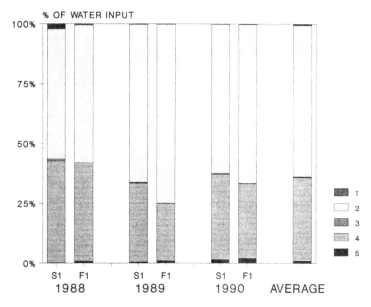

Figure 5. Water balances of test fields F1 and S1 as a percentage of the water input. 1, Surface run-off; 2, evapotranspiration including interception; 3, lateral drainage in topsoil; 4, discharge of drainage layer above liner; 5, discharge of underdrain below liner.

Evapotranspiration is a little higher on test field F1 (66% compared to 61% on S1) because due to the inclination the flat fields absorb more solar radiation while both fields are exposed in a northerly direction. On all six test fields average evapotranspiration is 65%, drainage 33%.

* There is little surface run-off under both slope conditions, though frequency and intensity are higher on the S-fields (0·8% of water input on the S-fields, 0·3% on the F-fields), but overall decreasing with time and growth of vegetation. Only very little erosion occurred during the first winter even though there was almost no vegetation on the fields during that time.
* Lateral flow within the topsoil only occurs on the S-fields. It only contributes 0·5% to the total balance.
* The widest range of flow intensities can be observed in the drainage layers. There is only little effect of slope concerning the total volume of flow (37% of the water input on S1 and 33% on

F1), but the flow intensity can be more than three times higher on the steep slope with a maximum flow rate of about 3 mm/h.

- The average liner leakage in both fields over the three years is 1% of the water input. It is a little higher on F1 (1·3%) than on S1 (0·7%). These figures, however, do not properly describe the efficiency of the liners, because the flow rates are not constant with time. The following chapter presents the results concerning the efficiency of the different liners and the processes that do occur.

Liner Leakage

Figures 6 to 9 show the amounts of water that have been collected below the different barrier systems on a weekly basis. The measurements started during construction, which ended on the steep fields in week No. 40 and on the flat fields in week No. 48 in 1987. The results are described separately for the different liner types.

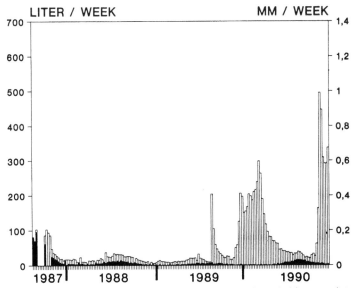

Figure 6. Water collected below the barrier systems of test fields S1 and S2 in liters/week and mm/week. White bars, test field S1 (compacted soil liner); black bars, test field S2 (composite liner: HDPE-liner on top of compacted soil liner).

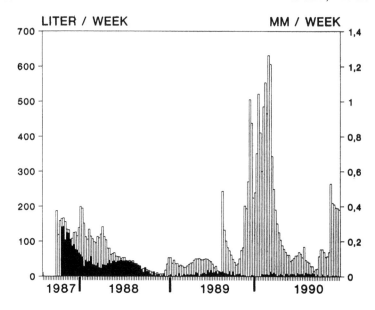

Figure 7. Water collected below the barrier systems of test fields F1 and F2 in liters/week and mm/week. White bars, test field F1 (compacted soil liner); black bars, test field F2 (composite liner: HDPE-liner on top of compacted soil liner).

Compacted soil liners (test fields F1, S1 and S3). In Figs 6 and 7, the flow rates on the fields S2 and F2 with a combination of HDPE and compacted soil liner (black bars) are compared with the flow rates of the respective fields S1 and F1 with only a compacted soil liner (white bars). During construction, the flow rates are similar and more or less constant underneath both liner types. The flow rates decrease after the end of construction. The decrease of flow is more significant below the composite liners of fields S2 and F2. This flow regime can be explained with pore water discharge out of the compacted soil liner due to consolidation during and after construction of drainage layer and topsoil. The black bars in Figs 6 and 7 therefore show the pore water flow out of the compacted soil liners under the impermeable HDPE-membranes while the white bars give the sum of pore water flow and leakage through the compacted soil liners of S1 and F1.

All the flow rates, including pore water discharge, show a low intensity in the beginning. The pore water discharge lasts for about a

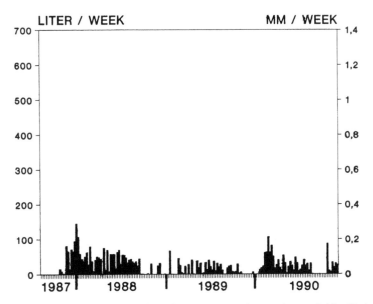

Figure 8. Water collected below the composite liner of test fields F3 in liters/week and mm/week (HDPE liner (not welded) on top of a compacted soil liner).

year on the flat fields. After that period the leakage increases at the end of 1988. A very sharp increase in the leakage of the compacted soil liners (F1, S1) occurred at the end of August 1989 after heavy rainfall. The same effect could be observed regarding the compacted soil layer on top of the capillary barrier in field S3 (Fig. 9, the lateral discharge in the capillary layer is identical to the leakage through the compacted soil layer). Before that event there had been no visible response of the discharges below the compacted soil liners to high flow rates in the drainage layer above the liners. This changed dramatically during the summer of 1989. From that time on, the discharges below the liners of F1, S1 and also S3 reacted within hours to changes in the drainage flow above the liners reaching high flow rates in winter time.

A tracer experiment carried out on field F1 (for details, see Melchior *et al.*, 1990) proved the formation of highly conductive macropores in the compacted soil liner during the summer of 1989. The soil hydrological data show a significant desiccation and strong,

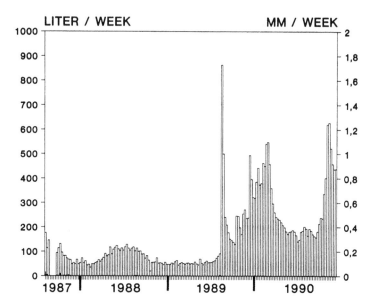

Figure 9. Performance of the extended capillary barrier design of test field S3. White bars, lateral drainage within the capillary layer; black bars, breakthrough of water across the layer boundary.

upward directed hydraulic gradients during the summer (see Melchior *et al.*, 1990).

Excavations of the soil liner outside the test fields, where it was protected by a HDPE membrane, showed in one-third of all inspected locations the presence of a soil structure with desiccation cracks at the surfaces of those liner lifts that were left unprotected against atmospheric influences for too long in times with high evaporation potential (high wind velocities and solar radiation). Furthermore, horizontal fissures between individual lifts were found. The compaction equipment (smooth vibrating rollers) was able to compact and mix material with a proper soil water content, but was unable to knead the dry surface of a lower lift and mix it with the fresh material of the next lift. These observations give way to the hypothesis that the desiccation cracks in the soil liners of the test fields F1, S1 and S3 were not completely formed in the summer of 1989 but where already initially present, they only had to be extended to become hydraulically effective.

The leakage rates through the compacted soil liners reach, with only short time lags, maximum values that stay constant over hours during periods with very high flow rates in the drainage layers on top of the liners. These maxima allow the calculation of the hydraulic conductivity k of the liners using the Darcy equation:

$$Q/A = i \times k \quad (\text{m}^3/\text{m}^2\text{s}) \tag{2}$$

where
  Q: volumetric flow rate (m^3/s),
  A: cross-sectional area of flow (m^2),
  i: hydraulic gradient dh/L: difference in pressure head (h) over the length of flow (L),
  k: saturated hydraulic conductivity (m/s); if conditions are unsaturated, k is a function of the soil water content.

Assuming the hydraulic gradient to be 1, the hydraulic conductivity of the soil liners are $2 \cdot 4 \times 10^{-9}$ m/s on F1, $2 \cdot 9 \times 10^{-9}$ m/s on S1 and $8 \cdot 9 \times 10^{-9}$ m/s on S3. This is much higher than German regulations for landfill liners usually allow. In case of S3 almost all the natural groundwater recharge (under central European climatic conditions) can potentially pass the compacted soil liner. The reasons for the actual liner leakage being much less than the potential leakage are evapotranspiration and good lateral drainage above the liner. For most of the time only limited amounts of water are available to percolate through the macropores.

Composite liners using a flexible membrane directly on top of a compacted soil liner (test fields F2, F3 and S2). The composite liners are much more effective. In winter when there are high discharges in the drainage layer above the liners no flow occurs below the liners. In summer, however, a few liters per week can be measured underneath the composite liner (Figs 6 and 7). Melchior *et al.* (1990) have shown that these discharges correspond with a change of temperature gradient in the compacted soil liner. In summer the temperature is higher at the surface of the soil liner than at the lower part. The thermal gradient for liquid and vaporous water movement is therefore, during that time, directed downward and produces lower matric potentials within the soil liners. This thermally induced desiccation is much less significant than the desiccation of the compacted soil liners of fields S1 and F1, where capillary rise of water leads to shrinkage of the liners. It is as yet unknown whether the balance of liquid and vaporous water movement over the year will cause shrinkage of the material of the mineral liners within the composite liner systems. The welded HDPE liners of fields F2 and S2 seem to be impermeable and prevent any capillary rise. Their long-term stability is not known. The discharges below the composite liner of test field F3 (Fig. 8), where the flexible membrane is not welded but is overlapping in the slope direction, are higher than on F2 and much lower than on F1. In addition to the discharges during the summer there are some events in winter that produce discharges. The total liner leakage over the year, including the water that moves out of the liner due to temperature gradients, however, is still very low (2·6 mm in 1990).

Extended capillary barrier (test field S3). The best results have been achieved with the extended capillary barrier system on S3 (Fig. 9). Nearly 100% of the water which infiltrates from the compacted soil layer above into the capillary layer moves laterally within the fine sand and does not pass the capillary barrier. Only during construction were some liters of water collected underneath the capillary barrier, that would have percolated into the disposed waste.

CONCLUSIONS

The results show that there are risks involved in relying on only one single barrier to prevent rainfall infiltration into landfills. The better strategy is to use a multibarrier concept to minimize the potential infiltration capacity into the waste. In the case of the compacted soil liners in test fields F1, S1 and S3, due to desiccation cracks, measurable leakage rates have been detected.

Evapotranspiration and lateral drainage play the key role in the water balance. Surface run-off does not, even on the steep slope, contribute much to the balance. Its optimization would only be possible by a complete sealing of the surface, which, for several reasons is not aimed at. Evapotranspiration alone cannot, under humid climatic conditions, prevent rainwater penetration into a landfill.

There are two possibilities to create lateral flow of water in combination with a liner system: in the traditional way, a permeable drainage layer is placed on top of a liner; secondly, the wick effect can be used to drain the water in the upper of two permeable layers (capillary barrier). In this case a liner is needed too. It does not have to be as efficient as in the first case and in contrast it is located above the drainage layer to limit the infiltration rate into the capillary layer.

Compacted cohesive soil liners (mineral liners) are very sensitive to shrinkage due to desiccation caused by a capillary rise of water into the topsoil during dry periods. In addition, they have the disadvantage that they are very difficult to construct because of the close control on the water content of the material that is needed and hard to achieve under always-changing weather conditions. A good deal of effort is necessary for testing material properties and construction technology before starting the construction and for quality assurance during construction. The investigated compacted soil liners failed within 2 years of their construction, even though the clay content was rather low, the clay mineralogical composition was not unusually sensitive to shrinkage and the construction technology and control were up-to-date.

Even if soil liners are protected against capillary rise of water by covering them with flexible membranes and if they are constructed in the best possible way, the results of the study show that a risk of desiccation remains due to thermally-induced liquid or vaporous water movement. Further work is necessary to identify and quantify the processes that produce the observed discharges below the composite liners of test fields F2 and S2 during the summer and to quantify the risk of shrinkage. The flexible membrane liners have seemed to work perfectly up to now, but their long-term stability is yet unknown. At the present state of knowledge it seems questionable whether or not the compacted soil liner would back up the plastic liner in the case of any future failure for whatever reason.

The extended capillary barrier design is very promising. Even though the compacted soil liner has lost a lot of its efficiency due to the desiccation cracks, the capillary layer is still able to drain off the infiltrating water totally.

Further investigations will show how the different systems perform on a long-term basis. On sites where suitable materials are available nearby, the extended capillary barrier might be the best and most cost-effective type of cap. Compared with compacted soil liners, a capillary barrier can be constructed more easily and under a broader range of weather conditions. It should be aimed at influencing the quality and the deposition of the waste in such a way that will avoid at least differential settling and will create a surface slope that will be suitable for capillary barriers. In 1991 an experiment was started to determine the maximum drainage length and the influence of slope, infiltration rate, subsidence and material properties on those systems in order to create design criteria for the practical use of capillary barriers.

ACKNOWLEDGEMENTS

This chapter is based on a study that is supported by the German Federal Ministry of Research and Technology (BMFT) and the City of Hamburg (FHH, Amt für Altlastensanierung der Umweltbehörde) under the Project No. 144035914. The authors are responsible for the content.

REFERENCES

Andersen, L. J., Clausen, E., Jakobsen, R. & Nilsson, B. (1988). Two-year water balance measurements of the capillary barrier test field at Bøtterup, Denmark. Preliminary results. *Proceedings of the UNESCO Workshop on the Impact of Waste Disposal on Groundwater and Surface Water*, ed. E. Rørdam & S. Vedby. National Agency of Environmental Protection, Denmark, 483 pp.

Cartwright, K., Larsen, T. H., Herzog, B. L., Johson, T. M., Albrecht, K. A., Moffet, D. L., Keefer, D. A. & Stohr, C. J. (1987). A study of trench covers to minimize infiltration at waste disposal sites. Final Report, US Nuclear Regulatory Commission, Washington, NUREG/CR-2478, Vol. 2.

Daniel, D. E. (1984). Predicting hydraulic conductivity of clay liners. *Journal of Geotechnical Engineering*, 110(2), 285–300.

Grube Jr., W. E., Roulier, M. H. & Herrmann, J. G. (1987). Implications of current soil liner permeability research results. *Proceedings of the 13th Annual Research Symposium on Land Disposal, Remedial Action, Incineration and Treatment of Hazardous Waste*, EPA Document No. EPA/600/9-87/015. US Environmental Protection Agency, Cincinnati, Ohio, pp. 9–25.

Hoeks, J., Ryhiner, A. H. & Van Dommelen, J. (1987). Feasibility study on landfill liners. Institute for Land and Water Management Research (ICW), Rep. No. 21. Wageningen, The Netherlands, 12 pp.

Melchior, S., Berger, K., Rook, R., Vielhaber, B. & Miehlich, G. (1990). Lysimeter and tracer studies on the efficiency of different landfill covers. *Zeitschrift der Deutschen Geslogischen Gesellschaft,* **141**, 339–47 (in German).

Warner, R. C., Wilson, J. E., Peters, N. & Grube Jr., W. E. (1984). Multiple soil layer hazardous waste landfill cover: design, construction, instrumentation and monitoring. *Proceedings of the 10th Annual Research Symposium on Land Disposal of Hazardous Waste,* Document No. EPA/600/9-84/007. US Environmental Protection Agency, Cincinnati, Ohio, pp. 211–21.

Wohnlich, S. (1987). The effect of remedial action on the water balance of landfills under special consideration of surface liners. Dissertation, University of Karlsruhe, Germany, 269 pp. (in German).

2.8 The 'Capillary Barrier' for Surface Capping

MICHEL BARRÈS & HUBERT BONIN

Bureau de Recherches Géologique et Minières, Department of Environment, BP 6009, 45060 Orleans Cedex, France

INTRODUCTION

Among the various techniques of confinement, the 'capillary barrier' seems to meet the basic criterion of durability. This technique, based on a two-layer system in which a layer of fine sand overlies a layer of gravel, has been successfully tested both on laboratory models and on site in small experimental trenches (Rançon, 1972). Full-scale experiments in the field are rare. However, an experimental site (15 × 100 m) at Botterup in Denmark should be mentioned where, since 1985, the real capacities of a capillary barrier have been studied by the Danish Geological Survey (Anderson & Christiansen, 1986). The materials used were specially washed and sieved, and the system is directly exposed to rain with no vegetation cover.

The study carried out by the Bureau de Recherches Géologiques et Minières (BRGM) at Chevilly is directed toward more applied objectives: to test a capillary barrier under the same conditions as a real waste disposal site respecting the management economics of such a site. The methods used to construct the site were therefore fairly simple, as far as concerns the materials used (raw quarry materials) and the excavation techniques (mechanical shovel).

THE 'CAPILLARY BARRIER' PRINCIPLE

This principle is based on the difference in grain size between two materials. In an unsaturated medium, the interface between a fine material and the coarser material functions as a 'capillary barrier' and water cannot pass from the fine medium to the underlying coarse medium due to the difference in capillarity, which is stronger than gravity.

Thus the water is in the first stage retained in the upper layer of fine sand. As long as the capillary attraction at the interface remains sufficiently strong, the hydraulic conductivity of the fine sand will be higher than that of the underlying gravel, thus limiting infiltration; if the fine sand is sufficiently well drained, the sand retains enough capillary to maintain this effect.

If the interface is horizontal, the infiltrated water will accumulate just above the interface without penetrating into the underlying gravel, until the force of gravity becomes greater than the forces of capillary pressure; under this condition the water will pass the interface and flow rapidly through the coarse material.

On the other hand, if the interface is an inclined plane, with a slope between 5° and 10°, less water will accumulate above the interface as it will flow laterally through the fine sand, and the penetration of water into the underlying layer will thus be limited. So simply draining the fine sand efficiently will give a watertight system made of natural materials. These 'capillary barrier' phenomena have been described during studies on laboratory models by Rançon (1972) and *in situ* in small experimental trenches by Rançon (1980). These studies show conclusively the efficiency of such a system.

THE EXPERIMENTAL SITE

Design and Construction of the System

The shape decided on for the capillary barrier was the dome. As already mentioned, it is essential that the flow rate allows water to drain laterally through the fine sand. This flow rate increases with the length of the system. However, the efficiency of the barrier over this length can be maintained by gradually increasing the hydraulic gradient, i.e. the slope of the barrier. In addition, the circular shape of the system provides a progressively larger drainage section as the size of the system is larger (which is not the case for a linear system). These two advantages can coexist only in a dome-shaped system.

The experimental system was built on the class 2 domestic waste

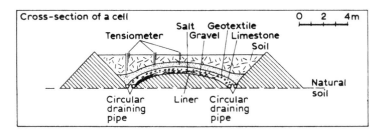

Figure 1. Experimental storage in a 'dry structure': cross-section of a cell.

disposal site at Chevilly (Loiret Department), about 10 km north of Orléans. The system comprises two containers inside which the waste is represented by a thin layer of soluble salts spread on an earth dome. Each dome is protected from water infiltration by means of a capillary barrier (Fig. 1).

The two containers, of slightly different dimensions, were placed side by side. The earth dome is 4 m in radius inside container 1 and 3·50 m in radius inside container 2.

At the bottom of the containers, the earth domes are covered by a watertight membrane (black BTP type plastic, 0·35 mm thick) upon which is spread a film of very soluble salts (0·35 cm of nitrates), simulating the waste to be protected beneath the barrier and acting as a tracer for any infiltrations. The capillary barrier placed upon this is made from ordinary natural materials having undergone no particular preparation. The lower layer, about 30 cm thick, is of rounded gravel (2–8 mm in diameter) coming directly from gravel pits. The entire perimeter of the gravel layer is drained by a tubular drain 50 mm in diameter.

The upper layer of the capillary barrier is made of untreated, uncompacted sand with a grain size not exceeding 0·25 mm. The barrier has a permeability of about 10^{-5} m/s.

The layer of sand is 30–40 cm thick and its perimeter is drained by a second drain that is situated parallel to, and outside, the first one. In order that the Fontainbleau sand does not penetrate into the underlying gravel layer, a Bidim U14 type geotextile was placed at the interface. The geotextile also made it considerably easier to put the fine sand in place.

Finally, the whole system was covered with a layer of soil (50 cm thick at the centre), which serves more than one purpose:

- it protects the fine sand from erosion,
- it provides a flat surface above the containers to avoid run-off.
- it acts as a buffer against sudden and heavy rain.

Equipment and Measurement Systems

The behaviour of each barrier is monitored by a network of tensi-ometers for measuring the water pressure in the fine sand. As the containers are exposed to natural precipitation, a rain gauge was placed between the two containers. In addition, a sprinkler enables simulation of exceptional precipitation and allows the study of the steady state behaviour of the capillary barrier.

RESULTS

Natural Rainfall Conditions

Monitoring under natural conditions took place over a period of one year, from July 1987 to July 1988, on a weekly basis. Data collection (flow rates and tensiometers) proved to be difficult.

The first recording phase (July–December 1987) is divided into two periods (see Fig. 2)

- The first phase (days 0–100), with little rain, includes long dry periods during which the efficiency of the barrier decreased substantially. however, the rain on the 55th day (26 mm) was entirely stopped by the barrier. The efficiency of the system therefore increases with the intensity of rainfall, which is contrary to the principle of the capillary barrier.
- the second period, with more frequent rain, shows a different behaviour of the system. The efficiency between two periods of rain is higher (between 90 and 100%). However, it falls to about 75% on days with over 10 mm rainfall.

The differing behaviour of the system is explained by the low degree of compaction of the fine sand layer, since it was emplaced without compaction. This resulted in a higher porosity and therefore reduced capillary force. This explains the reduced efficiency of the barrier after its installation.

The second period corresponds to times with heavier and more

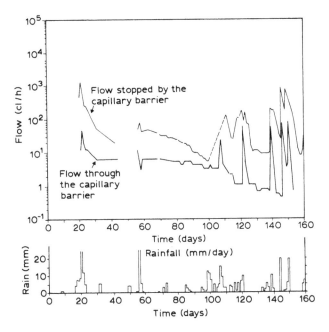

Figure 2. Monitoring under natural conditions (July–December 1987).

frequent rain, provoking natural compaction of the fine sand (the settling was observed on the site) and enabled the 'capillary barrier' effect to appear.

After emplacement of the fine sand it should be compacted to the optimum degree (to be determined in the laboratory), so that the capillary barrier operates from the time it is installed.

The data collected in June–July 1988 confirm the previous results. Except during periods of heavy rain, the efficiency of the capillary barrier is about 90%. It falls rapidly below 80% as soon as rainfall exceeds 10 mm/day. Nevertheless, it is difficult to assess efficiency as a function of daily rainfall. This is due to varying evapotranspiration rates and the hourly intensity of the rainfall. For this reason, in order to identify any relationships between efficiency of the barrier and the flow of water arriving at the barrier (the flow being defined as the sum of the flow stopped and that not stopped by the system), all the recordings of the efficiency and flow from containers 1 and 2 were correlated, eliminating the starting phase when the fine sand was poorly compacted. This kind of data evolution did not give convincing results since the values scattered to a high degree. This situation may be explained by interference phenomena:

- hysteresis effects;
- phenomena due to the difference in travel time of the water through the fine sand and the gravel (influenced by the respective water content of the material).

Detailed analysis of the efficiency of the capillary barriers is therefore difficult under natural conditions. In order to bypass certain uncontrollable factors, sprinkling with a constant flow rate was carried out on the site during the studies in 1987 and 1988.

Sprinkling

Three 24-h sprinkling sessions were carried out (container 1, on 9 and 10 July 1987, container 2 on 13 and 14 August 1987 and 29 and 30 June 1988). The intensity of the simulated rainfall was 4 mm/h for the first session, 10 mm/h for the second session and 15 mm/h for the third session.

The third sprinkling session (June–July 1988) was only carried out on container 2, and lasted 14 h 50 min at an intensity of about 15 mm/h. Dry periods after sprinkling were monitored for two days (Fig. 3).

Unfortunately, the sprinkling coincided with a storm (13·3 mm at 20.00 hours), which caused an immediate decrease in the effectiveness of the barrier (down to 35%).

This storm explains the first peak observed in the flow rate variation that was not stopped by the capillary barrier; the influence of the storm is also seen in the efficiency/flow rate diagram (Fig. 4), whose cycle begins with a sudden fall in efficiency (35%) and then rises again to about 50% for the rest of the sprinkling period. As soon as the infiltration is less than 200 mm/day the barrier is appreciably more efficient, rising above 90%. In spite of the interference caused by the storm, this diagram clearly shows the importance of the hysteresis which affects the relationship between efficiency and flow rate. These results are not surprising when the importance of the hysteresis is considered which already affects the curves characterising an unsaturated medium. In particular the curves showing permeability/

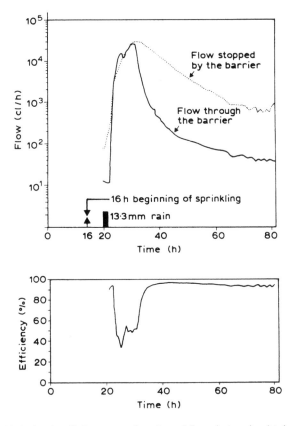

Figure 3. Variation in efficiency as a function of flow during the third sprinkling phase (container 2, June–July 1988).

water content and capillary attraction/water content have to be considered in this regard.

CONCLUSIONS

The objective of this experimental study was to test the practical application of capillary barriers to control water infiltration into landfills. From this study the following conclusions can be drawn:

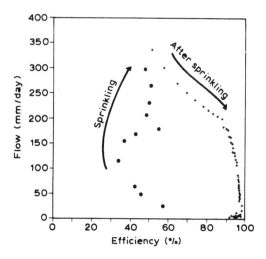

Figure 4. Variation in efficiency as a function of flow during the third sprinkling.

- A capillary barrier can be constructed at very reasonable costs by using ordinary techniques and ordinary standard materials.
- The barrier thus designed has good efficiency. Under normal weather conditions, the efficiency is always greater than 50%, and generally between 80 and 100%.
- The capillary barrier, due to the retention capacities of the fine sand, acts as a buffer and regulator during rainy periods; the flow arriving at the barrier never exceeds 10 mm/day.

The experiment also shows that the capillary barrier effect is not sufficient if water infiltration into a landfill has to be avoided. Its efficiency is low during rainy periods of more than 10 mm/day precipitation. Such a barrier is therefore only useful if it is combined with other lining methods.

On the other hand, other lining techniques have their weak points and may not be used on their own. So the concept of a *multilayer cover* is needed to satisfy all the criteria necessary for reliable and durable confinement.

The capillary barrier incorporated into such multilayer systems would, in addition, have several other advantages so long as it is placed underneath the other layer.

• It acts as a safety layer, in case problems occur with the first hydraulic barrier immediately above it. It is therefore more effective at this level since the flow rates to be disposed of would not be high. A local leak in the upper hydraulic barrier would not be effective since the water would disperse laterally in the fine sand layer of the capillary barrier. This second layer will retain the water with an efficiency of about 90%.

The underlying position of the capillary barrier is also logical because its essential advantage is its durability. It will also provide back-up sealing when maintenance work has to be done on the first hydraulic barrier.

• It is an ideal alarm system, thereby invoking the possibility of incorporating an effective alarm system to ensure the adequate functioning of the first upper liner. These alarm systems are increasingly being used. The properties of the materials used for capillary barrier render them particularly suitable for different alarm systems either by

—placing pressure sensors (tensiometers) in the fine sand, or
—by monitoring the flow rate from the drains.

• It enables gases to be extracted. The underlying gravel layer that is in direct contact with the waste enables any gas generated by the waste, as well as any toxic volatile substances, to migrate to the surface where they can be collected.

• It is an excellent base for certain artificial hydraulic barriers; in fact the fine sand comprising the upper part of the capillary barrier can be used as an appropriate basis layer for a plastic liner, a bituminous membrane, etc.

These properties are such that the capillary barrier should have a place in a multilayer type cover, since it adds a number of advantages to the final system, rendering it much more reliable. In such a context the 'capillary barrier' technique will be much appreciated. In the meantime, more experiments should be carried out to test such multilayer covers on a real waste disposal site.

REFERENCES

Andersen, L. J. & Christiansen, J. C. (1986). The capillary barrier: Part 1—Design construction and preliminary results of a full-scale test field. 14th

Nordic Hydrologic Conference—1986 (NHK-86), Reykjavik, 11th–13th August 1986.

Andersen, L. S., Clausen, E., Jakobsen, R. & Nilsson, B. (1988). Two-year water balance measurements of the capillary barrier test field at Botterup. Preliminary results, Geological Survey of Denmark.

Andre, J. R., Arnould, M., Billiotte, J., Deveughele, M. & Rousset, G. (1985). Une protection naturelle contre l'infiltration des eaux météoriques: la barrière capillaire principe—modélisation des transferts d'eau. Wiston Salem, Octobre 1985.

Bourgeois, M. (1986). Le concept de barrière capillaire—Etude par modèle numérique. Thèse de doctorat, Ecole normale supérieure des Mines de Paris.

Eagleman, J. R. & Jamison, V. C. (1962). Soil layering and compaction effects on unsaturated moisture movement. *Soil Science Society Proceedings,* pp. 519–22.

Frind, E. O., Gillham, R. W. & Pickens, J. F. (1978). Application of unsaturated flow properties in the design of geologic environments for radioactive waste storage facilities. Department of Earth Sciences, University of Waterloo, Ontario, Canada.

Nyhgar, J. W., Abeele, W., Hakonson, T. & Lopez, Z. A. (1986). Technology development for the design of waste repositories at arid sites: field studies of biointrusion and capillary barriers. Los Alamos National Laboratory, New Mexico.

Rançon, D. (1972). Structures sèches et barrières capillaires en milieux poreux—Application au stockage dans le sol. Centre d'Etudes Nucléaires de Cadarache, Rapport CEA-R-4310.

Rançon, D. (1981). Validité de la méthode neutronique pour les contrôles d'étanchéité sur de longues durées. Bulletin du CFHN, Novembre 1981, no. 10, pp. 54–61.

Rançon, D. (1980). Application de la technique des barrières capillaires aux stockages en tranchées. In International Atomic Energy Agency, Vienna, pp. 241–65.

Saillie, L. (1984). Some aspects of sandwiched soil cover design for toxic waste repositories. Department of Soil and Rock Mechanics, Royal Institute of Technology, Stockholm, Sweden.

2.9 Failure Mechanisms for Clay Cover Liners

CAROL J. MILLER & MANOJ MISHRA

Wayne State University, 2158 Engineering Building, Detroit, Michigan 48202, USA

INTRODUCTION

The primary objective of the cover liner design is minimization of moisture additions to the waste. In reality this objective is achieved only to a certain degree. Instead, the constructed liner is subject to numerous flow paths unintended in the original design.

This chapter reports and discusses the results of 4 years of field research on flow through landfill cover liners, attempting to resolve the failure modes for clay liners. The site of the research is a disposal facility near Detroit, Michigan. Both hazardous and non-hazardous landfill operations are conducted at this facility. Various failure mechanisms for cover liners, and the associated consequences, are described in relation to research at this particular facility.

COVER LINER DESIGN

Figure 1 presents a schematic of the cover liner for one cell at this facility. Above the solid waste is placed a layer of leveling soil, of several feet in thickness. A synthetic membrane liner, of 30 mil PVC, is placed above the leveling soil after all sharp objects, including stones and construction debris, are removed from the surface. A sand drain with perforated collection pipes covers the PVC liner. Above the sand drain is a five foot thick clay liner, placed in nine inch lifts and each

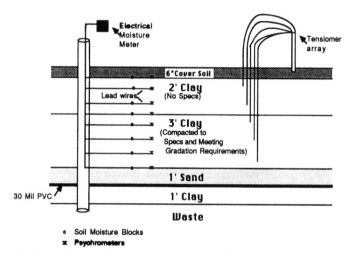

Figure 1. Cover liner and instruments at site of field work.

compacted using sheepsfoot rollers. Above the clay is a one foot thick layer of organic soil which is seeded to support vegetative growth.

The cover liner construction extended over two seasons. Two lifts of the clay were placed before the extreme winter weather halted construction operations. The remaining construction work was completed the following spring. This type of partial completion has ramifications when failure mechanisms are reviewed.

FIELD RESEARCH

The authors instrumented the cover liner of one of the hazardous waste cells (Fig. 1) at the facility for an investigation of the infiltration through the clay cover liner. During 1985 and 1986, data from gypsum blocks and psychrometers placed at various elevations within the liner were evaluated for determination of the variably saturated flow through the liner. Approximately 2% of the site precipitation passed through the clay cover liner, based on the gypsum block data. Later work indicated that a large portion of flow through the clay liner may have bypassed measurement by the gypsum blocks due to movement through large cracks or macro-pores (Miller, 1988).

Figure 2. Macro-pores in clay layer of landfill liner.

The existence of macro-pores in the cover liner were confirmed by field observations of the cover during the placement of the various lifts of clay. Figure 2 shows one example of the macro-pores formed by the crack network in the clay. The crack widths extended to one inch in several cases. These cracks form vertical channels, penetrating nearly the entire lift thickness. Depending on the duration of the atmospheric pressure and the local climate, the development of desiccation cracks can finally break the initially homogeneous compacted clay layer into individual clay peds. These peds are completely isolated from their surroundings by the cracks, and can be removed from the clay liners with little effort and negligible interaction with the surrounding soil. In such a physical state, the clay layer contains vertical channels distributed throughout. Such channels can significantly reduce the effective thickness of the compacted clay liner. In addition, they can quickly conduct a significant portion of the surface runoff directly to the lower lifts of the liner.

Shrinkage of the compacted clay during the desiccation process may actually reduce the hydraulic conductivity of the clay forming the individual peds. However, the channels formed around the peds by the dessiccation cracks increase the hydraulic conductivity of the compacted clay layer as a whole. As a result, during a rainfall event, the infiltration through the surface of the individual peds may be desirably low while the volume and rate of leakage through the whole layer may be unacceptable. In a situation like this, conventional laboratory tests, on small samples from individual peds, may indicate compliance with the regulatory requirements for hydraulic conductivity while in fact this is not the case.

Some investigators have suggested that the problem of clay cracks is limited in clay liners because of the multi-layer construction. They reason that the multiple layers will buffer lower clay from environmental effects and the propagation of cracks. However, when a cracked clay lift is overlaid by the subsequent lift of loose clay, not all of its cracks are completely filled up. In addition, the depth of the clay lift is greater than the length of the teeth of the sheepsfoot rollers used for the compaction. Therefore, the channels in the underlying old clay layer stand a good chance of remaining mostly undestroyed and unsealed. They are merely filled up by loose clay and remain highly permeable compared to the clay matrix surrounding them. In time, the new layer also develops desiccation cracks and the associated channel network. Now, the flow paths in the combination of the two layers appear similar to those in Fig. 3.

Many of the sources of crack formation are related to construction techniques. The compaction of a layer takes several days or weeks and it may not be immediately followed by the compaction of the next layer. During this time, the compacted clay layer is exposed to the

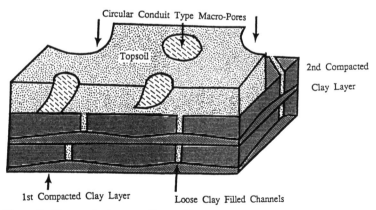

Figure 3. Flow paths in cover liner with macro-pores.

atmosphere. Consequently, it experiences the drying associated with wind exposure and temperature effects and subsequently develops desiccation cracks (see Fig. 2).

The formation of cracks in the clay creates the obvious problem of diminished containment characteristics. In addition, the cracks create a pathway for movement of landfill gases to migrate to the surface. This creates problems of air pollution and fire/explosion hazards. In addition, the cracks may cause placement problems for the cover synthetic liner.

At the site of the author's field study, large volumes of gas migrated upward through the clay cracks although gas collection vents had been installed. Figure 4 shows the expansion of the synthetic liner, ballooning through the sand drain, caused by the pressure of the gas migrating through the cracks of the underlying clay layer. In order to complete the placement of the synthetic liner and sand drain, these 'bubbles' must be punctured, allowing the gas to escape. The liner is then patched and covered as quickly as possible with sand bags to prevent the process from occurring again. The patches in the liner provide another possible failure mechanism for leakage.

Figure 4. Expansion of the synthetic liner due to gas migration through the cracks.

The topsoil layer is also subject to the development of macropores. The topsoil layer is densely populated by plant roots, and it provides a habitat to the worms and burrowing animals associated with the vegetation. This situation leads to the formation of the macro-pores that can be approximated by circular conduits. A significant fraction of the runoff generated on the landfill surfaces may be directly conducted to the topmost layer of the compacted clay by a network of such interconnected conduits.

NUMERICAL MODEL SIMULATION

Mathematical modelling of leakage through clay cover liners has been attempted by Remson *et al.* (1968), Fenn *et al.* (1975), Perrier and Gibson (1980), Skaggs (1980), Miller and Wright (1988), and Schroeder (1984). The first three of these researchers have used the water balance method, without considering the mechanism of the fluid flow through the porous media, to estimate the leakage. This approach has been found to result in gross underestimates of the field conditions (Gee, 1981; Gibson & Malone, 1982).

Skaggs (1980) improved over the earlier attempts by incorporating Darcy's equation in his model to compute the rate of moisture movement through the saturated porous media. Miller and Wright (1988) used a quasi-two dimensional, deterministic approach to model flow through the cover liner. Schroeder's (1984) model, renamed HELP (Hydrological Evaluation of Landfill Performance) and distributed by the US EPA, also uses this approach. However, the HELP model assumes the clay liner to be saturated, while Miller's model covers the full range of saturation. Miller's model has a drawback in that the surface hydrology must be solved a priori. HELP includes a surface hydrology component.

Each of these models assumes the clay liner to be a homogeneous mass of clay with uniform hydraulic properties. None account for degradation with time, i.e. the formation of cracks or other macropore channels due to desiccation, settlement, freeze/thaw, root penetration, etc. As such, it is likely that the models are most applicable to the early times after placement. For later times, the leakage estimates provided by these models are expected to underestimate the actual field condition.

CONCLUSIONS

The containment of landfill waste relies on the hydraulic characteristics of the liner system. One important failure mechanism for clay cover liners is the development of macro-pores, attributed to desiccation, freeze/thaw cycles, and animal activity. This type of failure can lead to additional problems with the synthetic liner and gas migration. Most important, however, is the additional volume of moisture that enters the waste through the macro-pore openings of the cover liner.

There are several public domain models currently available to simulate moisture transport through cover liners and leachate percolation through the waste. However, none of the existing models incorporate the potential failure of cover liners and the resulting leachate generation. Therefore, predictions from such models should be reviewed cautiously.

REFERENCES

Fenn, D. G., Hanley, K. J. & Degeare, T. V. (1975). Use of the water balance method for predicting leachate generation from solid waste disposal sites. US EPA, Report SW-168.

Gee, J. (1981). Prediction of leachate accumulation in sanitary landfills. Proceedings of the Gas and Leachate Management Conference, University of Wisconsin, Madison, Wisconsin, 2–4 November 1981.

Gibson, A. C. & Malone, P. G. (1982). Verification of the US EPA HSSWDS Hydrologic Simulation Model. In *Proc. of the 8th Annual Research Symposium on Land Disposal of Hazardous Wastes*. Rep. No. EPA-600/9-82-002.

Miller, C. J. (1988). Field investigation of clay liner performance. *Hazardous Waste and Hazardous Materials*, 5(3), 231–8.

Miller, C. J. & Wright, S. J. (1988). Application of variably saturated flow theory to clay cover liners. *Journal of Hydraulic Engineering of the American Society of Civil Engineers*, 114(10), 1283–99.

Perrier, E. R. & Gibson, A. C. (1980). Hydrologic simulation on solid waste disposal sites. US EPA, Report SW-868.

Peyton, R. L. & Schroeder, P. R. (1988). Field verification of HELP model for landfills. *Journal of Environmental Engineering, American Society of Civil Engineers*, 114(2), 247–69.

Remson, I., Fungaroli, A. A. & Lawrence, A. W. (1968). Water movement in an unsaturated sanitary landfill. *Journal of the Sanitary Engineering Division of the American Society of Civil Engineers*, 94(2), 307–17.

Schroeder, P. R., Gibson, A. C. & Smolen, M. D. (1984). The hydrologic evaluation of landfill performance (HELP) model: Documentation for Version I. US EPA Report EPA/530-SW-84-010, 2.

Skaggs, R. W. (1980). A water management model for artificially drained soils. North Carolina Agricultural Research Service, Technical Bulletin No. 267.

US EPA (1986). SOILINER Model—Documentation and user's guide. Report EPA/530-SW-86-006.

2.10 Geosynthetics for Surface Capping

SJOERD E. HOEKSTRA & HENNY C. BERKHOUT

Akzo Industrial Systems bv, Postbus 9300, 6800 SB Arnhem, The Netherlands

INTRODUCTION

Geosynthetics may be used in surface capping for various purposes: to improve gas removal, reduce infiltration of rain into the landfill, and improve the stability of capping constructions. These aspects are discussed in this chapter.

IMPROVED GAS REMOVAL

The differential pressure (ΔP_{max}) that can be applied to remove good-quality gas from a landfill is restricted by the oxygen (false air) intake through the surface. As the waste is deposited in layers, gas removal by vertical vents is much more effective because they link the individual layers. It has been found that ΔP_{max} for vertical vents cannot be higher than about 30–50 mbar, for horizontal systems only about 10–20 mbar (Tabasaran & Rettenberger, 1987). The radius of the area that can be degased by each vertical, perforated pipe is directly related to ΔP_{max} (Doedens & Weber, 1988). For this reason provision of an airtight capping system over a properly permeable layer to facilitate gas transport along the surface towards the vents will permit a high ΔP and consequently have a positive effect upon gas production.

Figure 1 (Doedens & Weber, 1988) plots the amount of good-quality gas Q_r that can be removed by a vent against the radius (R) of its exhaust area. Curves a, b and c show that any reduction in the impermeability of the capping system immediately necessitates a

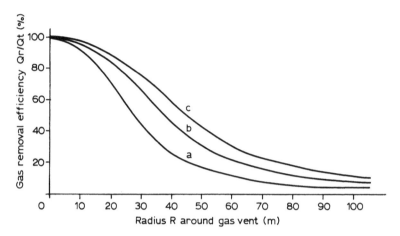

Figure 1. Gas removal efficiency in relation to the gas vent pattern for various systems (Doeden & Weber, 1988): (a) air-permeable cover; $\Delta P_{max} \leqslant 10$ mbar; (b) clay cover (fluctuating false air suction), $\Delta P_{max} \leqslant 30$ mbar; (c) airtight geomembrane cover (no false air), $\Delta P_{max} \leqslant 50$ mbar.

reduction in ΔP_{max}, and consequently lowers the efficiency of gas production.

For safety reasons, capping constructions possessing some air permeability require at the top of the vent a non-perforated pipe section about 2 m long. If an airtight geomembrane is incorporated in the capping construction, perforation of the entire vent is acceptable. Obviously this enhances the effectiveness of the vent.

FULL-SCALE EXPERIENCES

In Northeim county (Germany) at the Blankenhagen Landfill Site, each section of the tip for municipal waste that has reached the final height is covered. The capping construction is shown in Fig. 2. Depending upon the experiences gained in the coming years after the final closure of the site a second, definitive sealing layer will be added.

The gas removal system was designed by the Technical University of Hannover, which also monitors its performance. (Doedens & Weber, 1988). Since the waste is capped by an airtight system, the vents are perforated right to the top. Due to the permeable boundary with the section where waste is still being dumped, ΔP_{max} is

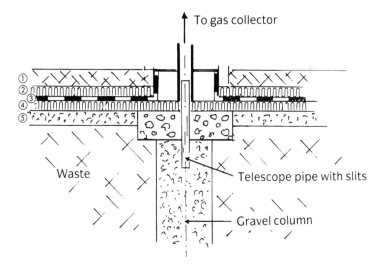

Figure 2. Capping construction near gas vent at Blankenhagen Landfill Site (Germany): 1, 0·5-m soil cover; 2, geosynthetic drain; 3, liner (2 mm); 4, geosynthetic drain; 5, 0·15-m sand.

temporarily limited to about 10 mbar. This reduces the obtainable gas quantity to about 70% of the generated volume. Of this volume, some 70% will flow through the vents; the remaining 30% arrives at the top and should flow through the permeable layer below the surface cap.

For economic reasons, the size of the highly efficient geosynthetic drain sections around each vent was limited to $R_1 = 10$ m. Over the remaining area ($R_2 - R_1 = 35 - 10$ m) the gas had to pass through a 0·15-m layer of sand. Across the sand layer there was a pressure drop of 17 mbar, while across the 10 m of geosynthetic drain layer a pressure drop of only 0·56 mbar was calculated. Since the gas removal system of the landfill operates at a ΔP_{max} of 10 mbar, obviously the flow rate of the gas is reduced by the sand layer.

DRAINAGE OF EXCESS RAINWATER

Whether the waste is capped by an impermeable geomembrane or by a mineral seal of low permeability, excess rainwater seeping through the soil cover has to be removed.

With a mineral seal the water-head formed after a heavy rainfall will cause water to seep through into the waste. Cracks in the sealing surface will make the removal of excess rainwater even more necessary. With a geomembrane liner a drainage layer is often needed to obtain enough friction for the soil cover. A 'conventional' drain consists of a gravel layer separated by non-wovens from adjacent layers which could otherwise cause clogging. If gravel is used on top of a geomembrane, additional geotextiles are needed to prevent the liner from being cut or punctured. The application of a much thinner but equally effective geosynthetic composite drain creates extra storage space which can be very valuable.

Non-woven Drain Layers

Investigations carried out at the Akzo Laboratories in Arnhem indicate that the use of a thick non-woven as a drain layer, or even one consisting of two layers with different pore sizes, cannot be recommended. In the apparatus schematically shown in Fig. 3, the ring-shaped drain sample is first compressed and then charged with a number of fluid batches that are forced radially through the material. These batches consist of water 'contaminated' with small amounts of clay or other fine particles. The effect of the passage of a clay suspension upon the transmissivity of a non-woven versus a geocomposite with an open three-dimensional drain core is shown in Table 1.

The experiments were carried out under the following conditions:

Liquid: $0 \cdot 1\%$ of clay ($85\% < 5\ \mu$m) in water.
Test cycle: $4 \cdot 5$ litres clay suspension $+ 4 \cdot 5$ litres water.
Load: $P = 80\ \text{kN/m}^2$.

The effect of the passage of water contaminated with small amounts of colloidal iron hydroxide was also measured in the same apparatus. The results are shown in Fig. 4. This graph illustrates that, already at a fairly low pressure of $13\ \text{kN/m}^2$, even heavy needle-punched non-wovens become seriously clogged by small amounts of very fine particles in a relatively short time.

Geosynthetic Filter–Core Composites

The drainage capacity of geosynthetic composites consisting of a three-dimensional, coarse, open drain core with one or two non-wovens is very satisfactory (Hoekstra & Berkhout, 1986). Such

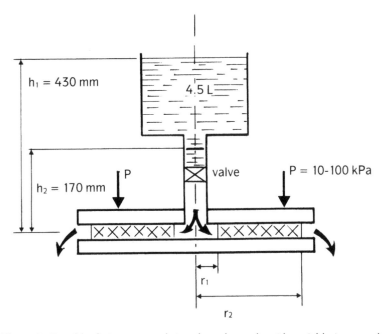

Figure 3. Portable drain tester and ring-shaped sample with variable inner and outer radii: $r_1 = 50-100$ mm; $r_2 = 125-300$ mm.

TABLE 1. Reduction of Transmissivity (θ) due to Clogging of a Drain by Clay Particles

Product	Transmissivity, θ $(m^2/s)^a$		
	At start	After 55 cycles	Reduction (%)
Open three-dimensional drain core	$10 \cdot 2 \times 10^{-5}$	$8 \cdot 0 \times 10^{-5}$	22
Non-woven drain (660 g/m^2)	$7 \cdot 7 \times 10^{-5}$	$1 \cdot 4 \times 10^{-5}$	82

^{a} Transmissivity, θ = drain capacity in m^3/s m at $i = 1$.

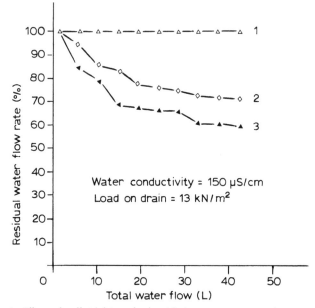

Figure 4. Effect of colloidal iron hydroxide in tapwater upon the transmissivity of two needle-punched non-wovens and an open drain core: 1, open drain core; 2, 750 g/m² non-woven; 3, 250 g/m² non-woven.

composites both save valuable storage capacity and effectively remove excess rainwater. The functioning of the non-woven covering the drain core will be further discussed here. It should:

- allow free passage of excess rainwater but prevent soil particles from clogging the drain core (filter function);
- provide enough friction to safely retain covering soil on slopes;
- possess enough stiffness to resist being pressed down unduly into the drain core by the covering soil particles.

During the past 20 years a great deal of research has been done to determine the performance of geotextile filters. Initially, such filters were mostly used in coastal protection constructions and shipping canals. Hence the primary research theme used to be the filter stability under dynamic waterflow conditions that favour needle-punched non-wovens. Rainwater seepage through the top cover of a landfill is fundamentally different.

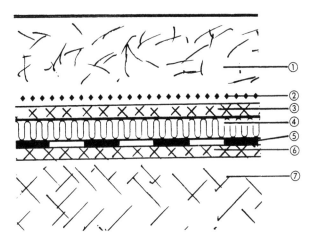

Figure 5. Capping construction at trial section of the Gerolsheim, landfill site, (Germany): 1, soil cover (0·55 m); 2, non-woven geotextile filter (125 g/m²); 3, open synthetic drain core; 4, geosynthetic drain; 5, HDPE liner (2·5 mm); 6, gas-conducting layer; 7, Waste.

The Institute for Land and Water Management (Stoving Centre) in Wageningen showed that under such conditions the soil structure has a decisive impact on the risk of clogging (Stuyt & Oosten, 1987). A special study was made at the Gerolsheim landfill site, Germany (Muth, 1988/1990). The construction of the trial section is shown in Fig. 5.

Over a 5-year period the thermally bonded filter geotextile and the soil on top of the filter were periodically sampled. For all filter samples both the weight increase due to soil particles that could not easily be washed out and water permeability were measured. The results can be summarized as follows (see Fig. 6):

- Most soil particles are trapped in the first rainy month and add to the filter 40–100% of its own weight of 125 g/m².
- The residual permeability of about 1×10^{-3} m/s stayed unchanged over the 5 years of the investigation and remained more than 3,000 times higher than the permeability of the adjacent soil, given $k_f = 3 \times 10^{-7}$ m/s.

Some of the non-woven geotextile filter samples were further tested at Stoving Centre (Wageningen). By means of high-frequency vibra-

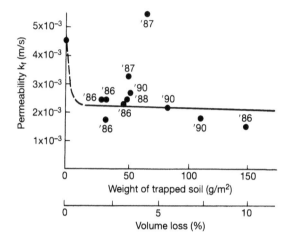

Figure 6. Relation between permeability k_f at $i = 1$ and soil uptake in a Colback filter fabric (125 g/m²) over a 5-year period.

tion, the trapped soil particles were removed from the filter, screened in a particle size analyser and counted by computer (see Table 2). The results give rise to the following observations:

- Only some 3% of the particles in the filter are larger than 100 μm whereas the soil directly above the filter contains 40%.

TABLE 2. Particle Size Distribution of (a) Soil Trapped in the Non-woven Geotextile Filter and (b) That of the Covering Soil

Particle size (μm)	Composition of soil trapped in the filter (%)	Composition of soil directly above the filter (%)
0–5	—	20
5–10	6·6	10
10–16	39·3	5
16–35	25·5	51
35–50	9·6	4
50–75	8·3	3
75–100	7·7	3
>100	3	40

• In the non-woven geotextile no particles smaller than 15 μm are found although the soil directly above contains 20%.

Since these and other investigations showed that no soil particles remain in the drain core, this leads to the following tentative conclusion:

• Only particles from a very thin layer of soil adjacent to the filter are transported by water. Particles up to 5 μm pass through the filter, while those over 100 μm remain above it.

STABLE CAPPING CONSTRUCTIONS

For optimum usage of the allotted space in a landfill, the slopes of the impoundment should be rather steep. This requires extra attention for the stability of the capping construction and the execution of friction tests. Friction tests indicate that if mineral layers are chosen for impermeabilization the stability of the construction is only slightly reduced by the incorporation of a geosynthetic drain. When placed under a certain load, the non-woven filter apparently adapts snugly to the contours of the mineral seal. This results in a coefficient of friction between geotextile and soil that is almost equal to that of the mineral layer itself. Obviously the same holds for the transition non-woven and soil cover. When geomembranes are used, notably in the wet state, a satisfactory stability of the soil cover upon the smooth surface of the liner is much more difficult to obtain.

Incorporation of a Geotextile Composite

At the Rahmedetal landfill site (Germany), compensation for the lack of friction between geomembrane and soil cover is provided by the introduction of a geotextile composite consisting of a fabric with sufficient tenacity combined with a synthetic grip layer. This construction is schematically shown in Fig. 7. Since no frost-sensitive mineral materials are used for impermeabilization, a thick insulating cap is not necessary. The soil cover, which should of course be stable in itself, is

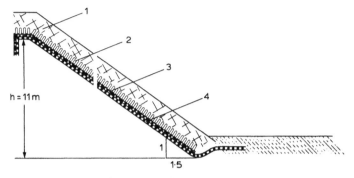

Figure 7. Construction of a stable soil cover on a geomembrane on a steep slope: 1, 20-cm soil cover; 2, grip layer; 3, high-tenacity fabric; 4, geomembrane.

firmly anchored in the geocomposite. The part of shear force along the slope that is not compensated by friction is absorbed by the fabric. Depending upon the type of polymer, a certain percentage of the fabric strength can be utilized permanently to hold the cover in place.

Since 1984 the construction has performed according to plan: the geomembrane is protected from atmospheric deterioration, and every spring wild flowers bloom on the grassy slopes.

CONCLUSIONS

- In many cases the use of a geotextile composite with a highly permeable drain core for gas removal below an airtight surface cap can provide benefits including a higher gas production and/or a larger waste storage capacity.
- Non-wovens are not suitable for in-plane water transport. Geotextile composites with a separate filter and drainage function should be preferred. Such drains also protect the geomembrane surface liner from mechanical damage and provide extra waste storage volume.
- High-tenacity fabrics in combination with a three-dimensional, synthetic grip layer offer a solution for stability problems encountered in the construction of soil covers upon steep slopes. Such geotextile composites are particularly valuable if geomembranes are used for impermeabilization.

REFERENCES

Doedens, H. & Weber, B. (1988). Degasification of a capped landfill site for domestic refuse. In *Proc. ISWA 5th Int. Solid Wastes Conf.*, *Vol. II*, ed. L. Andersen & J. Møller. Copenhagen, Denmark, pp. 69–75.

Hoekstra, S. E. & Berkhout, H. C. (1986). Geotextile/'Geospacer'. Composites for environmental projects. In *Proc. 3rd Int. Conf. on Geotextiles*. Vienna, Austria, 343–7.

Muth, W. (1988/1990). Filter stability of Enkadrain geocomposites, landfill Gerolsheim, Rep. No. 8812-AKZOG and 8812-AKZOG/E90. Fachhochschule Karlsruhe, Germany.

Stuyt, L. C. P. M. & Oosten, A. J. (1987). Mineral and ochre clogging of subsurface land drainage systems in The Netherlands. *Geotextiles and Geomembranes*, 5, 123–40.

Tabasaran, O. & Rettenberger, G. (1987). Design data for degasification systems. *Handbook for Waste Disposal 1/87*. Erich Schmidt Verlag, Berlin, Germany.

2.11 Design Considerations of a New Liner System over an Existing Landfill

UMESH DAYAL, JOHN M. GARDNER & ERIC D. CHIADO

Almes & Associates Inc., Box 520, Pleasant Valley Rd, Trafford, Pittsburgh, Pennsylvania 15085, USA

INTRODUCTION

Many municipal solid waste landfills are nearing capacity or are being forced to close due to more stringent regulations. Concurrently, regulatory agencies are applying more pressure to limit development of new sites. Consequently, many site operators are considering expansions over the top of and/or adjacent to an existing landfill facility. However, this presents several technical issues relating to the integrity and stability of the liner and cap systems along the interface (or overlap) between existing and future refuse.

Proper design of these liner and cap systems requires the establishment of several geotechnical parameters. One of the key issues is to estimate post-capping total and differential settlement of the existing underlying refuse due to its own weight, decomposition of the existing waste, surcharging by the additional new refuse and the resultant strain of the new liner and cap systems. Additional relevant parameters that must be identified and determined include shear strength of the refuse, frictional resistance between the various geosynthetic components and their overall individual and combined behavior with respect to shear, tensile and compressive loadings.

As an example of a vertical landfill expansion design, a case study is presented in this chapter. The landfill is located in Conemaugh Township, Somerset County, Pennsylvania, USA. This chapter evaluates several specific design issues related to the vertical expansion of this site with emphasis placed on providing simple reliable methods for designing the liner and cap systems.

BACKGROUND

The landfill currently serves as the disposal site for municipal, commercial and some select industrial wastes. The landfill was nearing its permitted capacity and in order to continue landfilling operations Chambers proposed a partial closure of the facility. This partial closure plan would cover the majority of the existing landfill with a final cover system to prevent the infiltration of precipitation into the refuse and the remaining existing portion would be overlain with a new liner system to permit vertical expansion. A typical cross section of the vertical expansion is shown in Fig. 1.

As a consequence of evolving state regulatory policy, the interface between the existing refuse and the future waste had to be closed and lined with a double liner system prior to placement of any new (overlying) refuse. There was no specific policy that required a cap system separate from an overlying double liner system on this type of interface. As such, the interface between the existing and future waste was designed using a synthetic double liner system. A cross section of the liner system along the interface is shown in Fig. 1.

The incorporation of a synthetic liner system on a slope overlying existing refuse made it necessary to consider design aspects for the liner system that may not be commonly required when designing landfills on stable soil or rock slopes. Analysis and design of this landfill facility identified the following critical issues concerning vertical landfill development:

- Settlement of the existing waste could be excessive during and after future (overlying) waste placement, which could introduce strains in the interface liner system that could potentially rupture the liner. A reliable method for estimating settlement of waste was required and a determination of the effect of this settlement on liner integrity/stability, and a method for controlling this impact was required.

- The existing waste slope along the interface tended to be steep, typically 3 horizontal to 1 vertical (3H:1V). Placement of a liner system and protective cover over this slope had the potential to introduce additional stresses and strains in the liner system components. These stresses and strains had to be estimated and accounted for, so as to insure internal stability of the liner system.

- The number of geosynthetic components proposed for the liner

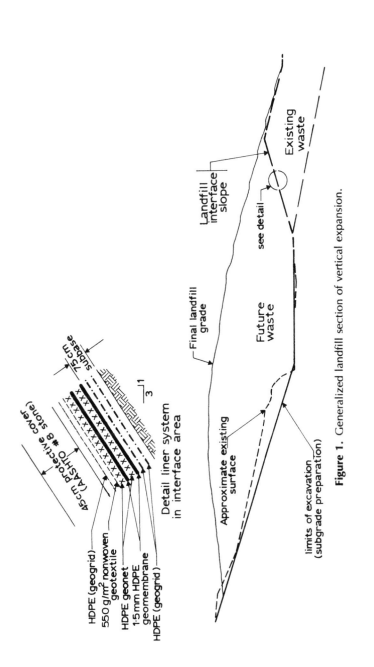

75 cm subbase

Detail liner system
in interface area

HDPE (geogrid)
550 g/m² nonwoven geotextile
HDPE geonet
1·5 mm HDPE geomembrane
HDPE (geogrid)

45cm protective cover) (AASHTO #8 stone)

Landfill interface slope

see detail

Existing waste

Final landfill grade

Future waste

Approximate existing surface

limits of excavation (subgrade preparation)

Figure 1. Generalized landfill section of vertical expansion.

system resulted in a number of interfaces along which frictional resistance would be minimal. A determination of the location and magnitude of critical friction angles in the liner system was required.

- Finally, different components used in the liner system could tend to behave individually under load, such that the liner system may not have reacted to loadings as a single unit. The component or components governing the overall behavior of the liner system had to be determined, and its properties under load were applied in subsequent analyses of liner integrity and stability.

LANDFILL SETTLEMENT

Total settlement is comprised of: (1) settlement due to the weight and decomposition of the existing refuse itself; and (2) the weight of (future) superimposed surcharge loading from additional overlying new refuse. As a consequence of the many unknown and highly variable properties of waste, no precise analytical method is available to predict settlement. A review of available literature shows that there are three primary methods for estimating settlement of waste. These methods are conventional consolidation theory, the rate process model and Maxwell type model. The later two models require the determination of several consolidation coefficients of the material, which can be determined experimentally. The limited amount of field measurements of waste settlement have shown that the conventional consolidation method predicts the settlement of waste relatively accurately and as such, was used in the present analysis. Additionally, localized settlement was also addressed to account for the condition of localized decomposition of the existing waste, which is not a function of surcharging by the new overlying waste.

Total Settlement

Perhaps the earliest and most commonly referenced model of waste settlement is that attributed to Sowers (1972). Sowers indicated that waste is extremely compressible and that consolidation involves several mechanisms: (1) mechanical consolidation; (2) biochemical decomposition and decay; (3) collapse of hollow bodies; and (4) ravelling or sloughing of fines into the voids between larger irregular particles. Sowers notes the similarity of the consolidation process in waste to that in peats, which have been more widely studied than waste. The total waste settlement comprises an initial, rather rapid, primary consolidation followed by a secondary compression which can continue for years and can be a significant portion of the (total) ultimate settlement. It is noted that primary and secondary consolidation processes in a typical geotechnical sense involve obviously different processes than the primary/secondary phases discussed herein. However, the two-phase process (rapid and time dependent compression) appears to be common to both soils and waste. Load-related compressibility of waste (primary consolidation) can be expressed as:

$$\Delta H = \frac{C_c H}{1 + e_0} \log \frac{\sigma'_0 + \Delta \sigma'}{\sigma'_0} \tag{1}$$

where: ΔH = settlement; e_0 = initial void ratio; H = existing waste thickness; C_c = compression coefficient; σ'_0 = initial effective vertical stress; $\Delta \sigma'$ = change in vertical effective stress produced by surcharge loads.

The compression index (C_c) varies considerably for any value of void ratio, depending upon the organic content of the refuse. For low organic content waste, the lower bound value of C_c is approximately equal to $0.15e_0$, and for high organic content waste the upper bound value of C_c is approximately equal to $0.55e_0$.

The secondary consolidation of waste is also a function of void ratio and can be expressed as:

$$\Delta H = \frac{\alpha H}{1 + e_0} \log \left(\frac{t_2}{t_1} \right) \tag{2}$$

where; α = coefficient of secondary consolidation; t_1 = time of initial (primary) consolidation; t_2 = time of secondary consolidation; α, analogous to C_c, is related to the fill environment and typically is given as: $\alpha = 0.03e_0$ (anaerobic; poor decomposition); $\alpha = 0.09e_0$ (aerobic; good decomposition).

The total settlement is obviously the sum of the primary and secondary consolidations as defined above by eqns (1) and (2). The use of the above equations necessarily involves either the measurement or assumption of the initial void ratio (e_0) in order to calculate the respective compression coefficients, C_c and α. In addition, estimates of the overburden load (in this case, the new overlying waste) must also be made to determine σ'_0 and $\Delta\sigma'$ in eqn (1).

Additional modifications to these original equations have been made in recent years with the majority attempting to limit or better define the assumptions needed for the variables. However, in the majority of cases where field measured settlements are compared with predictive models, the original model by Sowers provides reasonable estimates of waste settlement. Although Sowers' method involves estimating the initial void ratio and the existing effective stress state as well as the type of decomposition to be expected, it does allow for the adequate determination of approximate conditions for design purposes. All of the other settlement models require field or laboratory testing in order to determine various consolidation parameters. In addition, it should be noted that Sowers' model has been referenced in whole, or in part, in numerous field studies of waste settlement and the agreement has generally been good.

Because of the difficulty in predicting future waste conditions, both upper and lower bound values of C_c and α were used to estimate the range of magnitude of settlements likely to occur along the interface between existing and future waste. Primary settlement of the existing waste was calculated assuming that the future waste surcharge would be applied instantaneously. However, for calculation of secondary consolidation, a 10-year time period was used to ensure that the majority of secondary consolidation was included. These two settle-ments were added together to determine the theoretical future settlements of the existing waste, which would occur after the closure of the existing waste and application of future waste. Settlement was estimated along several cross sections through the interface. Areas of the interface where the magnitude of settlement was equal were connected with a line to produce an 'iso-settlement' contour (refer to Fig. 2). Figure 2 shows the 'iso-settlement' contour map (each line represents a line of equal maximum settlement) prepared for the interface of the vertical expansion.

When the 'worst case' conditions (i.e. largest settlement, high organic material-aerobic decomposition) were investigated, the maxi-mum calculated total settlement was about 6·3 m. Conservatively, zero settlement was assumed at points beyond the influencing zone of the new waste (i.e. toe and top of the interface). This resulted in a maximum differential settlement of about 6·3 m over an approximate distance of 748 m, producing a resultant strain of 0·9% in the liner

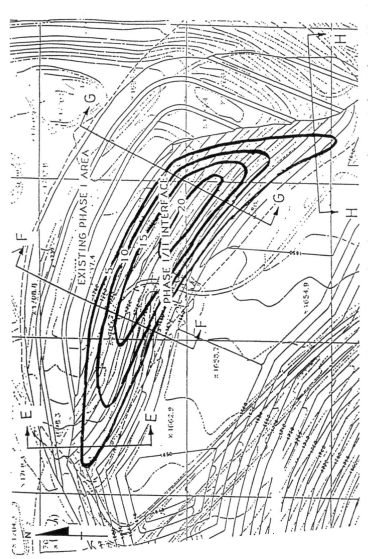

Figure 2. Estimated total settlement; 'iso-settlement' contour lines for maximum settlement (in feet; 1 foot ≅ 30 cm).

system along the interface. It should be realized that this strain, due to overall settlement, will not occur until closure of the new portion of waste over this interface, since the calculations were performed using the final thicknesses of new waste.

The liner system on the interface was capable of accommodating approximately 10·0% strain. Therefore, the maximum strain in the liner system on the interface, attributable to overall settlement of the existing waste (i.e. 0·9%) was well below the design strain of the liner system, and it was concluded that potential overall settlement of the sub-base would not damage the liner system.

Localized Differential Settlement

Aside from the overall settlement that could occur along the interface, localized pockets of excessive settlement in the existing waste could also occur. It is probable that the localized depressions would be temporary, since the surrounding waste would eventually settle. However, these temporary conditions would represent a worst case scenario, and the liner system had to be designed to accommodate these conditions. To prevent excessive elongation of the liner components due to localized settlement, geosynthetic reinforcement (geogrids) were incorporated into the design of the interface liner system sub-base. The reinforcement is located in the sub-base of the interface liner system (refer to Fig. 1) and is there solely to limit deflection of the overlying liner system.

The design for localized settlement conservatively assumed that localized settlement would take the form of 1·8-m diameter approximately-circular void (rusted refrigerator assumption) of infinite depth. Furthermore, maximum strain in the lining system was limited to the allowable strain of the geomembrane liner, which was 10%. The following methodology is based on reinforced earth theory coupled with soil arching theory, which is summarized in Giroud *et al.* (1988). Terzaghi's equation for soil arching was used to determine the pressure exerted on the liner system due to the future waste overlying the interface liner and (depression) void, and is given as follows:

$$p = 2\gamma_{\mathrm{AVG}} r (1 - e^{-0\cdot 5H/r}) \tag{3}$$

where: p = pressure exerted in the liner system; γ_{AVG} = average unit weight of material above the void; r = radius of void; H = height of new waste above the void.

This vertical pressure is transferred to the geosynthetic reinforcement layer in the form of tensile stresses. Tensile stresses that could be generated in the reinforcement are estimated as follows (Giroud *et al.*, 1988):

$$T = pr\Omega \qquad (4)$$

where: T = tensile stress in reinforcement geogrid; Ω = dimensionless factor (other parameters defined previously).

The dimensionless factor (Ω) is a function of the strain that is allowed to be applied to the reinforcement (this relationship can be found in Giroud *et al.*, (1988)). A value of 0·98 for Ω, which limited reinforcement strain to 4%, was used in the design. Strain in the reinforcement due to localized settlement was designed to be limited to 4% so that the strain from overall settlement of the existing waste (previously presented as 0·9%) as well as the strain from potential mine subsidence (not discussed in this paper) could be superimposed on the liner system without overstraining it.

Table 1 contains the results of the localized settlement analysis performed to determine the tension in the reinforcement resulting from allowing a strain of 4%. As can be seen, the tension generated in the reinforced sub-base stabilizes beyond a certain new waste height, which is attributable to the arching effect over the void. The calculated maximum tension of 1680 kg/m was used to choose a geosynthetic reinforcement which possessed a long-term tensile strength of at least 1680 kg/m at less than 4% strain.

TABLE 1. Tension in Liner System on Interface due to Localized Settlement

Height of new waste (m)	*Tension in reinforcement due to localized settlement of existing waste (kg/m)*
3·0	1455
6·0	1635
12·0	1680
15·0	1680
22·5	1680
30·0	1680
42·0	1680

LINER STABILITY

Aside from settlement of the existing waste and its effect on the stability and integrity of the interface liner system, the other major areas of concern evaluated herein include:

1. overall slope stability of the new waste with its effect on the interface liner system; and
2. slippage between various geosynthetic components within the interface liner system.

Overall Slope Stability

The overall slope stability of the new waste, existing waste and interface liner system was investigated using the PCSTABL5 slope stability computer program. The overall stability of the landfill depends on the shear strength properties of the soils, waste, and geosynthetic componentes used in its construction. The design parameters and shear strength properties used in the overall slope stability analyses are given in Table 2 and are based on laboratory tests and manufacturer's recommendations (refer to Fig. 1 for the interface liner system and the various components described in Table 2).

The interface liner system represents the most critical plane in the landfill since it contains the weakest planes, with regard to stability. A number of the liner component interfaces listed above possess friction angles of approximately 12°. Low values such as these indicate that slippage would likely propagate along one or more of these interfaces in a deep-seated failure. By envisioning these weak interfaces as providing little to no resistance, it is conceivable that whole-body rotation of the new waste could occur and was considered in this analysis as a worst case scenario.

Based on the above discussion, the following minimum factors of safety (FS) were obtained from the PCSTABL5 slope stability analyses:

FS (circular failure surface) = 1·53
FS (sliding block failure surface) = 1·74

The calculated factors of safety indicated that the overall stability of the landfill was adequate, and that a deep-seated catastrophic failure was unlikely.

TABLE 2. Slope Stability Design Parameters

Design parameters	
Unit weight of waste	$= 830 \text{ kg/m}^3$
(Approximate weight of refuse including daily and intermediate cover soil)	
Shear strength properties of refuse	
Effective angle of internal friction ($\bar{\phi}$)	$= 25°$
Effective cohesion ($\bar{c}$)	$= 0$
Shear strength properties of protective cover material	
Moist unit weight	$= 2000 \text{ kg/m}^3$
Effective angle of internal friction ($\bar{\phi}$)	$= 31°$
Effective cohesion ($\bar{c}$)	$= 0$
Shear strength properties of sub-base material	
Moist unit weight	$= 2300 \text{ kg/m}^3$
Effective angle of internal friction ($\bar{\phi}$)	$= 31°$
Effective cohesion ($\bar{c}$)	$= 0$
Geosynthetic interface friction values	
Friction angle between non-woven geotextile filter and protective cover (δ_{fc})	$= 25°$
Friction angle between non-woven geotextile filter and high density polyethylene (HDPE) drainage net (δ_{gd})	$= 30°$
Friction angle between HDPE drainage net and HDPE drainage (δ_{dd})	$= 12°$
Friction angle between drainage net and smooth HDPE geomembrane (δ_{dp})	$= 12°$
Friction angle between HDPE geomembrane and sub-base (δ_{ps})	$= 30°$

Interface Liner System Stability

The analysis of liner stability on the interface was performed using the following methodology proposed by Giroud and Beech (1989). The tension that is generated in the liner system can be approximated by the following equation:

$$\alpha = \frac{\gamma_c T_c^2}{\sin 2\beta} \left[\left(\frac{2H_c \cos \beta}{T_c} - 1 \right) \frac{\sin(\beta - \phi_{im})}{\cos \phi_{im}} - \frac{\sin \phi_{cm}}{\cos(\beta + \phi_{cm})} \right] \quad (5)$$

where: α = tension in liner system; γ_c = unit weight of protective cover; H_c = height of protective cover; T_c = thickness of protective cover; β = slope angle; ϕ_{cm} = friction angle of protective cover that is mobilized (ϕ_c/FS); ϕ_{im} = critical friction angle of liner system that is mobilized (ϕ_i/FS); ϕ_c = angle of internal friction of protective cover, and ϕ_i = critical friction angle of the liner system.

The slope of the interface is approximately 33% ($\beta = 18\cdot4°$) and the most critical angle of friction is 12°.

To prevent slippage, the liner system must be anchored to provide stability. Because the system must be anchored, applied loads from the protective cover and future waste will result in tensions being induced in the liner system. The liner system must be designed to carry this tension with some factor of safety.

The tension created by these loads will be resisted by the incorporation of one layer of geosynthetic reinforcement (i.e. geogrid) directly between the interface liner system and the protective cover (refer to Fig. 1).

The maximum length of the interface is approximately 128 m. This length was used to determine the required strength of the geosynthetic reinforcement and staging of the protective cover. Tension in the liner system can be limited by placing the protective cover in increments. Design of the protective cover increment length was based on the following assumptions: (1) the maximum strain the liner system can handle before it yields is governed by the yield strain of the primary and secondary geomembranes; (2) for 1·5 mm HDPE geomembrane strain at yield is approximately 10%; (3) the maximum strain due to trough, sinkhole subsidence and overall waste settlement equals 5·7%; (4) the strain that can be safely carried by the liner system due to tension caused by the weight of protective cover and future waste is: 10% less 5·7% = 4·3%; (5) the long-term tensile strength of the geosynthetic reinforcement (uniaxial geogrid) at 4% strain was approximately 2100 kg/m, and (6) the tensile strength of the other liner system components was neglected.

Based on these assumptions, the maximum additional tension that could be introduced into the liner system without straining it beyond 10% was 2100 kg/m. The maximum height of the protective cover that can be placed before exceeding 2100 kg/m is 3·25 m (determined by solving eqn (5) for H_c). Thus the maximum permissible length of the first increment of protective cover along the 3H:1V interface slope was approximately 10·5 m. By completely lining the interface slope with the geosynthetic liner system and then placing no more than 10·5 m of protective cover along the slope, the strain will be kept below 10% at a maximum, and the liner system will be stable.

The length of subsequent increments of protective cover will depend on the height of future waste placed adjacent to the protective cover. The future waste will act as a buttress, which will result in reducing tension in the liner. The subsequent increments of protective cover length were maintained at 10·5 m to limit tension in the liner system to 2100 kg/m.

SUMMARY AND CONCLUSIONS

Because of the escalating cost of the new development and scarcity of suitable new landfill sites, vertical expansion of existing landfill sites appears to be very attractive. However, this presents several complex geotechnical issues such as: (1) estimating total, differential and localized settlement and the resultant strain; (2) stability of the liner system; and (3) stability of various geosynthetic units within the liner system.

A case study on the vertical expansion of a landfill site located in Pennsylvania, USA is discussed. Settlement of the existing waste would be due to its own weight, decomposition of the waste and surcharging from future refuse. A literature search was conducted relating to waste settlement and the mathematic model proposed by Sowers (1972) was found to be most appropriate for this site as other models require several other parameters for estimating settlement that are not readily available. Aside from the overall settlement, the condition of excessive localized settlement was also evaluated.

Several alternative methods for lining the interface between existing waste and future waste have been evaluated. It is recommended that three layers of high strength geogrid reinforcements are used to ensure the integrity of the other geosynthetic liner components. The analysis has shown that the proposed geosynthetic units will:

- conform to the settled ground shape for overall settlement and differential settlement, and the maximum strain in the geosynthetic units due to overall existing waste settlement is minimal;
- bridge a localized depression of infinite depth and up to a maximum diameter of 1·80 m without failure to the geosynthetic units;
- allow flow of leachate through the leachate collection/detection

system along the interface—this is because the overall slope of the calculated settled surface of the existing refuse is fairly uniform;
* withstand potential tensile stresses which may develop due to the critical governing friction angle(s) within the interface liner system;
* maintain liner tension to within acceptable strains because incremental construction of the protective cover on the interface liner system, followed by future waste placement acts as an effective buttress.

REFERENCES

Giroud, J. P. & Beech, J. F. (1989). Stability of soil layers on geosynthetic lining systems. Geosynthetics 1989 Conference, San Diego, California.
Giroud, J. P., Bonaparte, R., Beech, J. F. & Gross, B. A. (1988). Load carrying capacity of soil layer supported by a geosynthetic overlying a void. International Geotechnical Symposium on Theory and Practice of Earth Reinforcement, Fukuoka, Japan.
Sowers, G. F. (1972). Foundation engineering in waste disposal fills. Proceedings of 21st Annual Conference on Soil Mechanics and Foundation Engineering, University of Kansas, Technical Publication No. 2, Kansas, USA.

2.12 Construction of Composite Lining Systems Under Unfavourable Weather Conditions

ERNST BIENER & TORSTEN SASSE

UMTEC—Ingenieurgesellschaft für Abfallwirtschaft und Umwelttechnik, Stresemannstrasse 52, D-2800 Bremen 1, Germany

INTRODUCTION

Construction of a lining system in unfavourable weather conditions may cause deterioration such as shrinkage, cracking and mineral lining softening as well as extreme wave formation in plastic liners. These quality reductions can be prevented by design and construction management measures and by specific considerations on the selection of materials. These measures are outlined in this chapter. The first part of the chapter deals with general requirements regarding

—design measures
—selection of materials
—construction management measures

In the second part these requirements are explained in detail, and two waste disposal site projects are taken as examples.

GENERAL REQUIREMENTS

As a premise, an indication of the basic requirements regarding the design and construction of base sealing systems is provided. This should be strictly observed when waste disposal sites are constructed under unfavourable weather conditions.

Design Measures

In order to limit the negative effects of heavy rainfall during the construction phase, the provision of suitable drainage conditions is of the utmost importance and adequately dimensioned drains should be provided. In addition to the observance of minimum gradients, a minimum slope of 3% is required for rapid drainage of plain surfaces, especially during the construction phase. Free drainage, possibly in combination with provisional rain-retaining basins are preferable to pumping. As a general rule, drainage operation during the construction phase should be taken into account during design.

Selection of Materials

During dry periods, mineral liner materials with a high percentage of fine particles and in particular, with a large clay fraction, tend to shrink and crack. This phenomenon cannot always be fully prevented on site, in spite of the taking of suitable comprehensive precautions such as temporary coverage by means of foils and artificial moistening (see also Chapter 3.3). Regarding the emplacement and welding of plastic liners, such precautions unfortunately cause problems since they aggravate the timetable of the welding. On the other hand, cover foils to reduce the drying-out phenomena have proven to be very effective and should be available on every waste disposal construction site.

Sealing materials with a smaller amount of fine materials are characterized by a less distinct shrinkage and cracking behaviour, but due to their minor cohesion value, tend to develop marked erosion phenomena.

In addition to the above-mentioned factors, the selection of materials should also be based on accessibility, susceptibility to frost, and general material handling. Furthermore, manufacturing aspects should be respected; in cases of doubt, further clarification can be gained from the installation of pilot test areas.

The use of mineral sealing materials with a clay content of 20–40% and, for reasons of grading, a considerable percentage of sand and fine gravel (up to a total of approximately 20%) is recommended. In addition, a proportional silt content should be included. The sealing materials should be in the medium range of plasticity since materials with a narrow range of useful water content are weather-dependent. Positive experiences were gained at a waste disposal site where a bentonite–sand mixture was used which neither shrank, cracked nor formed erosion grooves. This is described in further detail later in the chapter.

Construction Management Measures

With regard to material quality and the construction of a liner, a wide range of quality improvements are possible and in many cases necessary. In particular, the use of a stationary mixing plant should be considered in which, contrary to the mixed-in-place procedure, the mineral sealing material can be produced under controlled and initially weather-independent conditions. Such plants, first used to add fine materials, e.g. bentonites, provide the following performance range:

- preparation of the basic material (if a natural soil is the basis of the mixture)
- crushing of particles
- accurate dosing of the mix ingredients
- regulation of water content
- production of a homogeneous weather-independent total mixture
- immediate reaction to changed conditions

These requirements are met preferably by plants operating discontinuously which should be equipped with the following:

- feeding device (e.g. box hopper)
- material crushing facilities (e.g. roller crusher)
- weightbatcher to weigh all ingredients to be mixed
- electronically controlled water feeding
- pan-type mixer with high homogeneity effect (e.g. double-shaft mixer)
- central, computer-aided and controlled supervision unit

In order to ensure adequate homogeneity, the actual mixing time should not be less than 1 min. The minimum output of such a plant should not fall short of 25 m^3/h.

Satisfactory results have been achieved by similar plants in recent years, e.g. in the case of the surface sealing of the Gerolsheim hazardous waste landfill (Rettenberger *et al.*, 1980).

There is a wide field of application for these preparation plants. At the Inden waste disposal site, in a continuously-operating machine, a highly plastic clay (clay content of approximately 70%) was mixed with approximately 5–10% water in order to obtain a material that could be handled satisfactorily (Sasse, 1989).

Mixing plants for mineral sealing materials thus provide advantages even when no fine materials need to be added. They are generally technically superior to the mixed-in-place technique and should be used increasingly in the future (Fig. 1).

Figure 1. Stationary mixing plant.

EXPERIENCE IN PRACTICE

The requirements outlined above are specified and described in further detail in the following paragraphs: these regard the mono-landfill of Berlin-Kladow and the hazardous waste disposal site of Billigheim, both in Germany.

Mono-Landfill Facility of Berlin-Kladow

Before 1988, in Berlin, large quantities of coal have been stored periodically as a reserve. In order to be able to use this area for the building of industrial enterprises, the coal (approximately 500 000 t) had to be removed to an active sandpit. In order to protect underground water supplies, the coal should be hydraulically insulated on all sides by means of a lining system.

The main difficulty was the extremely short time available for design and construction. Only six months in the period winter–spring were available for excavation of the sandpit (approximately 570 000 m³), construction of the base and slope liners and storing the coal.

The coal disposal site was therefore planned according to strict requirements with special consideration to be given to the expected weather conditions.

Lining system. The lining system of the deposit, planned as a pit landfill (total depth approximately 17 m), consists of base sealing above the highest ground water level, slope and surface sealing (see Fig. 2).

The landfill area is almost rectangular (300 m × 200 m). The total area was subdivided into five sectors each of equal dimensions (approximately 60 m × 200 m). Thus hydraulic separation and independent operation in the individual sectors during construction was made possible.

With regard to rainfall and seepage water run-off, the base sealing of the individual sectors is roof-shaped (Fig. 2) with a minimum slope of 3%. In the 'roof ridge', drainage trenches are positioned to drain the collected water to the pump.

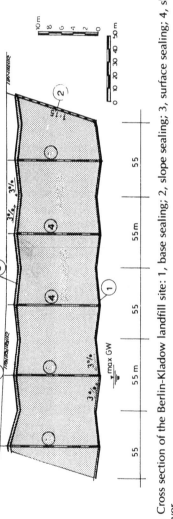

Figure 2. Cross section of the Berlin-Kladow landfill site: 1, base sealing; 2, slope sealing; 3, surface sealing; 4, shaft axes; 5, top cover.

Figure 3. Stacked drainage trenches.

The drainage trenches consist of wire mesh cages, filled with coarse crushed rock. They can be prefabricated and are set up in rows, cage by cage (Fig. 3). In winter this method affords considerable operational advantages.

The surface sealing is designed in a similar way to the base liner.

For base, side and surface, different lining systems were chosen

BASE SEALING

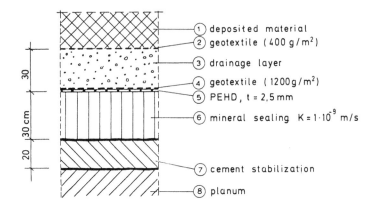

BERLIN KLADOW - LANDFILL

Figure 4. Bottom barrier system of the Berlin-Kladow landfill site: 1, deposited material; 2, HDPE geotextile (400 g/m²); 3, drainage layer gravel, 5/32 mm; 4, HDPE geotextile (1200 g/m²); 5, HDPE geomembrane (2·5 mm); 6, mineral sealing (kmin = 1×10^{-9} m/s); 7, cement stabilization; 8, existing soil.

according to their individual functions and hydraulic conditions. The design of the bottom barrier system is represented in Fig. 4.

The top cover system which has a long-term safety function, was designed as follows:

- 65 cm top soil
- 15 cm cement stabilized sand (qu, min = 5 MN/m²)
- 20 cm sand filter layer
- drainage geotextile mat (d = 20 mm)
- plastic liner (HDPE modified)
- mineral liner (50 cm, kmin = 1×10^{-9} m/s)
- geotextile (400 g/m²)
- deposited material

Contrary to common solutions, in this case a drainage mat and cement-stabilized sand were used. The cement-stabilized layer was designed to protect the liner system from damage. The drainage mat drains off leachate migrating through the top soil. The hydraulic efficiency of the drainage mat was proven under a load of 15 m sand.

Construction of the disposal site. After a rough surface preparation had been carried out, the underground was cement-stabilized, resulting in better accessibility, especially during rainfall. The cement was added to the soil by means of rotavators and compacting was achieved by means of medium-heavy smooth-wheel rollers. After having compacted the first of the five sectors, (each $10\,000\,m^2$ in size), compacting was performed in the next field; simultaneously, construction of the mineral sealing started in the already-compacted field.

A purpose-installed mixing plant was used to produce mineral sealing material; it was used to mix medium sand, high-swelling sodium bentonite and water to obtain a homogeneous material. When compared to the mixed-in-place technique, the mixing plant not only assures a higher quality but morevoer represents a lower weather risk. Furthermore, the sealing material can be produced under unfavourable weather conditions and can be stored temporarily.

The bentonite-sand mixture, already used in positive experiences under similar conditions (Sasse, 1988), has a relatively low sensitivity to weather conditions, and therefore met the requirements particularly well. The related special properties are as follows (Fig. 5):

- satisfactory accessibility to material

Figure 5. Placement of the mineral liner.

- low erosion tendency
- very moderate tendency to dry, shrink and crack
- low potential regarding softening for the already-installed sealing material
- relative frost resistance of material

In view of the unfavourable season for construction, special tests were conducted to investigate resistance to frost. Usually, frost severely attacks natural sealing materials containing a high portion of fine particles. The sealing material used in this case (bentonite–sand mixture) was resistant to several frost cycles; a standard frost–thaw interaction test was performed to investigate deformation, compaction and permeability properties. Thus, in terms of practical site management, negligible frost impacts (e.g. at night) could be accepted without hesitation.

Subsequently the plastic sealing linings were placed on the smoothly-rolled mineral sealings, and then welded. It should, however, be mentioned that the lining material of modified HDPE is principally easier to weld than a pure HDPE material. In combination with higher flexibility, especially at lower temperatures, the customer rated these advantages higher than better permeability and increased chemical resistance of the pure HDPE lining.

In order to protect the plastic sealing linings from damage, a protective geotextile (specific weight $1200 \, \mathrm{g/m^2}$) was emplaced. Drainage trenches were then installed, drainage layers placed in the front-dump mode of operation and concurrently a separating geotextile and material to be deposited in layers of at least 70 cm were put into position. The placement machines invariably moved on a $1 \cdot 0$-m-thick cushion.

The division of the foundation into five a hydraulically-independent sectors had a positive effect on the construction and operation sequence. Even under the most adverse weather conditions when work on the actual sealing system had to be stopped, other tasks in surrounding fields could be carried out. However, owing to the in-line structure of the construction site, the division in separate fields cannot compensate for long periods of adverse climatic conditions.

When planning the sequence of operations, special attention had to be given to the run-off of rainfall waters during the individual construction phases. Together with the described gradient conditions, the circumstances were thus provided to reduce the adverse effects of rainfall.

The design, construction and site management measures outlined above, in connection with a very dry and mild winter, meant that construction made unusually quick progress. Three months after commencement of site installation, site filling was started. Over the following three months the construction of 45 000 m² mineral sealings, the placement of plastic sealing linings and drainage layer, as well as the construction of 10 000 m slope sealings were completed. In the subsequent six months a total of 640 000 t coal were deposited. The surface combination sealing (55 000 m² including drainage layer and spade stop), was likewise constructed in an extremely short period of time (4 months).

Quality assurance. A quality assurance programme was prepared to aid the construction of the disposal site. This programme considered all components of the barrier system and determined the scope of quality assurance for internal and external supervision, as well as the quality requirements for individual components. The contracting firm provided a fully-equipped soil mechanical on-site laboratory in which all necessary field tests, including permeability tests, were performed. Thus, all test results from internal supervision were rapidly elaborated and were immediately available for practical work.

Hazardous Waste Disposal at Billigheim Landfill

The Billigheim landfill is situated about 20 km north of Heilbronn in Baden-Wurttemberg and is installed in a former clay pit. The work at the second construction section (60 000 m² area) was started in summer and completed in winter.

Measures taken during the construction phase of the mineral sealing. Accurate examination of the design of the combination sealing system (Fig. 6) reveals the central significance of measures aimed at protection from unfavourable weather conditions, taken when the Billigheim hazardous waste disposal site was extended. From the point of view of site management, construction of the 3·0-m-thick mineral sealing was an effective challenge.

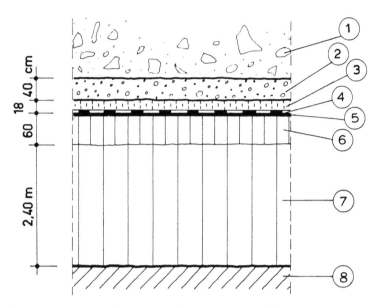

Figure 6. Formation of the base sealing system of the Billigheim waste disposal site: 1, deposited material; 2, drainage layer; 3, mineral protective layer; 4, geotextile; 5, plastic sealing linings HDPE (2·5 mm); 6, mineral sealing (3 layers of 20 cm; $k_{min} = 1 \times 10^{-10}$ m/s); 7, mineral sealing (11 layers of 22 cm; $k_{min} = 1 \times 10^{-9}$ m/s); 8, base.

The suitability of an existing natural medium-plastic clay soil with an approximate clay content of 30% was analysed. The placement and compaction of the 14 layers was carried out by rotary laser-controlled push dozers, sturdy rotavators for homogenizing, a centrifugal harrow, graders, high-speed compactors, sheepsfoot and smooth-wheel rollers (Fig. 7).

Although the clay sealing material used meets the German standards of the TA-Abfall (BMU, 1990) regarding soil-mechanical characteristic values, unfavourable weather conditions, especially rainfall, impair the sequence of construction. This applies in particular to the mineral sealing which may be washed out and softened by rainfall. On the other hand, attention should also be paid to the drying-out of the clay soil. These difficulties were overcome by the installation of the following comprehensive on-site facilities to protect from adverse weather conditions:

Figure 7. Construction of the mineral sealing material.

- roofed-over storage for depositing of mineral sealing of material in adverse climatic conditions;
- semi-automatic sprinkler system for dry weather;
- semi-automatic device for placing impermeable foil to protect from rainfall and drying-out.

These measures enabled a more rapid response than usual to weather changes. It should, however, be pointed out that the measures mentioned above should be selected individually, case by case. The experience in Billigheim showed that the protective foil was well suited to preventing the drying-out effect and made the sprinkler system largely unnecessary.

Application of foil against heavy rainfall proved efficient in the prevention of erosion groove formation in the mineral sealing material. The foil should be positioned as quickly as possible over the entire area. A semi-automatic unwinding device was installed on site for the test. The foil should be weighted against wind suction and checked regularly for efficient functioning (Fig. 8). The protective foil used should not be thin PVC construction foil but a strong tension material for multiple usage in order to prevent the production of waste.

In order to dry out wet mineral sealing materials, the contractor

Figure 8. Installation of the protective foils.

attempted to equip a centrifugal harrow with an additional heater blower. Thus, horizontal centrifugal soil harrowing was combined with concurrent blowing-in of hot air. Parallel measurements confirmed however that, owing to water evaporation (approx. 2300 kJ/kg), considerable quantities of energy were required to reduce water content. Alternatively, during the excavation of soil, transportation and intermittent storage is recommended to counteract a high water content; this can be obtained by spreading and drying of the sealing material on suitable free areas.

Plastic lining sealing operations. Plastic lining sealing operations are similarly subjected to weather conditions:

- placement can be hindered according to wind conditions;
- welding activities have to be interrupted during rainfall;
- in the case of high outside temperatures and intense sunshine, plastic liners tend to form intense waves;
- temperature changes have a considerable impact on the strength properties of thermoplastic materials (i.e. HDPE) and welded joints.

Furthermore problems are created by a moist base as placement of the plastic lining is hindered and welding operations complicated. This reveals moreover that preventive measures to protect against adverse weather conditions must be carefully evaluated, as provided for in German directives (BMU, 1990).

Quality assurance of the plastic liner was carried out. Several aspects, especially temperature impact on welding, are dealt with in this chapter since they are directly correlated to weather influence.

Documentation supplied for approval by the manufacturer of the plastic liners (Niederberg-Chemie, 1988, unpublished) filled an entire file and contained a comprehensive description of the product including specifications, description of placement and welding methods, representation of the welding seam test, placement plans including structural details, test reports on the suitability of the linings, as well as material and welding samples. The plastic sealing linings were exclusively made of HDPE, vestolene A 3512 R type, with 2% soot content.

The linings were welded on site, applying the hot wedge technique. In preliminary tests at the lining manufacturer's the effects of different welding parameter adjustments were investigated, covering:

• speed
• preheating temperature
• hot wedge temperature
• contact pressure of the hot wedge
• welding roller pressure

On-site temperature fluctuations were simulated by means of air-conditioning equipment. This preliminary investigation showed that welding parameters depended on the temperature of the linings. These were documented in operating instructions concerning the individual welding units and were the basis of on-site welding operations. Figure 9 shows an example of operation range with regard to welding speed versus temperature.

Similar instructions were developed for welding seam tests; tests to check air-pressure were chosen according to lining temperature in view of the fact that polyethylene, at high temperatures, loses strength considerably and the deformation module decreases. At a lining temperature of 50°C, the yield point and Young's module will amount to only approximately 70% of respective values obtained at 20°C.

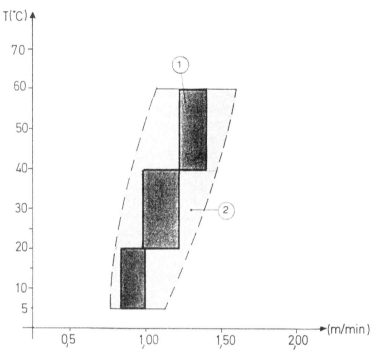

Figure 9. Range of operation (welding speed, m/min) for HK 3 welding unit: 1, range of operation; 2, admissible range.

Tests performed at constant pressure, frequently used in the past, do not of course comply to the above. As a decrease of Young's module, via deformation effects, increases seam load, the test pressures at higher temperatures were reduced non-linearly in relation to the decrease in strength, as follows:

Surface temperature	Test pressure
5–20°C	5 bar
20–35°C	4 bar
35–50°C	3 bar

This technique has been adequately proven in construction site

Figure 10. Welding of plastic liners (including continuous measuring and indication of the decisive welding parameters).

practice (Fig. 10). With regard to lining temperature measurements, non-contact temperature measuring instruments equipped with immediate temperature indication are suitable.

FINAL REMARKS

Performance requirements as described are mandatory in construction work and should be treated accordingly in the tender documents. As a rule, each constructor has particular views as to the meaning of preliminary remarks such as 'all weather protection measures shall be included into the prices per unit'. Moreover, such a definition generally violates the principle of contracting regulations that all performance requirements must be fully specified, reliable and should be definitely calculable for each bidder.

These items should be specified in such a way that the performance quantities to be expected are clearly defined and the economic value may be calculated and compared. The integration of such items in the bill of quantities guarantees that protection measures are implemented on competitive conditions with regard to the disposal site project.

This chapter outlines how the quality of disposal site systems may be further improved even when constructed under unfavourable weather conditions. This does not mean, however, that such projects should be carried out with the same priority in winter months. If overall conditions are the same, a base barrier system for a waste disposal site constructed in summer will be of superior quality. The authors wish to underline this fact in order to avoid the occurrence of situations whereby discussion on improvement of minor details is started in July and thereby delays the beginning of construction by months.

All reliable time estimations should, however, include an adequate number of bad weather days which may cause an interruption of all plant and equipment operations. The measures presented in this chapter, which refer mainly to soil engineering measures but indicate also impact on the placement of the plastic sealing linings, may help to reduce the number of such days.

REFERENCES

BMU (Bundesminister für Umwelt, Naturschutz und Reaktorsicherheit) (1990). Technical directive on the deposit, chemical/physical and biological treatment and burn up of especially supervision-requiring waste. In *Administrative Regulation of 17/12/1990 to Amend the 2. Administrative Regulation of the Waste Law* (*Technical Directive Waste*). Section I. Gemeinsames Ministerialblatt 41st Annual Vol., No 35, Carl Heymanns Verlag KG; Bonn (in German).

Rettenberger, G., Sasse, T. & Urban, S. (1988). Conception on the surface sealing system of the hazardous waste disposal site of Gerolsheim. *Stuttgarter Berichte zur Abfallwirtschaft*, Vol. 29. Erich Schmidt Verlag, Bielefeld (in German).

Sasse, T. (1988). Construction of combination sealing systems. Seminar documentation on construction and operation techniques regarding waste disposal sites, 17 November 1988. Technische Akademie, Wuppertal (in German).

Sasse, T. (1989). Experiences gathered in constructing combination sealing systems. Seminar documentation on construction and operation techniques regarding waste disposal sites, 12–13 October 1989. Technische Akademie, Wuppertal (in German).

3. PROPERTIES AND QUALITY CONTROL OF CLAY MATERIALS

3.1 Effect of Organic Liquids on the Hydraulic Conductivity of Natural Clays

ROBERT M. QUIGLEY & FEDERICO FERNANDEZ

University of Western Ontario, Faculty of Engineering Science, Ontario, Canada N6A 5B9

INTRODUCTION

This chapter presents a discussion of the influence of a variety of organic liquids on the hydraulic conductivity, k, of a natural, water-compacted clay. The clay tested was a weathered near-surface soil from southern Ontario, Canada, which contained ~20% quartz and feldspar, ~10% carbonate, ~50% illite, ~8% chlorite and ~12% smectite. The $<2\,\mu$m fines content amounted to ~50%. The 'extra' clay minerals are believed to be present in sand and gravel sized shale fragments in this waterlaid till.

A constant flow rate permeameter adapted for constant head testing was used as described in detail by Fernandez and Quigley (1985, 1990). The dielectric constant, ε, of the organic liquids ranged from 80 (polar water) to 32 (ethanol) to 2 (cyclohexane, toluene, dioxane). Both water-soluble (hydrophilic) and water-insoluble (hydrophobic) organic liquids were used. A range of static effective stresses from zero to 160 kPa was also employed to clearly illustrate the beneficial effects of stress in inhibiting large increases in k caused by pure soluble organics at low stresses.

This chapter is based on publications by Fernandez and Quigley (1985, 1988, 1991) and Quigley and Fernandez (1989, 1990, 1992). Dry clay–organic mixtures are not discussed since they represent an artificial, highly flocculated system that should not be typical of a barrier clay (Quigley & Fernandez, 1990). It is noted, however, that water contaminated with either soluble organic liquids or inorganic salts should *never* be used as a molding fluid, since the flocculated soils generate hydraulic conductivity values up to 1000 times greater than those for water-dispersed clays.

All information in this chapter pertains to dispersed, water-compacted clays.

PERMEATION BY PURE INSOLUBLE ORGANICS

Constant Head Tests

An unflawed, water-compacted clay resists penetration by insoluble, hydrophobic organic liquids due to a small pore size and very strong interparticle surface tension of water at the clay/organic interface. As shown by the constant head tests in Fig. 1, pure cyclohexane undergoes a step-wise penetration of the clay without breakthrough, up to a static total head of $\sim3\cdot46$ m (solid curve). At this head, $k = 0$ even though cyclohexane occupies about 17% of the total pore volume of $21\cdot9$ ml. For these tests, there was zero applied vertical effective stress on the fixed-ring sample.

Also illustrated in Fig. 1 is a curve for cyclohexane with 5% soap (polyvalent phosphate) added as a *surfactant*. The presence of soap serves to destroy the above-mentioned surface tension between the two liquids, resulting in easy penetration by the cyclohexane. The curve is slightly nonlinear indicating a slight increase in k as the head increases. At the end of testing, $k \approx 1\cdot2 \times 10^{-9}$ cm/s indicating a competent clay in spite of cyclohexane penetration. Since surfactants are probably present in most toxic organic wates, it is suggested that we should not rely on high breakthrough pressures of a good barrier clay to completely retain insoluble organic liquids.

Mutually soluble, semi-polar, organic liquids (such as the alcohols and acetone) render insoluble organics, water-soluble over certain concentration ranges (Fernandez & Quigley, 1989; Quigley & Fernandez, 1990*b*). These materials also permit easy penetration and unexpected increases in k by insoluble organic liquids.

Constant Flow Rate Tests

At low rates of forced flow, q, our constant flow rate permeameter, generates relatively low values of gradient, i, and measured hydraulic

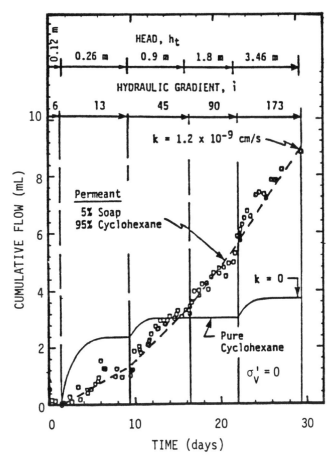

Figure 1. Constant head tests for insoluble cyclohexane penetration in water-compacted clay both pure and in the presence of 5% surfactant (soap) (after Quigley & Fernandez, 1989).

conductivities for pure insoluble organics that are close to or less than those measured for water. Increasing the flow rate (or gradient) to some critical value causes an abrupt enlargement of the flow channels and abrupt increases in the measured k. These features are illustrated in Fig. 2. For the curve shown, k started to increase from values close to those for water ($\sim 1 \times 10^{-8}$ cm/s) at a critical gradient of ~ 185. At the final gradient of ~ 430, k_f was 150×10^{-8} cm/s; much higher than

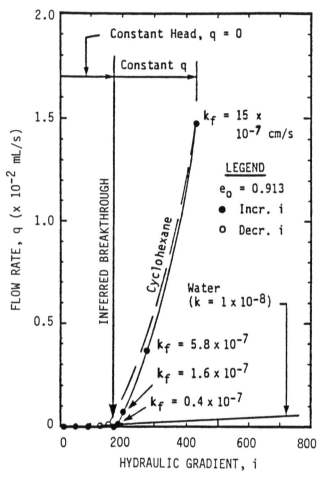

Figure 2. Flow rate versus gradient for insoluble pure cyclohexane flow through water-compacted brown silty clay. Note that five constant head and four constant flow rate values are plotted (after Quigley & Fernandez, 1989).

for water due to forced enlargement of channels and microfissures in the soil. For reference, the linear q versus i plot that yields $k = 1 \times 10^{-8}$ cm/s for water is also shown.

The most important aspect of the plot, however, is that if the flow rate curve is extrapolated back to zero flow, it intercepts the x-axis at a gradient inferred to be the breakthrough gradient. Using this important feature of constant flow rate testing, breakthrough gradients (or heads) may be determined on a clayey soil at different void ratios and different effective confining stresses just by increasing the flow rate in increments during a k-test.

Summary plots of flow rate versus hydraulic gradient are presented in Fig. 3 for six such tests. The four dashed curves were obtained on the test clay compacted to four different void ratios and tested at $\sigma'_v = 0$ kPa. The two solid curves were obtained at $\sigma'_v = 40$ and 160 kPa after wet compaction at a void ratio of about 0·90.

The four dashed curves show extreme nonlinearity of q versus i and indicate increasing breakthrough gradients as the void ratio decreases. This is completely consistent with the concept of increasing strength of capillary water as the pore size of a soil decreases. In all cases where the final pore fluid was analyzed at the end of testing, the insoluble organic occupied only 6–10% of the pore space, confirming its restricted passage along selected flow channels.

The two curves for cyclohexane flow under confining pressures of 40 and 160 kPa yield much higher breakthrough gradients at void ratios similar to those tested at $\sigma'_v = 0$ kPa. For example, at a final void ratio of 0·71, breakthrough gradients of ~410 and ~1500 were obtained at 0 and 40 kPa respectively. This indicates that the presence of effective stress on a linar system is an important ally in protecting the integrity of clayey barriers against non-aqueous phase liquid (NAPL) penetration.

This is further demonstrated by the plot of breakthrough gradient versus final void ratio plotted in Fig. 4. The gradients or heads required for breakthrough at $\sigma'_v = 0$ kPa are obviously much lower than those required if stresses are in place on the barrier clay.

In all of the tests reported in Figs 3 and 4, the soils were compacted by kneading at water contents above Harvard miniature optimum. For this clayey test soil, the Harvard miniature $\gamma_d(\text{max})$ and ω_{opt} values are very close to standard Proctor values. The soils are thus believed to have been relatively dispersed and free of large compaction-induced fractures and inter-clod pores that are very difficult to eliminate during dry-of-optimum compaction.

The hydrophobicity causing the above-described phenomena is illustrated pictorially in Fig. 5. Hydrophobic xylene placed in contact with clay particles suspended in water remains completely inert on a glass slide (Fig. 5(a)). If, however, water is placed next to a

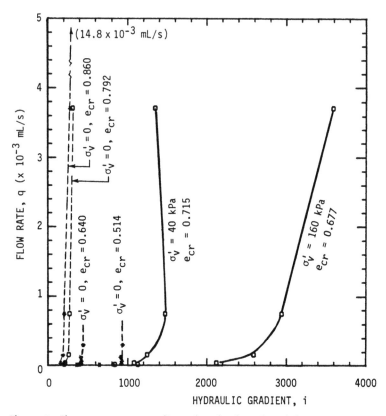

Figure 3. Flow rate versus gradient plots for forced cyclohexane permeation through water-compacted clay: e_{cr} = void ratio at which breakthrough is inferred to have occurred.

suspension of clay in xylene, the clay particles are observed to literally 'fly' towards the water, crossing the interface to produce a clay–water suspension.

Early work or reviews by Fernandez and Quigley (1985), Mitchell and Madsen (1987), Foreman and Daniel (1986) and Acar *et al.* (1985) indicated that forced permeation of insoluble organics through water-wet clays did not increase k beyond that for water. The above work for fixed-wall testing, however, clearly shows that testing at very high heads or very high flow rates may well generate anomalously high

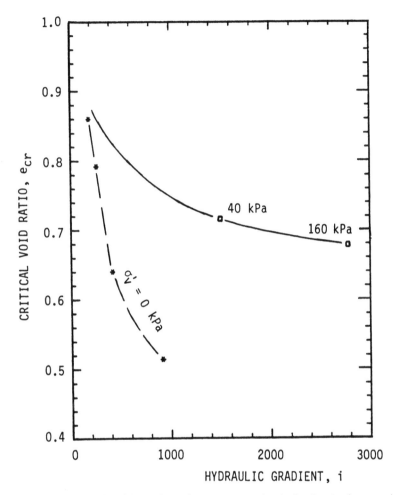

Figure 4. Inferred breakthrough gradient versus void ratio for Sarnia clay tested at σ_v'(static) = 0 kPa (four data points) and $\sigma_v' = 40$ and 160 kPa.

k values by a process of channel expansion which produces a nonlinear q versus i relationship.

The breakthrough resistance of clays to hydrophobic liquids was effectively employed by Lambe (1956) to store oil in water-compacted surface reservoirs in Venezuela.

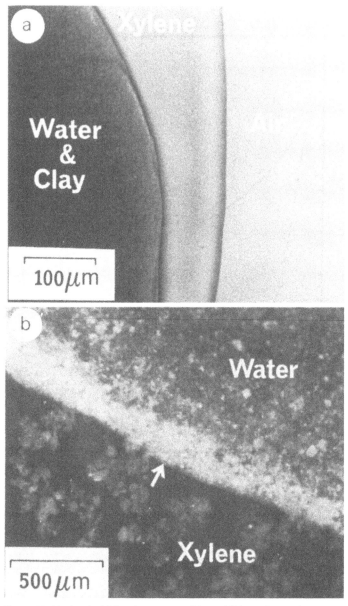

Figure 5. Hydrophobicity demonstrated by addition of (a) xylene to a suspension of 0·5 g clay in 30 ml of water, and (b) addition of water to a similar clay/xylene suspension (source: Fernandez & Quigley, 1985).

PERMEATION BY MIXTURES OF WATER-SOLUBLE ETHANOL AND MUNICIPAL SOLID WASTE LEACHATE

Most water-soluble organic liquids are polar in nature, making them completely soluble in polar water. Ethanol, having a dielectric constant of 25, is such a compound, and is the organic under discussion in this chapter. Some non-polar aromatic compounds such as dioxane are also soluble in water. These materials frequently have oxygens substituting for one or more of the carbons in the benzene ring enabling a hydrogen-to-oxygen linkage that produces complete solubility. For studies of dioxane behavior towards water-wet clays, the reader is referred to Fernandez and Quigley (1988, 1991).

A summary plot showing the hydraulic conductivity of water-compacted clay to ethanol/leachate mixtures is presented in Fig. 6. The mixtures vary in composition from 100% commercial ethanol to 100% municipal solid waste (MSW) leachate. For all test samples, an initial k was obtained for water (solid circles) prior to organic liquid permeation (solid triangles). Two sets of tests are presented; one for zero static effective stress on the samples ($\sigma'_v = 0$ kPa) and one for σ'_v (static) = 160 kPa.

Tests at Zero Effective Stress

Six tests are presented at $\sigma'_v = 0$, at which the reference value for water (k_w) was about 6×10^{-9} cm/s. Permeation with 'pure' MSW leachate (left axis) produced a slight decrease in k. Permeation with 'pure' ethanol (actually 95% ethanol plus 5% water, right axis) produced a 100-fold increase in k to $\sim 6 \times 10^{-7}$ cm/s. This very large increase in k is caused by ethanol entry into the double layers of the clay causing them to contract because of a reduction in dielectric constant from 80 (water) to 32 (commercial grade ethanol). At a constant void ratio, this increases the relative volume of free pore space in the sample producing the large increase in k. It is also possible that some side wall leakage develops as a result of sample shrinkage. As will be discussed later, very small confining stresses on a triaxial k-test sample would wipe out these large increases in k so that they can only be seen in fixed ring tests at σ'_v (static) = 0 kPa.

For all mixtures of ethanol and leachate up to $\sim 70\%$ ethanol,

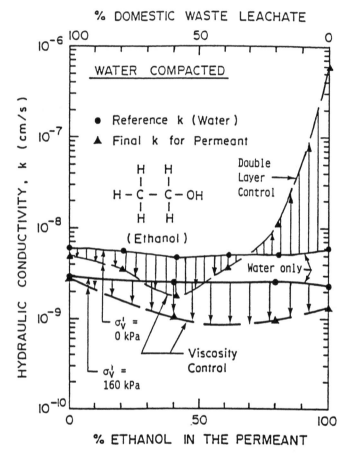

Figure 6. Hydraulic conductivity of water-compacted clay permeated with mixtures of ethanol and municipal solid waste leachate. Upper curves at σ_v'(static) = 0 kPa. Lower curves at σ_v'(static) = 160 kPa (source: Quigley and Fernandez, 1992).

significant decreases in k relative to water permeation were observed. These decreases are believed to be directly caused by increases in viscosity which reach a maximum at approximately 50% ethanol. This coupled with ethanol exclusion from the double layers where the cations prefer to hydrate with more polar water molecules results in the decreases in k. For a more thorough discussion of these phenomena, the interested reader is referred to Fernandez and Quigley (1988).

Tests at 160 kPa Effective Static Stress

Also presented in Fig. 6 are data points for k-tests run at $\sigma_v' = 160$ kPa. The reference k_w values are below those for $\sigma_v' = 0$ kPa yielding initial values of $\sim 3 \times 10^{-9}$ cm/s. The three data points for 40%, 80% and 100% ethanol all indicate a decrease in k. In all three cases, the permeant viscosity is greater than that for pure water and the presence of σ_v'(static) = 160 kPa on the samples results in significant consolidation and 'healing' of the samples during flow. These and other tests at 20, 40, 80 and 120 kPa all show similar consolidation and healing effects which are illustrated in Fig. 7.

The Beneficial Role of Stress

Summary plots of hydraulic conductivity and settlement versus static vertical effective stress are presented in Fig. 7(a) and (b). All tests were run first with water followed by commercial (95% pure) ethanol. The reference water curve shows a decrease in k from $\sim 7 \times 10^{-9}$ cm/s at $\sigma_v' = 0$ kPa to $\sim 2 \times 10^{-9}$ cm/s at 160 kPa. At $\sigma_v' = 0$ kPa, ethanol permeation causes a 100-fold increase in k to $\sim 6 \times 10^{-7}$ cm/s as illustrated previously in Fig. 6. The curve defined by the open circles clearly indicates that for static stresses greater than ~ 25 kPa present on the water-wet barrier, ethanol permeation results in *decreases* in hydraulic conductivity. In other words, very low levels of pre-damage stress may serve to completely prevent the damaging increases in k noted for $\sigma_v' = 0$ kPa. The settlement associated with ethanol permeation is quite large (1·5 mm or $\sim 7\%$ of the initial sample volume at 160 kPa) and is caused by both seepage stresses and chemical consolidation which is currently the subject of further research.

The application of stress after a test barrier has been damaged is much less effective in healing the clay. As shown by the post-damage stress curve (3) in Fig. 7(a), $\sigma_v' \approx 100$ kPa is required to heal the clay and the resulting settlements appear to be significantly less than those for the pre-damage stress application (Fig. 7(b)).

A complicating factor is the presence of seepage drag stresses on the

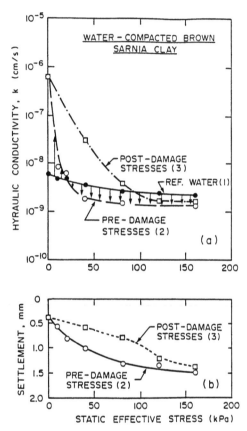

Figure 7. (a) Hydraulic conductivity and (b) settlement versus static effective stress level for ethanol permeation through water-wet clay (after Quigley & Fernandez, 1990).

samples during permeation which varies from 0 at the top of the barrier to $\mathcal{I}_{max}$ at the bottom. To illustrate the significance of this factor, the curves for reference water (1) and pre-damage stresses (2) are replotted in Fig. 8 with $\mathcal{I}_{max}$ included. The two curves labelled σ'_v(top) in the lower left part of the figure are the same as those plotted in Fig. 7 and again show that a static effective stress of only 20 kPa on the water-wet clay is enough to prevent any increases in k caused by reaction with the permeating ethanol.

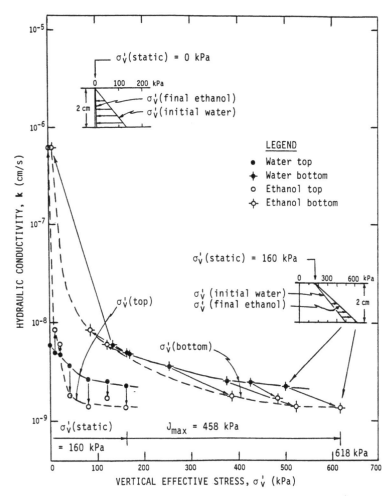

Figure 8. Hydraulic conductivity versus vertical effective static and seepage stresses for water-compacted clay permeated with commercial ethanol.

The two curves labelled σ_v'(bottom) represent a combination of static and seepage generated effective stresses at the bottom of the 2-cm-thick barrier. This set of curves shows a crossover point at ~120 kPa which means that the average stress required on the barrier to prevent any increases in k is about 60 kPa.

Two effective stress diagrams for the actual 2-cm-thick test specimens are included in Fig. 8 for σ'_v(static) = 0 and 160 kPa. At zero static stress (top diagram), the seepage stresses in the sample *decrease* dramatically as ethanol replaces the water and k increases dramatically from 7×10^{-9} cm/s to 6×10^{-7} cm/s. At 160 kPa static stress, the seepage stresses in the sample *increase* significantly as ethanol replaces the water and k decreases from $2 \cdot 2 \times 10^{-9}$ cm/s to $1 \cdot 4 \times 10^{-9}$ cm/s.

SUMMARY AND CONCLUSIONS

Permeation of liquid hydrocarbons through water-wet barrier clays may have either negligible or very damaging effects. While much work remains to be done, some conclusions may be drawn from the work presented herein.

Insoluble Organics

- Water-compacted clayey barriers resist penetration by insoluble organic liquids (NAPLs) due to very strong surface tension effects at the organic-soil/water interface. Even after breakthrough, k remains very low due to lyophobicity effects and flow restricted to macropores. Flaws in a barrier such as compaction- or shrinkage-induced fractures create higher velocity flow channels and much higher values of k.
- Use of excessive inlet heads or gradients may cause very high apparent k-values in the laboratory related to flow channel expansion that is completely non-representative of field conditions.
- Surfactants and mutually-soluble organic association liquids destroy surface tension effects, permitting easy entry of insolubles. This problem has received very little study, but work to date suggests that little reliance should be placed on breakthrough resistance in the field unless high stress levels are present.

Soluble Organics

- Water-soluble organics may cause huge increases in hydraulic conductivity at concentrations greater than 70% in low stress environments. In high stress environments, the increases appear to be eliminated by a combination of seepage stresses and chemical consolidation. Contraction of the double layers surrounding clay particles appears to be the cause of the large increases in k associated with 'pure product' permeation.

- At concentrations below 70%, decreases in k are observed for soluble organic permeation, even at low stress levels. These decreases are related to large increases in viscosity that reach a maximum at ~50% concentration for most soluble organics.
- Pre-damage effective stresses on a water-wet barrier are very beneficial in preventing the increases in k associated with concentrated organic permeation. For soluble organics with intermediate dielectric constant (alcohols with $\varepsilon = 25$ to 30), relatively low stress levels of 50–100 kPa may be adequate to prevent damage. For soluble organics with a very low dielectric constant (dioxane with $\varepsilon \approx 2$; Fernandez & Quigley, 1991), much higher stress levels of 200 kPa plus may be required.
- Post-damage application of stresses appears to be much less effective in healing a damaged barrier than pre-damage stresses are in preventing damage.

ACKNOWLEDGEMENTS

The research summarized in this chapter has been carried out using research funds to the first author from the Natural Sciences and Engineering Research Council of Canada and the Ontario Ministry of the Environment.

REFERENCES

Acar, Y. B., Hamidon, A., Field, S. D. & Scott, L. (1985). The effects of organic fluids on hydraulic conductivity of compacted kaolinite. In *Hydraulic Barriers in Soil and Rock, ASTM STP 874* ed. A. I. Johnson, R. K. Frobel, N. J. Cavalli and C. B. Pettersson. American Society for Testing and Materials, Philadelphia, pp. 171–87.

Fernandez, F. & Quigley, R. M. (1985). Hydraulic conductivity of natural clays permeated with simple liquid hydrocarbons. *Canadian Geotechnical Journal*, **22**, 205–14.

Fernandez, F. & Quigley, R. M. (1988). Viscosity and dielectric constant controls on the hydraulic conductivity of clayey soils permeated with water-soluble organics. *Canadian Geotechnical Journal*, **25**, 582–9.

Fernandez, F. & Quigley, R. M. (1991). Controlling the destructive effects of clay–organic liquid interactions by application of effective stresses. *Canadian Geotechnical Journal*, **28**, 388–98.

Foreman, D. & Daniel, D. E. (1986). Permeation of compacted clay with organic chemicals. *ASCE Journal of Geotechnical Engineering*, **112**, 669–81.

Lambe, T. W. (1956). The storage of oil in an earth reservoir. *Journal of Boston Society of Civil Engineers*, **43**, 179–241.

Mitchell, J. K. & Madsen, F. T. (1987). Chemical effects on clay hydraulic conductivity. In *Proceedings, Specialty Conference on Geotechnical Practice for Waste Disposal '87*, ed. R. D. Woods, American Society of Civil Engineers, Geotechnical Special Publication 13, pp. 87–116.

Quigley, R. M. & Fernandez, F. (1989). Clay/organic interactions and their effect on the hydraulic conductivity of barrier clays. In *Proc. International Symposium on Contaminant Transport in Groundwater*, ed. H. E. Kobus & W. Kinzelbach. A. A. Balkema, pp. 117–24.

Quigley, R. M. & Fernandez, F. (1990a). Hydrocarbon liquids and clay microstructure. In *Microstructure of Fine-Grained Sediments: From Mud to Shale*, ed. R. H. Bennett, W. R. Bryant & M. H. Hulbert. Springer-Verlag, New York, pp. 469–74.

Quigley, R. M. & Fernandez, F. (1992). Organic liquid interactions with water-wet barrier clays. In *Proc. International Conference on Subsurface Contamination by Immiscible Fluids*. A. A. Balkema, pp. 49–56.

3.2 Diffusive Transport of Pollutants through Clay Liners

R. KERRY ROWE

Department of Civil Engineering, Geotechnical Research Centre,
The University of Western Ontario, London, Ontario, Canada N6A 5B9

INTRODUCTION

The evaluation of the suitability of a clay liner has often focused on the hydraulic conductivity. In some jurisdictions a clay liner is required to have a hydraulic conductivity, k, of less than 10^{-7} cm/s while in other cases, a much tighter specification of hydraulic conductivity of less than 10^{-8} cm/s has been required (Lahti *et al.*, 1987). The objective of controlling the hydraulic conductivity is clearly one of limiting advective contaminant transport (i.e. the movement of contaminants with moving water) through the liner. However, despite more than a decade of research and the existence of good supporting field data, it is only recently that it has been generally recognized that there is a second contaminant transport process which will occur even through a very low hydraulic conductivity clay liner: that process is chemical diffusion. As will be demonstrated in this chapter, diffusion may be the dominant contaminant transport mechanism in a well-constructed clay liner. Furthermore, contaminants can escape from a waste disposal site, by diffusion through a liner, even if water flow in the liner is into the landfill.

DIFFUSION AS A TRANSPORT MECHANISM

Diffusion involves the movement of contaminants from points of high chemical concentration (e.g. a landfill) to points of lower chemical concentration (e.g. an underlying aquifer or drainage system). The driving force for diffusive transport is a concentration gradient (or, more precisely, a chemical potential gradient). The mass flux, f, (e.g. the mass of contaminant transported through a unit area of liner in a unit time) can be given by a slightly modified version of Fick's first law:

$$f = -nD_e \frac{\partial c}{\partial z} \tag{1}$$

where

f is the chemical mass flux $[ML^{-2}T^{-1}]$,

n is the porosity of the clay liner (dimensionless),

D_e is the diffusion coefficient in the clay $[L^2T^{-1}]$,

c is the solute concentration in the pore fluid at a depth z $[ML^{-3}]$,

z is the point of interest $[L]$ and

$\partial c/\partial z$ is the concentration gradient (i.e. the change in concentration with position).

The negative sign arises from the fact that contaminants move from high to low concentration and hence the gradient $\partial c/\partial z$ will be negative.

The effective diffusion coefficient D_e will depend on numerous factors including the valence and ionic radius of the contaminant species, the temperature, the type of clay, the pore size and pore size distribution (for example, see Quigley *et al.*, 1987*a*). Attempts are often made to separate the chemical and soil components by decomposing the effective diffusion coefficient into two components:

$$D_e = \tau D_0 \tag{2}$$

where

τ is called a tortuosity factor and is intended to represent the effect of the soil upon the effective rate of diffusion,

D_0 is the free solution diffusion coefficient (e.g. the diffusion coefficient in pure water) and is intended to represent the chemical aspects (i.e. effect of temperature, ionic radius, valence, etc.)

The diffusion of chemicals at infinite dilution (i.e. in pure water) has been extensively examined in the chemical literature and values for the

TABLE 1. Tracer Diffusion Coefficient, D_o, of Selected Species at Infinite Dilution, 25°C (modified from Lerman, 1979; Quigley *et al.*, 1987*a*)

Cation	D_o (m²/annum)	Anion	D_o (m²/annum)
H^+	0·294	OH^-	0·166
Li	0·032 5	Cl^-	0·064
Na^+	0.041 9	HS^-	0.054 5
K^+	0.061 8	SO_4^{2-}	0.033 7
NH_4^+	0.062 4	N_2^-	0.060 2
Mg^{2+}	0.022 2	NO_3^-	0.060 0
Ca^{2+}	0.025 0	HCO_3^-	0.037 2
Mn^{2+}	0.021 7	CO_3^-	0.030 1
Fe^{2+}	0.022 6	PO_4^{3-}	0.019 2
Cu^{2+}	0.023 1	CrO_4^{2-}	0.035 3
Zn^{2+}	0.022 5		
Cd^{2+}	0.022 6		
Pb^{2+}	0.024 8		

free diffusion coefficient D_o have been published in the literature as shown in Table 1. The values given in Table 1 are at 25°C. At lower temperatures, the diffusion coefficient can be lower than that at 25°C. Since there are a number of complicating factors associated with changing temperature, the effect of temperature is often best considered by performing tests at or near the expected field temperature.

As discussed by Quigley *et al.* (1987*a*), the 'tortuosity' factor τ is a complex factor which incorporates components including decreased fluidity related to adsorbed double-layer water, an electrostatic interaction factor, and a physical tortuosity factor. Furthermore, it must be recognized that in real situations contaminant species do not move in isolation but rather, they move along with other contaminant species. For example, an anion, such as chloride (Cl^-), does not move alone but must have some cations moving with it to maintain electroneutrality. Thus, the movement of a species such as chloride will depend on the availability and mobility of associated cations. This means that, in addition to the factors listed above, the migration of even a 'conservative species' such as chloride (which does not interact directly with the clay) will depend on the type of clay and the chemical characteristics of the pore fluid and the leachate.

It must be admitted that diffusion is a complicated process; diffusion of leachate through soil is even more complicated and there are still unanswered questions. As a result, it is difficult to calculate a diffusion coefficient in soil from fundamental principles. Fortunately, however, an alternative simpler 'engineering approach' tends to provide quite acceptable results. This approach involves estimating the effective diffusion coefficient directly by simple laboratory tests using the soil and a leachate as close as practical to that to be used in the field. Often, the diffusion coefficient so determined can then be used to estimate contaminant migration through liners. This approach will be discussed in the following sections:

EQUATIONS GOVERNING ADVECTIVE–DIFFUSIVE TRANSPORT

It is well recognized (for example, see Freeze & Cherry, 1979) that contaminant transport through soils can be generally modelled using the advection–dispersion equation which, for one-dimensional conditions, can be written as

$$n\frac{\partial c}{\partial t} = \left(nD\frac{\partial^2 c}{\partial z^2} - nv\frac{\partial c}{\partial z}\right) - \rho K_d \frac{\partial c}{\partial t} \qquad (3)$$

where

c is the concentration at depth z at time t [ML^{-3}],
n is the porosity (dimensionless),
D is the coefficient of hydrodynamic dispersion [L^2T^{-1}],
v is the average linearized groundwater (seepage) velocity [LT^{-1}],
ρ is the dry density of the soil [ML^{-3}]
K_d is a distribution or partitioning coefficient [L^3M^{-1}] and
v_a $= nv$ is the Darcy or discharge velocity (flux) [LT^{-1}].

Equation (3) is based on considerations of conservation of mass and simply states that the increase in contaminant concentration within a small volume of soil is equal to the increase in mass due to advective–diffusive transport minus the mass of contaminant removed from solution by what are loosely referred to as 'sorption processes.' Here, it is assumed that the sorption processes are linear and can be presented in terms of a partitioning or distribution coefficient K_d (see Freeze & Cherry (1979) for more details regarding 'sorption processes'). The product ρK_d is a dimensionless measure of the amount of sorption that is likely to occur. A contaminant species is said to be 'conservative' if there is no sorption (i.e. $\rho K_d = 0$).

The coefficient of hydrodynamic dispersion is often decomposed into two components

$$D = D_e + D_m \qquad (4)$$

where D_e is the effective diffusion coefficient (as previously discussed) and D_m is the coefficient of mechanical dispersion. The processes of diffusion and dispersion are quite different. However, for most practical purposes, they can be modelled in the same way and, hence, are lumped together as the coefficient of hydrodynamic dispersion.

When dealing with contaminant migration through intact caly, it has been found that diffusion is the dominant process (e.g. see Gillham & Cherry, 1982; Rowe, 1987; Rowe *et al.*, 1988) and to all practical purposes

$$D \simeq D_e \qquad (5a)$$

In aquifers, the opposite tends to be true and dispersion tends to dominate. It is often convenient to model dispersive processes as a linear function of velocity (Bear, 1979):

$$D \simeq D_m = \alpha v \qquad (5b)$$

where α is the dispersivity [L]. The dispersivity α tends to be scale dependent and is not a true material property (see Gillham & Cherry, 1982).

DETERMINATION OF DIFFUSION COEFFICIENTS

The diffusion coefficient, the partitioning or distribution coefficient (expressed in terms of the dimensionless product ρK_d) and the effective porosity can be estimated from laboratory tests. For example, Rowe *et al.* (1985, 1988) have developed a technique that allows the determination of these parameters using 'undisturbed' samples of the proposed liner material (be it a natural clay depoist or a compacted clay liner).

In the proposed test, an undisturbed sample soil is placed in a column and the leachate of interest is placed above the soil. Contaminant is then permitted to migrate through the specimen under the prescribed head (which may be zero). The volume of leachate above the soil will normally be selected to be sufficiently small to allow a significant drop in concentration of contaminant within the source

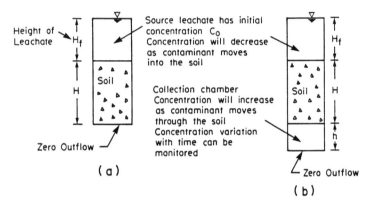

Figure 1. Schematic of test used to determine diffusion and distribution coefficients: (a) zero flux at base of the soil; (b) migration into a collection chamber (source: Rowe, 1988. Reproduced with the permission of the *Canadian Geotechnical Journal*).

solution (typically the height of leachate in the column above the clay will range from 0·03 to 0·3 m). This drop in concentration with time should be monitored.

A number of possible boundary conditions at the base of the sample may be considered. If the test is to be conducted with advective transport through the specimen, then a porous collection plate can be placed beneath the sample and the effluent collected and monitored (see Rowe *et al.*, 1988). If there is no advective flow, then two other base boundary conditions may be considered.

Firstly, the base could be an impermeable plate (see Fig. 1(a)), in which case the base boundary condition is simply given by

$$f_b(t) = 0 \qquad \text{for all } t \tag{6}$$

where f_b is the mass flux at the base of the soil column $[ML^2T^{-1}]$.

The second alternative is to have a closed collection chamber (reservoir) similar to that for the leachate but initially having only a background concentration of the contaminant of interest (see Fig. 1(b)). Thus, as contaminant passes through the soil, it accumulates (and can be monitored) in this collection chamber.

Suppose that the volume of source solution (leachate) is equal to $A \cdot H_f$ where A is the plan area of the column and H_f is the 'height of the leachate' in the column (e.g. see Fig. 1). Then at any time t, the mass of any contaminant species of interest in the source solution is equal to the concentration $c_t(t)$ in the solution multiplied by the volume of solution (assuming here that the solution is stirred so that $c_t(t)$ is uniform throughout the solution). The principle of conservation of mass then requires that at this time t, the mass of contaminant in the source solution is equal to the initial mass of the contaminant minus the mass which has been transported into the soil up to this time t. This can be written algebraically as

$$c_t(t) = c_0 - \frac{1}{H_f} \int_0^t f_t(\tau) \, d\tau \qquad (7)$$

where

$c_t(t)$ is the concentration in the source solution at time t_1 [ML³],

c_0 is the initial concentration in the source solution ($t = 0$) [ML³],

H_f is the height of leachate (i.e. the volume of leachate per unit area) [L] and

$f_t(\tau)$ is the mass flux of this contaminant into the soil at time τ [ML²T⁻¹].

Contaminant is allowed to migrate from the source chamber through the soil and, if present, into the collection chamber. If no additional contaminant is added to the source chamber, then the concentration of contaminant will decrease with time as mass of contaminant diffuses into the soil (see Fig. 2). The rate of decrease can be controlled by the choice of the height of leachate, H_f. Conversely, as contaminant diffuses into the collection chamber (Fig. 1(b)), the increase in mass gives rise to an increase in contaminant concentration in this reservoir (see Fig. 2) which can be modelled by an equation analogous to eqn (7):

$$c_b(t) = c_b + \frac{1}{h} \int_0^t f_b(\tau) \, d\tau \qquad (8)$$

where

$c_b(t)$ is the concentration in the collection reservoir at time t [ML³],

c_{b0} is the initial concentration of the species in the collection reservoir (which may be zero) [ML³],

h is the volume of the collection reservoir per unit area (see Fig. 1) [L] and

$f_b(\tau)$ is the mass flux into the collection reservoir at time τ [ML²T⁻¹].

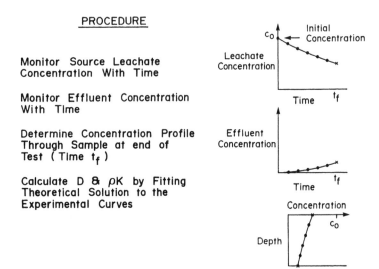

Figure 2. Experimental procedure used to determine the diffusion coefficient *D* and distribution coefficient K_d from the test shown schematically in Fig. 1(b) (source: Rowe, 1988. Reproduced with the permission of the *Canadian Geotechnical Journal*).

The rate of decrease in concentration in the source and, where appropriate, the increase in the collection chamber should be monitored with time. At some time t_f, the test is terminated and the concentration profile through the soil sample may be determined (see Fig. 1). Assuming linear sorption, theoretical models can then be used to estimate the parameters *n*, *D* and ρK_d by solving eqns (4), (7) and either (6) or (8). This theoretical analysis has been described in detail by Rowe and Booker (1985a, 1987) and has been implemented in the computer program POLLUTE (Rowe & Booker, 1983–90). This approach permits very accurate calculation of concentration in only a few seconds on a microcomputer and hence is well suited for use in the interpretation of the results of the column tests. Applications of this technique has been described in detail by Rowe *et al.* (1988) and Barone *et al.* (1989).

To illustrate the application of this approach and the need for caution in determining parameters, Fig. 3 shows results obtained for potassium migration in two diffusion tests on the same clay where two different contaminant sources were used. The first source was a raw

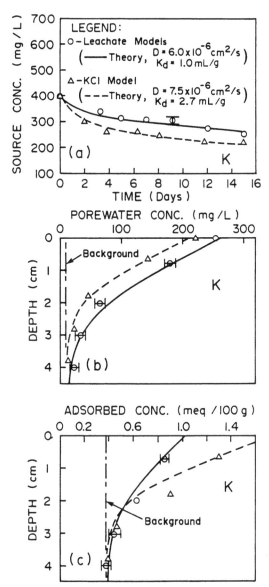

Figure 3. Migration data for potassium: (a) variation in source concentration with time; (b) pore water concentration versus depth at the end of the test—15 days; (c) absorbed concentration versus depth at the end of the test (source: Barone *et al.,* 1989. Reproduced with the permission of the *Canadian Geotechnical Journal*).

leachate whose chemical composition has been described by Barone *et al.* (1989) and which had significant concentration of Na^+, K^+, Ca^{2+} and Mg^{2+} with Cl^- being the primary anion. The second source solution was a single salt solution consisting of K^+ and Cl^- alone.

Figure 3(a) shows the observed decrease in concentration with time in the source solution as potassium diffuses into the clay. Figures 3(b) and 3(c) show the pore water concentration versus depth and the adsorbed concentration versus depth, respectively, at the termination of the test. The experimental results presented in Fig. 3 represent the average of two identical tests. The results of duplicate tests were in good agreement with the range of values being shown by the 'error bar' in Fig. 3 (where no range is shown, the difference in results was negligible).

The results presented in Fig. 3 show two things. Firstly, for a given soil and leachate, it was possible to obtain good fits between the experimental data and the theoretical model (which is based on eqn (3)); and hence both $D = D_e$ and ρK_d can be determined. It can, in fact, be shown that there is a unique combination of D and ρK_d which will give a good fit to the data in Figs. 3(a)–(c). Secondly, it is evident that, for potassium, both D and ρK_d are dependent on the chemistry of the source solution. As discussed by Barone *et al.* (1989), there is a good physiochemical explanation for this, related to competition for sorption sites, the desorption of Mg^{2+} and Ca^{2+} from the clay as K^+ is sorbed and the need to maintain electroneutrality. Thus, for example, the sorption (i.e. ρK_d) of K^+ for the single salt source is substantially higher than for the leachate source since in the former case there is no competing species for sorption sites in the source while in the leachate case K^+ must compete with high concentrations of Na^+, as well as Ca^{2+} and Mg^{2+} for sorption sites, and hence less potassium is sorbed. It is also noted that the diffusion coefficient is not the same. Even for a so-called conservative species (i.e. one which does not adsorb, $\rho K_d = 0$) such as chloride, the diffusion coefficient can vary depending on the source leachate (in this case by 25%) since the requirement that electroneutrality be maintained links the migration of Cl^- to the migration of cations in the pore fluid of the clay.

The key point to draw from Fig. 3 and the foregoing discussion is that diffusion coefficients and sorption parameters should be obtained on the proposed liner soil using a source leachate as similar as possible to that expected in the field situation.

Fortunately, despite the potetial variation in diffusion coefficients due to soil and leachate composition, the range of variation is relatively small compared to parameters such as hydraulic conductivity. In engineering terms, diffusion is in fact a very predictable process and can be predicted using simple models for time spans of thousands of years (for example, see Desaulniers *et al.*, 1981).

By comparing the results obtained from tests with advective transport ($v_a \leq 0.035$ m/annum) and pure diffusion tests, Rowe *et al.* (1988) have demonstrated that for the clayey soil they examined, mechanical dispersion is negligible. This is consistent with empirical/theoretical expectations discussed by Rowe (1987) and the resulting conclusions summarized in the following section.

Tests such as those described in this section can also be used to estimate the effective porosity. Thus, for a conservative species (e.g. Cl^-), one can adjust both the effective porosity or diffusion coefficient when attempting to fit the theory to the experimental data. Using this approach, Rowe *et al.* (1988) demonstrated that for the clay they were studying, the effective porosity was not significantly different from the porosity simply based on water content.

RELATIVE IMPORTANCE OF DIFFUSION AND ADVECTION

To illustrate the relative importance of diffusion and advection as transport mechanisms, consider a clay liner of thickness $H = 1.2$ m with very simple boundary conditions—a constant source concentration c_0 at the surface and a high flushing velocity at the base (such that all mass exiting the liner is immediately removed) so there is no concentration buildup at the base.

Figure 4 shows the variation in exit flux of chloride (i.e. the mass exiting the liner per unit area per unit time) for the case of pure diffusion ($D = 0.018$ m²/annum, $v_a = 0$), pure advection ($D = 0$, $v_a = 0.006$ m/annum) and advective–diffusive transport ($D = 0.018$ m²/annum, $v_a = 0.006$ m/annum). Coincidentally, in this example the maximum flux of 6 g/m²/annum is identical for both the pure diffusion and pure advection cases. Conventional calculations performed neglecting diffusion and assuming plug flow ($D = 0$, $v_a = 0.006$ m/annum), would suggest that no contaminant would escape into the aquifer until the seepage front arrived at the base after 75 years. However, diffusion is important and due to diffusion alone

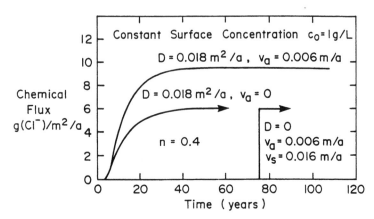

Figure 4. Effect of assumptions concerning the mechanisms for contaminant transport on the chemical flux exiting from a 1·2-m-thick clayey liner (after Rowe, 1987).

an exit flux exceeding 10% of the maximum flux would be expected after only 5 years. Indeed, the maximum flux of 6 g/m²/annum would be attained after only 50 years (compared to 75 years for plug flow). Consideration of both advection and diffusion gives a substantially higher flux at any time with the peak flux being 55% higher than the peak flux obtained by considering either diffusion or advection independently.

One can readily estimate the flux through the liner for the limiting conditions of pure diffusion and pure advection, viz.

$$f = -nD\frac{\partial c}{\partial z} = nDc_0/H \qquad \text{(for } v_a = 0\text{—pure diffusion)} \qquad (9a)$$

$$f = nvc_0 = v_a c_0 \qquad \text{(for } D = 0\text{—pure advection)} \qquad (9b)$$

where H is the thickness of the liner and c_0 is the constant leachate concentration (i.e. 1·2 m and 1 g/liter respectively in this case). One can obtain an estimate of the relative importance of the two transport mechanisms by determining the actual flux (based on solving eqn (3) for a given velocity v_a and then dividing this by an estimated flux based on eqn (9) (using whichever of eqns (9a) or (9b) gives the larger number). The results of this calculation are given in Fig. 5.

Diffusion is the dominant mechanism for Darcy velocities less than 2×10^{-4} m/annum and for these cases advection can be neglected.

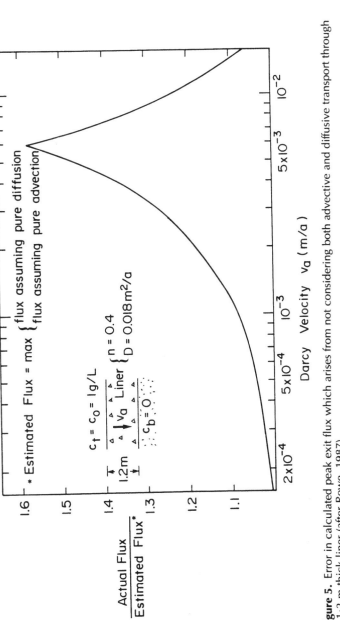

Figure 5. Error in calculated peak exit flux which arises from not considering both advective and diffusive transport through a 1·2-m-thick liner (after Rowe, 1987).

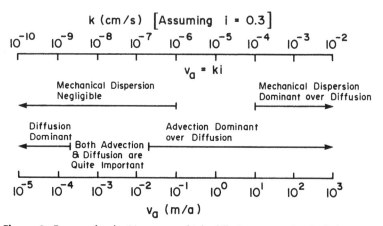

Figure 6. Range of velocities over which diffusion or mechanical dispersion controls the coefficient of hydrodynamic dispersion *and* the range of velocities over which diffusion or advection dominates the peak exit flux through a 1·2-m-thick clayey liner (see text for details) (after Rowe, 1987).

Advection dominates over diffusion for Darcy velocities greater than 0·02 m/annum in the case considered here. Both advection and diffusion play a very important role for intermediate velocities. In this case, the maximum interaction occurs for $v_a = 0·006$ m/annum. In many practical situations involving clayey barriers the hydraulic conductivity and gradient will give Darcy velocities in the ciritical range where advection and diffusion are important. For example, a hydraulic conductivity of 10^{-7} cm/s and gradient of 0·2 gives $v_a = 0·006$ m/annum. Similarly, a hydraulic conductivity of 2×10^{-8} cm/s and gradient of 1·0 gives 0·006 m/annum.

Figure 6 summarizes the range of velocities in which diffusion and advection have the dominant effect for this problem. Also shown is the range of velocities over which the coefficient of hydrodynamic dispersion is controlled by diffusion or mechanical dispersion as determined by Rowe (1987) based on the empirical relationship established by Perkins and Johnston (1963).

COMPARISON BETWEEN OBSERVED AND CALCULATED FIELD PROFILES

The Confederation Road landfill near Sarnia, Ontario is one of the best-documented case histories where the migration of contaminants

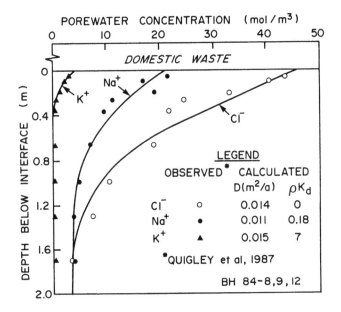

Figure 7. Comparison of observed and calculated migration profiles beneath the Confederation Road landfill after 16 years migration.

has been monitored in an in-situ natural clay barrier. The migration has been carefully monitored over a 16-year period by researchers at The University of Western Ontario (Goodall & Quigley, 1977; Crooks & Quigley, 1984; Quigley & Rowe, 1986; Quigley *et al.*, 1987*b*; Yanful *et al.*, 1988).

Figure 7 shows the observed concentration profiles for chloride (Cl^-), sodium (Na^+) and potassium (K^+) which were obtained in 1984 (after about 16 years migration). It is evident from Fig. 7 that after 16 years, chloride (a conservative species) has migrated about 1·3 m below the waste (see also Chapter 2.1). The cations Na^+ and K^+ are being retarded by interaction (cation exchange) with the clay, the more highly sorbed K^+ having migrated less than 0·4 m. An investigation of the migration of the heavy metals copper, lead, zinc and iron (Quigley *et al.*, 1987b; Yanful *et al.*, 1988) has shown that they have been restricted to the upper 0·1–0·2 m of the clay in this same time period.

Laboratory tests were performed to determine diffusion and partitioning coefficients for the three species Cl^-, Na^+ and K^+ (Rowe *et al.*, 1988). The diffusion coefficients that were obtained at laboratory temperature were adjusted for field temperature (for example, see Crooks & Quigley, 1984; Quigley *et al.*, 1987a) and the concentration profiles were calculated and plotted (Fig. 7). Allowing for inevitable experimental scatter, it can be seen that the agreement between observed and calculated profiles is quite encouraging. (A more detailed discussion of modelling at this landfill is given in Quigley & Rowe (1986).)

MASS OF CONTAMINANT AND LEACHATE COLLECTION

For waste disposal sites such as municipal landfills, the mass of any potential contaminant within the landfill is finite. The process of collecting and treating leachate involves the removal of mass from the landfill and hence a decrease in the amount of contaminant that is available for transport through the liner and into the general groundwater system. Similarly, the migration of contaminant through the barrier also results in a decrease in the mass available within the landfill. For a situation where leachate is continually being generated (e.g. due to percolation through the landfill cover), the removal of mass by either leachate collection and/or contaminant migration will result in a decrease in leachate strength with time (i.e. there will be a decrease in concentration similar to that observed in the laboratory test described in a previous section, see Rowe 1991). A detailed discussion of this is beyond the scope of this chapter. However, it should be noted that the potential impact of diffusion-controlled contaminant transport through a liner is highly dependent on the mass of contaminant available for transport into the liner.

The mass available for transport (i.e. excluding that collected by a leachate collection system) can be represented in terms of an equivalent height of leachate H_f in a similar manner to that adopted in the analysis of laboratory tests as described in an earlier section. Readers interested in the determination of the equivalent height of leachate, H_f, are referred to Rowe (1988). It is also noted that recent theoretical models also allow direct modelling of the removal of mass by a leachate collection system, for example see Rowe and Booker

(1988; 1990). In thse models, the total mass of contaminant and the volume of leachate collected per annum are inputs to the analysis, where the total mass is represented by a 'reference height of leachate' H_r. This approach is discussed by Rowe (1991).

DIFFUSION IN THE OPPOSITE DIRECTION TO ADVECTION

A parametric evaluation of the mass flux through a 1·2-m-thick liner for a given diffusion coefficient and a range of advective velocities given in Fig. 5 (and discussed previously) shows that diffusion can be important for low Darcy velocities (i.e. $v_a \leq 0·02$ m/annum) where advection and diffusion are in the same direction (i.e. both out from the waste and into the liner). If the direction of diffusive transport is the same as the direction of advective flow then it will increase the amount of contaminant transported and decrease the time it takes for contaminant to move to a given point away from the source. However, diffusion can also occur in the direction opposite to advective transport. For example, if the concentrations of contaminant in the landfill exceed those beneath the liner, then there will be diffusive transport out of the liner. If the water pressures at the base of the liner exceed those on the top of the liner, then groundwater flow will be into the landfill. Under these circumstances it is still possible for contaminant to escape from the landfill and the level of impact will depend on the chemical concentration gradient across the liner and the magnitude of the inward Darcy velocity. The impact can be calculated and a number of landfills have been designed to minimize impact by controlling the inflow of water to the landfill (this type of design is sometimes referred to as a 'hydraulic trap'; see also Chapter 2.6).

To illustrate the potential for contaminant migration in the opposite direction, consider a liner with thickness H and the boundary conditions of constant source concentration c_0 in the landfill and a high flushing velocity at the base of the liner (such that all mass exiting of the liner is immediately removed) so that there is no concentration buildup at the base. For these conditions, there is a steady-state concentration profile in the liner given by Al-Niami and Rushton (1977):

$$\frac{c}{c_0} = \frac{\exp\left[-v(H-z)/D\right] - 1}{\exp\left[-vH/D\right] - 1} \quad \text{for } v < 0 \tag{10}$$

where

c is the concentration at depth z [ML^{-3}],

c_0 is the source concentration $(x = 0)$ [ML^{-3}],

v is the linearized groundwater velocity taken as negative for inward flow [LT^{-1}],

D is the effective diffusion coefficient [L^{2}T^{-1}] and

H is the thickness of the liner [L].

because of the boundary condition that $c = 0$ at $z = H$, eqn (10) necessarily implies zero concentration at the base; however, this does not imply that no contaminant can escape from the liner. On the contrary, one can determine the flux into the liner from eqn (10) and it is given by

$$\frac{f_b}{c_0} = \frac{-nv}{\exp[-vH/D] - 1} \quad \text{for } v < 0 \tag{11a}$$

where f_b is the mass flux out of the liner [ML^{-1}T^{-1}] and all other terms are as previously defined. It is noted that for $v = 0$, eqn (11a) transforms to

$$\frac{f_b}{c_0} = \frac{nD}{H} \quad \text{for } v = 0 \tag{11b}$$

Figure 8 shows the calculated steady-state contaminant flux passing

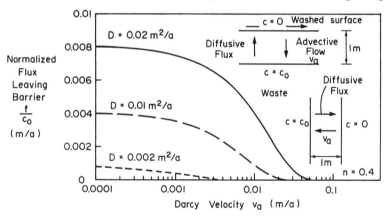

Figure 8. Diffusive flux of contaminant into an aquifer against an inward flow at a Darcy velocity v_a for steady state conditions (source: Rowe, 1988. Reproduced with permission of the *Canadian Geotechnical Journal*).

into an aquifer beneath a 1-m-thick clayey barrier for a range of inward Darcy velocities and assumed diffusion coefficients.

The flux f has been divided by the initial source concentration and the resulting normalized flux f/c_0 has units of velocity. (This may, in fact, be thought of as the equivalent outward velocity of contaminant migration which occurs due to the outward diffusive transport in opposition to inward flow.) The effective diffusion coefficient for chloride often lies in the range $D = 0.01$–$0.02\,\mathrm{m^2/annum}$. For these conditions, Fig. 8 shows that the inward Darcy velocity would have to exceed 0.025 and $0.05\,\mathrm{m/annum}$ for $D = 0.01$ and $0.02\,\mathrm{m^2/annum}$ respectively before the outward flux was reduced to negligible levels for a 1-m-thick liner.

PREDICTION OF CONTAMINANT IMPACT DUE TO DIFFUSION-CONTROLLED MIGRATION

The examples considered in Figs 5 and 8 illustrate that there can be contaminant transport due to diffusion for small Darcy velocities ($v_a \leq 0.02\,\mathrm{m/annum}$), even when the velocity is in the opposite direction to that of diffusive transport ($v_a < 0$). Both of these cases considered extreme boundary conditions involving a constant source inside the liner and zero base concentration outside the liner. In many practical situations the mass of contaminant will be finite and so the source concentration will decrease with time. Similarly, in many cases the base flushing velocity will not be infinite and there will be some increase in base concentration with time. Physically, this 'base' layer may be a natural aquifer or an engineered drainage system.

There are a wide variety of techniques available for the analysis of contaminant impact due to migration through a liner. These include finite element, finite difference, finite layer, boundary element and analytic solutions. Of these, the finite element programs are most general. For many practical problems, however, techniques such as the finite layer method provide a very attractive alternative for use in practical design situations. Finite layer techniques have been described by Rowe and Booker (1985*a,b,* 1987) and are available as computer programs for $1\frac{1}{2}$D (POLLUTE: Rowe & Booker, 1990) and 2D (MIGRATE: Rowe & Booker, 1988) conditions.

For situations involving a clayey liner overlying a drainage layer

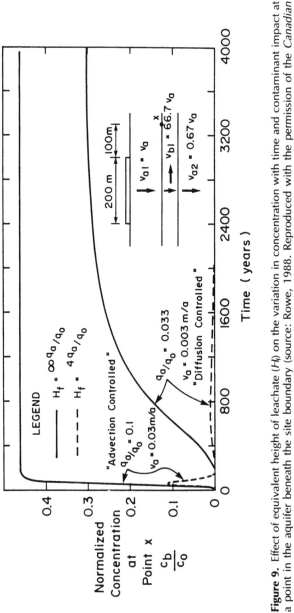

Figure 9. Effect of equivalent height of leachate (H_f) on the variation in concentration with time and contaminant impact at a point in the aquifer beneath the site boundary (source: Rowe, 1988. Reproduced with the permission of the *Canadian Geotechnical Journal*).

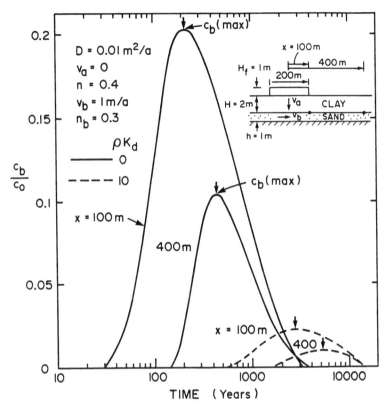

Figure 10. Attenuation of the contaminant due to diffusion into the clay adjacent to the aquifer and due to sorption (source: Rowe & Booker, 1985*b*. Reproduced with the permission of the *Canadian Geotechnical Journal*).

which can be pumped or overlying a thin natural aquifer (see Figs 9–11), a reasonable initial estimate of contaminant impact can be obtained in seconds using a microcomputer and $1\frac{1}{2}$D finite layer programs such as POLLUTE (Rowe & Booker, 1985*a*). The designation of these programs as $1\frac{1}{2}$D is intended to indicate that they consider one-dimensional transport down into the upper aquifer (or drainage layer) and also approximately take account of lateral migration within the aquifer. These techniques will often provide a reasonable estimate of both the magnitude of the peak impact beneath the landfill and the time at which this occurs.

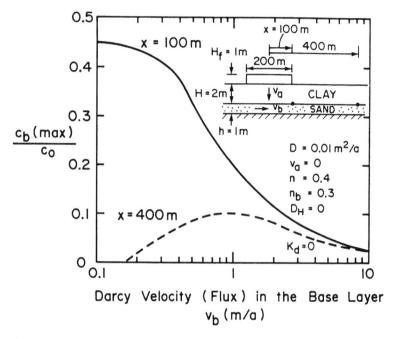

Figure 11. Variation in peak concentration at two points ($x = 100$ m, 400 m) as a function of the Darcy velocities in the aquifer, v_b (source: Rowe & Booker, 1985*b*. Reproduced with the permission of the *Canadian Geotechnical Journal*).

A more rigorous solution of this problem can be obtained using a full 2D analysis (e.g. program MIGRATE). A comparison of the two approaches has been given by Rowe and Booker (1985*b*). The 2D approach allows one to consider multiple layers (e.g. see Fig. 9) and impact at points outside the landfill. The full 2D analysis can be readily performed on a microcomputer but it does involve substantially more computation than the $1\frac{1}{2}$D analysis. Before performing 2D analyses, a $1\frac{1}{2}$D analysis should always be performed to estimate the likely magnitude of impact and its time of occurrence beneath the landfill; the 2D program can then run for appropriate times using the $1\frac{1}{2}$D results as a reference. One of the advantages of finite layer techniques over conventional finite elements methods is that it is not necessary to determine solutions at times prior to the time period of interest (which is usually the time period when peak impact will occur). Thus it is unnecessarily wasteful of the engineers (and the computers) time to evaluate solutions at times well before (or after) the peak impact will occur. As noted above, this can be avoided by using a $1\frac{1}{2}$D program to determine the time period where attention should be focused.

To illustrate the application of some of the concepts discussed in the previous sections, consider a landfill with dimensions 200 m × 1250 m ($A_0 = 25$ ha) separated from a thin underlying aquifer (1 m thick) by a 2-m-thick clayey till liner. The groundwater flow is in the direction of the shorter side (i.e. 200 m). The thin aquifer is assumed here to be underlain by an additional 10 m of clay till and a second 2-m-thick aquifer; however, the primary impact will clearly be on the upper thin aquifer so attention wil be focused on this. Figure 9 shows concentration at the downgradient monitoring point 'x' for an infinite mass of contaminant ($H_f = \infty q_a/q_0$: i.e. constant source concentration) and a finite mass of contaminant ($H_f = 4q_a/q_0$ m: where $q_a = \bar{q}/c_0$, the normalized average flux into the liner and q_0 is the average infiltration through the landfill cover—see Rowe (1988) for more details) for assumed downward Darcy velocities, v_a, of 0·03 m/annum and 0·003 m/annum. If one assumes that the concentration in the source remains constant for all time (i.e. $H_f = \infty$) then a 10-fold decrease in Darcy velocity v_a only reduces the peak concentration by about 35% from $0·46c_0$ to $0·3c_0$. However, when one considers the finite mass of contaminant (specifically $H_f = 4q_a/q_0$ m) then this 10-fold decrease in Darcy velocity gives rise to a more than 10-fold decrease in peak concentration from $0·11c_0$ to $0·01c_0$. The corresponding increase in the time required to reach this peak was from a little over 60 years to about 700 years. For a Darcy velocity of 0·003 m/annum as considered here, assuming a constant source concentration would result in an overestimate of the peak concentration by a factor of 30 if the mass of contaminant corresponds to $H_f = 4q_a/q_0$ m (e.g. 1 million tonnes of waste over a site of area 25 ha at an initial source concentration of 2000 mg/liter and assuming the contaminant of interest represents 0·2% of the waste).

Attenuation

Figure 9 shows results for a conservative contaminant species ($\rho K_d = 0$) which does not interact with the soil. Many contaminant species will interact and will be removed from solution by processes

such as cation exchange (e.g. metals) or by partitioning with the
organic matter in the soil (e.g. organic contaminants such as benzene,
toluene, etc.). Figure 10 shows the normalized variation in concentration in an aquifer with time at two locations (i.e. at the downgradient
edge of the landfill: $x = 100$ m, and 300 m downgradient of the
landfill: $x = 400$ m) for both a conservative species ($\rho K_d = 0$) and a
moderately-sorbed contaminant species ($\rho K_d = 10$). These results are
for a finite mass of contaminant ($H_f = 1$ m) and hence the concentration at any point increases to a peak value and then subsequently
decreases. As might be expected, the peak value is reached at the
downgradient edge of the landfill ($x = 100$ m) before it is reached at
the point $x = 400$ m. This delay is largely due to diffusion of
contaminant from the sand layer into the clay. This reduces the mass
of contaminant in the aquifer and results in a decrease in concentration as can be appreciated by comparing the peak impact at the points
$x = 100$ m and 400 m. It should be noted that these results were
obtained for the case where there was no advection through the clay
liner (i.e. $v_a = 0$). All the contaminant which reaches the aquifer does
so by the process of molecular diffusion.

Effect of Base Velocity

The results presented in Fig. 10 were obtained for a specific value of
the horizontal Darcy velocity within the aquifer ($v_b = 1$ m/annum).
This parameter is important; it is also difficult to determine in
practice. What *can* be determined is a reasonable estimate of the range
in which the velocity is expected to lie. Under these circumstances,
finite layer techniques can be easily used to determine the effect of this
uncertainty upon the expected impact. For example, Fig. 11 shows
the peak concentrations obtained at $x = 100$ and 400 m for analyses
performed for a range of base velocities v_b.

Beneath the edge of the landfill ($x = 100$ m), the maximum concentration decreases montonically with increasing base velocity due to
the consequent increased dilution of the contaminant in high volumes
of water. However, at points outside the landfill area, there is a critical
velocity which gives rise to the greatest 'maximum' concentration. As
indicated by Rowe and Booker (1985*b*), this situation arises because
of the interplay of two different attenuation mechanisms. The first of
these, diffusion into the surrounding clayey soil, is dependent on the

time required to reach the monitoring point. Generally, the lower the velocity v_b, the more time there is for contaminant to diffuse from the aquifer into surrounding clay and hence the lower the maximum concentration in the aquifer at points away from the landfill. Thus, diffusion brings contaminant down from the landfill to the underlying aquifer; however, once the contaminant moves out from beneath the landfill, diffusion has a beneficial effect of removing mass from the aquifer and hence giving rise to attenuation of contaminant concentrations. The second mechanism, dilution, involves decreasing contaminant concentration due to higher volumes of water (i.e. higher v_b).

An important practical consequence of the foregoing is that its not necessarily conservative to design only for the maximum and minimum expected velocities in the aquifer. In performing sensitivity studies, sufficient analyses should be performed to either determine the critical velocity or, alternatively, to show that the critical velocity does not lie within the practical range of velocities for the case being considered.

CONCLUSIONS

This chapter has reviewed transport process and has concluded that diffusion is an important (often critical) mechanism controlling contaminant migration in well-designed modern landfills with clayey barriers. Advective–diffusive transport can be readily modelled on a microcomputer using techniques such as the finite layer method. Parameters such as the diffusion coefficient and distribution coefficient can also be determined using these modelling techniques combined with laboratory column tests. These tests should be performed using the soil to be used in the liner and a leachate which is considered to be representative of that expected in the field.

Examination of actual diffusion profiles beneath existing landfills shows that while diffusion does occur, it is slow and predictable. Furthermore, considerable attenuation of heavy metals and organic chemicals can occur due to interaction with the clay.

The process of diffusion of chemical species through a clay liner is very complicated; however, experience to date would suggest that by using sound engineering judgement together with relatively simple modern laboratory and computer modelling techniques, safe landfills can be designed using clayey liners.

REFERENCES

Al-Niami, A. N. S. & Rushton, K. R. (1977). Analysis of flow against dispersion in porous media. *Journal of Hydrology*, **33**, 87–97.

Barone, F. S., Yanful, E. K., Quigley, R. M. & Rowe, R. K. (1989). Effect of multiple contaminant migration on diffusion and adsorption of some domestic waste contaminants in a natural clayey soil. *Canadian Geotechnical Journal*, **26**, 189–98.

Bear, J. (1979). *Hydraulics of Groundwater*. McGraw-Hill, New York.

Crooks, V. E. & Quigley, R. M. (1984). Saline leachate migration through clay: A comparative laboratory and field investigation. *Canadian Geotechnical Journal*, **21**(2), 349–62.

Desaulniers, D. D., Cherry, J. A. & Fritz, P. (1981). Origin, age and movement of pore water in argillaceous quaternary deposits at four sites in Southwestern Ontario. *Journal of Hydrology*, **50**, 231–57.

Freeze, R. A. & Cherry, J. A. (1979). *Groundwater*. Prentice-Hall Inc., Englewood Cliffs, New Jersey.

Gillham, R. W. & Cherry, J. A. (1982). Contaminant migration in saturated unconsolidated geologic deposits. Geophysical Society of America, Special Paper 189, pp. 31–62.

Goodall, D. E. & Quigley, R. M. (1977). Pollutant migration from two sanitary landfill sites near Sarnia, Ontario. *Canadian Geotechnical Journal*, **14**, 223–36.

Lahti, L. R., King, K. S., Reades, D. W. & Bacopoulos, A. (1987). Quality assurance monitoring of a large clay liner. In *Proceedings of ASCE Specialty Conference on Geotechnical Aspects of Waste Disposal, '87*, ed. R. D. Woods. Ann Arbor, pp. 640–54.

Lerman, A. (1979). *Geochemical Processes—Water and Sediment Environments*. Wiley-Intersciences, New York.

Perkins, T. K. & Johnston, D. C. (1963). A review of diffusion and dispersion in porous media. *Society of Petroleum Engineering Journal*, **3**, 70–84.

Quigley, R. M. & Rowe, R. K. (1986). Leachate migration through clay below a domestic waste landfill, Sarnia, Ontario, Canada: Chemical interpretation and modelling philosophies. In *ASTM Special Publication on Industrial and Hazardous Waste STP 933*, ed. D. Lorenzen, R. A. Conway, L. P. Jackson, A. Hamza, G. L. Perke & W. J. Lacy. pp. 93–103.

Quigley, R. M., Yanful, E. K. & Fernandez, F. (1987a). Ion transfer by diffusion through clayey barriers. In *Proc. ASCE Specialty Conference on Geotechnical Aspects of Waste Disposal '87*. Ann Arbor, pp. 137–58.

Quigley, R. M., Fernandez, F., Yanful, E., Helgason, T., Margaritis, A. & Whitby, J. L. (1987b). Hydraulic conductivity of a contaminated natural clay directly below a domestic landfill. *Canadian Geotechnical Journal*, **24**, 377–83.

Rowe, R. K. (1987). Pollutant transport through barriers. In *Proceedings of ASCE Specialty Conference, Geotechnical Practice for Waste Disposal '87*, ed. R. D. Woods. Ann Arbor, pp. 159–81.

Rowe, R. K. (1988). Contaminant migration through groundwater: The role of modelling in the design of barriers. *Canadian Geotechnical Journal*, **25**(4), 778–89.

Rowe, R. K. (1990). Contaminant impact assessment and the contaminating lifespan of landfills. *Canadian Journal of Civil Engineering* **18**, 244–53.

Rowe, R. K. & Booker, J. R. (1985*a*). 1-D pollutant migration in soils of finite depth. *Journal of Geotechnical Engineering, ASCE*, **111** (GT4), 479–99.

Rowe, R. K. & Booker, J. R. (1985*b*). 2D pollutant migration in soils of finite depth. *Canadian Geotechnical Journal*, **22**(4), 429–36.

Rowe, R. K. & Booker, J. R. (1987). An efficient analysis of pollutant migration through soil. In *Numerical Methods in Transient and Coupled Systems*, ed. Lewis, E. Hinton, P. Bettess & B. A. Schrefler. John Wiley & Sons, New York, pp. 13–42.

Rowe, R. K. & Booker, J. R. (1988). MIGRATE: Finite layer analysis program for 2D analysis. Users Manual, Rep. No. GEOP-1-88. Geotechnical Research Centre, The University of Western Ontario.

Rowe, R. K. & Booker, J. R. (1990). POLLUTE v5—1D pollutant migration through a non-homogeneous soil: Users manual. Geotechnical Research Centre Research Report, GEOP-1-90, Faculty of Engineering Science, University of Western Ontario, London, Canada.

Rowe, R. K., Caers, C. J., Booker, J. R. & Crooks, V. E. (1985). Pollutant migration through clay soils. *Proceedings of the 11th International Conference on Soil Mechanics and Foundation Engineering*. San Francisco, pp. 1293–8.

Rowe, R. K., Caers, C. J. & Barone, F. (1988). Laboratory determination of diffusion and distribution coefficients of contaminants using undisturbed soil. *Canadian Geotechnical Journal*, **25**, 108–18.

Yanful, E. K., Nesbitt, H. W. & Quigley, R. M. (1988). Heavy metal migration at a landfill site, Sarnia, Ontario, Canada—I: Thermodynamic assessment and chemical interpretations. *Applied Geochemistry*, **3**, 523–33.

3.3 Moisture Behaviour of Soil Liners and Subsoil beneath Landfills

ULRICH HOLZLÖHNER

Bundesanstalt für Materialforschung und -prüfung (BAM), Unter den Eichen 87, D-1000 Berlin 45, Germany

INTRODUCTION

The 'combination liner' is a special design of a composite liner in which a synthetic membrane is placed directly on top of a mineral liner. Because of its advantages, the combination liner has become the standard lining system in Germany (August, 1985). As the synthetic membrane seals the mineral liner from the drainage water, the possibility arises that the mineral liner could desiccate. This gave the impetus to investigations into the moisture movement in soil liners and subsoil beneath landfills.

THEORETICAL ASPECTS

Figure 1 shows a schematic picture of a landfill with a bottom liner and the ground water table underneath. As the side lengths of the base area of the landfill are very long compared to the distance to the ground water table, moisture movement can be assumed to be one-dimensional, i.e. only the 'middle zone' of the landfill (Fig. 1) need be considered. Figure 2 shows a soil prism of the middle zone subjected to the following conditions: overburden pressure due to the waste load on top, temperature decrease from top to bottom, and connection to ground water at the bottom. The presence of a synthetic membrane on top of the mineral layer means that there is no flux of water from the top.

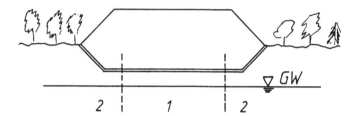

Figure 1. Landfill: 1, middle zone; 2, fringe area.

It is known from tests with confined soil samples that due to the temperature gradient the moisture content increases at the cold side and decreases at the warm side (Habib, 1957; Evgin & Svec, 1988).

The mechanism to be expected may be described as follows (see Holzlöhner, 1988). The temperature gradient and gravity drive the soil water towards the colder ground water. This movement is counteracted by the suction that the soil produces with increasing desiccation. The mineral layer could display cracks just at that time when the synthetic membrane might have become pervious due to ageing. This desiccation cracking could also occur with liner systems composed only of mineral layers because parts of the base can be dry from time to time.

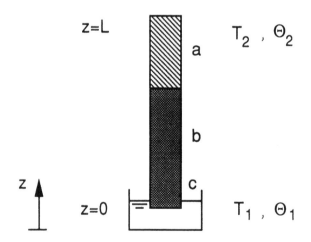

Figure 2. Soil prism used for modelling heat and moisture transport beneath landfill: a, liner; b, subsoil; c, ground water; T, temperature; Θ, moisture.

SUCTION AND OVERBURDEN PRESSURE

Water loss from the mineral liner may occur even without a temperature gradient. If a soil is sealed completely against precipitation and evaporation at the surface, a suction distribution develops that corresponds to the respective distance to the ground water table (Hartge, 1978). The suction determines the moisture content of the given type of soil: the greater the suction, the smaller the moisture content. If the mineral liner is placed with a higher moisture content than that corresponding to the distance to ground water, the mineral liner will lose water. The overburden pressure due to the waste acts similarly: the soil diminishes its volume, and pore water is pressed out. In a sample of saturated clay under isotropic confining pressure the pore water is loaded as if a suction of the same value is applied (Soeiro, 1957). Therefore it is possible to approximatively evaluate the relationship between suction and water content by oedometer tests. Figure 3 shows such results for a heavy clay with Atterberg limits $w_p = 33\%$, $w_L = 86\%$. The suction is presented in pF, which is the logarithm of the negative pressure head in centimetres of water. Figure 3 includes values determined by the suction plate as usual. The differences between the results can be attributed to the fact that in the oedometer cell lateral stress is smaller than vertical stress.

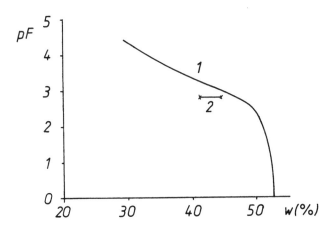

Figure 3. Suction versus water content determined by (1) oedometer and (2) suction plate.

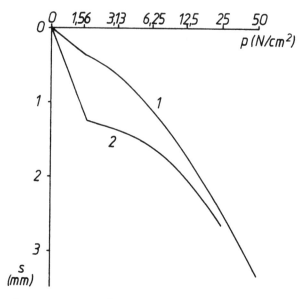

Figure 4. Compression versus load in oedometer tests: 1, no suction; 2, suction of 6 N/cm² applied.

Figure 4 shows the similar action of confining load and suction. The suction of 6 N/cm² applied in test 2 compresses the sample similarly to a load. With increasing load, the load–compression curve approaches that of test 1 without suction.

With respect to soil fabric, the type of load application is important: suction without confining pressure may produce cracks in the soil, whereas drainage by compression in the oedometer produces no cracks. This one-dimensional deformation takes place in the soil liner beneath the waste. If the distance to ground water is only a few metres, the overburden pressure of the landfills of today is far higher than the suction. Thus, no cracking will occur if there is no temperature gradient. This is even more true with increasing stability of the soil skeleton.

SOIL COLUMN TESTS

Soil columns of the type shown in Fig. 2 were investigated. Polyvinyl chloride tubes of 15 cm in diameter and 90 cm in length were filled

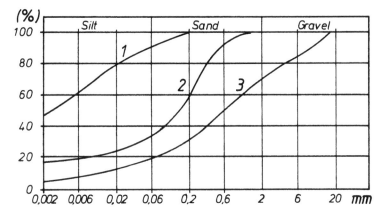

Figure 5. Particle size distribution of investigated liner materials: 1, heavy clay; 2, sand–silt mixture; 3, well-graded soil.

with soil material: a 15-cm upper layer of mineral lining material overlying 'subsoil'. Three liner materials were used (the particle size distributions of which are given in Fig. 5): the already-mentioned heavy clay, a sand–silt mixture and a Fuller-curve graded gravel–sand–silt mixture. The last two soils have been upgraded by adding 1% and 2% bentonite, respectively. Two subsoil materials were used, coarse and medium sands. All the soil materials were compacted by means of a Proctor device.

The clay was put in place with the natural water content; the water content of the other two soils were to the wet side of the Proctor optimum. The subsoil sands were emplaced in a moist state simulating a short distance to 'ground water'.

After being filled, the tubes were brought into an upright position where the lower ends were in a water bath at 10°C. Through a filter, this ground water was in connection with the pore water of the subsoil. The absolutely-sealed upper ends of the test facility were heated to 30°C, representing the temperature of the leachate in landfills. The tubes were insulated against loss of heat.

The aim of the tests was to measure the variation of the moisture content with time. As sampling would have disturbed the transport processes within the soil column, a non-destructive method was applied. The moisture content was determined by measuring the dielectric constant of the soil. The measuring device designed for this purpose consists of a rigid ring of polyvinyl chloride. Two rings of sheet copper, 10 mm wide, are fixed at the inner side of the rigid ring at a distance of 12 mm from one another. The device can simply be slid over the tubes. A 60-MHz voltage is applied at the copper rings, which act as a capacitor. The torus-like electric field extends through the plastic tube into the fringe area of the soil column. As the dielectric constant of water is about 20 times that of polyvinyl chloride or of dry soil material, the pore water strongly affects the capacity of the capacitor. The effect is displayed by an electric circuit. This measurement is carried out and the tubes are weighed from time to time.

Preliminary tests using smaller test tubes (30 cm in length and 5 cm in diameter) showed no desiccation; drying occurred in all the bigger tubes. Figure 6 shows the results for the three liner materials which have been placed on coarse sand subsoil. The clay and the sand–silt mixture shrank to such an extent that the soil column separated from the top and sides of the tube. Once a coherent air-filled gap exists, desiccation can proceed faster because of unhindered air movement. It is doubtful, however, if such cracks or gaps really develop in mineral liners, as discussed earlier.

The fuller-curve graded soil also dried to some extent but no cracks occurred. This soil obviously has a stable soil skeleton. Only the voids are filled with clayey material. This type of soil is more likely to regain its former impermeability when remoistened than the other two soils investigated.

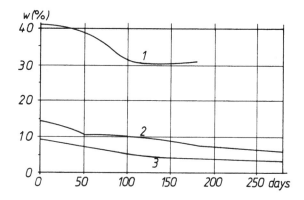

Figure 6. Decrease of water content due to temperature gradient. 1, Heavy clay; 2, sand–silt mixture; 3, well-graded soil.

The tests described above do not respect the overburden pressure due to the waste. As has been already discussed, this pressure might prevent the generation of cracks that might occur due to desiccation. The pressure could not be applied to the soil columns because the load would soon be influenced by the tube walls. Therefore thin samples will be investigated in further tests. The base of the cell will be a ceramic plate making possible the application of suction. Thus, soil behaviour under more realistic moisture and load conditions can be studied.

MOISTURE EQUILIBRIUM DUE TO TEMPERATURE AND GRAVITY

In the state of equilibrium the problem is stationary, i.e. the time variable can be omitted. Thus, flux q is influenced by volumetric moisture content Θ, temperature T, hydraulic head H, hydraulic conductivity K and the two transport coefficient D_0 and D_T (see Hillel, 1980). Flux q has the dimension of velocity, Θ is dimensionless.

$$q = -D_0 \nabla \Theta - D_T \nabla T - K \nabla H \tag{1}$$

As already pointed out, moisture movement beneath the landfill will be described as one-dimensional. In the case of the combination liner the synthetic membrane seals the top surface completely. For these conditions eqn (1) becomes

$$0 = -D_0 \frac{d\Theta}{dz} - D_T \frac{dT}{dz} - K \tag{2}$$

Figure 2 shows the considered system. For simplification it is assumed that the temperature variation is linear:

$$T(z) = T_1 + \frac{T_2 - T_1}{L} \cdot z \tag{3}$$

Thus, eqn (2) becomes:

$$\frac{d\Theta}{dz} = -\frac{D_T}{D_0} \frac{T_2 - T_1}{L} - \frac{K}{D_0}$$

$$\Theta = \Theta_1 - \left(\frac{D_T}{D_0} \frac{T_2 - T_1}{L} + \frac{K}{D_0} \right) \cdot z \tag{4}$$

Neglecting the effect of gravity yields

$$\Theta_1 - \Theta = \frac{D_T}{D_0}(T_2 - T_1)\frac{z}{L} \tag{5}$$

which means for the warm side $z = L$:

$$\Theta_1 - \Theta_2 = \frac{D_T}{D_0}(T_2 - T_1) \tag{6}$$

Three conclusions can be drawn

- The moisture difference $\Theta_1 - \Theta_2$, which describes desiccation, depends on the ratio of the transport coefficients.
- The moisture difference depends on the temperature difference and not on the temperature gradient.
- The moisture distribution depends not on z but on z/L.

These statements are still valid when the transport coefficients are taken as functions of volumetric moisture content Θ. In this case, as a model the soil has to be cut into slices within which the coefficients are assumed to be constant. The calculation proceeds from the lowest slice upwards. The assumed temperature distribution (eqn (3)) may iteratively be improved by considering the dependency of heat conductivity on moisture content and by solving the stationary heat conduction problem.

If the effect of gravity is included, the solutions depend not only on z/L but on z and on L (see eqn (4)). Note that the coefficient ratio K/D_0 is the reciprocal of the derivative of the suction head versus moisture content function (Hillel, 1980). The calculation proceeds as described above.

THEORETICAL PREDICTIONS: POSSIBILITIES AND LIMITATIONS

The moisture distributions measured with the relative short soil columns can be transferred to thicker layers by relating the depth variable z to total thickness L as shown in eqn (5). The tests give the lower limit of the desiccation effect, because the influence of gravity increases with the thickness of the layer. The results as described above also show that the calculation may include gravity and other effects as the dependency of the diffusion coefficients on moisture content.

The diffusion coefficients consist of two parts corresponding to vapour (v) and liquid (l) phase transport (Nielsen *et al.*, 1972).

$$D_0 = D_{0v} + D_{0l}$$
$$D_T = D_{Tv} + D_{Tl}$$ (7)

Temperature gradient induced diffusion is assumed to take place mainly in the vapour phase. With decreasing moisture content, this kind of transport increases. Therefore desiccation due to temperature is especially important at low moisture contents (see also Cary, 1965; Nielsen *et al.*, 1972). Note that the calculations above only require the ratio of the total diffusion coefficients.

All the reflections are based on the assumption that temperature-induced transport is proportional to temperature gradient. Although Cary (1965) uses a linear relationship between the flux and temperature gradient (Fig. 7), the extrapolated lines do not meet the abscissa at the origin. Therefore for small temperature gradients the moisture transport might be less than predicted by linear theory. Such small gradients often occur beneath landfills. In the opinion of the author, conclusions based on tests with higher gradients and linear theory most probably are conservative estimations regarding the occurrence of desiccation.

As an example, the digging up of the landfill Geldern-Pont should be mentioned. After 10 years, no desiccation below the synthetic membrane could be detected (Düllmann 1987, pers. comm.). More

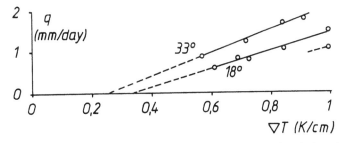

Figure 7. Water flux versus temperature gradient at a suction head of 2·45 m of water for average temperatures as indicated (after Cary, 1965).

experience of that kind of investigation is needed, where the specific situation is described in detail.

DAMAGE AND PREVENTIVE MEASURES

The problem of desiccation generally depends on the following conditioning factors:

- material of the mineral liner
- material of the subsoil
- overburden pressure
- distance to ground water
- temperatures of landfill base and subsoil.

Damage has occurred if the desiccation has produced cracks that do not vanish in case of remoistening.

The purpose of the above-described investigations is of course to apply the results to practice in order to avoid damage due to desiccation. This also applies to the construction and to the operation of the landfill. Since it could be shown during the investigations that a high temperature gradient results in extreme desiccation, high temperatures at the base of the landfill should be avoided. This can be achieved—at least to a certain degree—by classifying and conditioning the waste and by depositing inert material immediately above the base lining.

An important factor of influence is the depth of the ground water table. If it is not more than 3 m beneath the base liner, desiccation is not assumed to cause damage (Holzlöhner, 1988). The subsoil material affects desiccation because of its conductivity properties: transport of moisture in the vapour phase is easier in partly saturated sands than in clays. Temperature-induced desiccation may therefore be diminished by replacing pervious sands by clay or clay-like materials.

The following options may be respected during the design of the mineral liner:

- wetting the mineral liner from time to time. It is however, questionable if such a system will work for decades;
- enveloping the mineral liner by upper and lower synthetic membranes;

- addition of chemicals which bind the water more strongly to the soil material;
- selection of lining materials to minimize shrinkage, such as suitably graded soil or dry mineral material.

CONCLUSIONS

The desiccation of mineral layers in the base lining system of landfills may influence the permeability of the liner due to cracking. This problem must be investigated in each case. Desiccation may also occur if the soil liner is not covered by an overlying synthetic membrane. The main factor of influence is the temperature difference between the landfill base and the subsoil. A deep ground water table also is disadvantageous in this regard. The most effective preventative measure is the use of mineral lining materials that do not display cracks when drying out. Other properties must not be influenced due to desiccation, e.g. the flexibility of the soil liner. The proper construction of long-lasting lining systems requires the simultaneous consideration of various aspects. This problem will be investigated in the German BMFT research programme, 'Liner systems of landfills'.

ACKNOWLEDGEMENT

This research project was supported by the Federal Ministry of the Environment, Germany, whose help is greatly appreciated by the author.

REFERENCES

August, H. (1985). Untersuchungen zum Permeationsverhalten kombinierter Abdichtungssysteme. Mitteilungen des Instituts für Grundbau und Bodenmechanik, TU Braunschweig, Heft 20, pp. 205–19.

Cary, J. W. (1965). Water flux in moist soil: thermal versus suction gradients. *Soil Science*, **100**(3), 168–75.

Evgin, E. & Svec, O. J. (1988). Heat and moisture transfer characteristics of compacted Mackenzie silt. *Geotechnical Testing Journal*, **11**(2), 92–9.

Habib, M. (1957). Thermo-osmose. Annales de l'Inst. du Bâtiment et des Travaux Publics, No. 110, pp. 130–6.

Hartge, K. H. (1978). Einführung in die Bodenphysik. Enke-Verlag, Stuttgart.

Hillel, D. (1980). Fundamentals of Soil Physics. Academic Press, Orlando.

Holzlöhner, U. (1988). Das Feuchteverhalten mineralischer Schichten in der Basisabdichtung von Deponien. Müll und Abfall, 20(7), 295–302.

Nielsen, D. R., Jackson, R. D., Cary, J. W. & Evans, D. D. (1972). Soil Water. Am. Soc. of Agronomy and Soil Science Soc. of America, Madison, Wisconsin.

Soeiro, M. (1957). Succion. Annales de l'Inst. du Bâtiment et des Travaux Publics, 110, 121–9.

3.4 Effects of Leachate on the Permeability of Sand–Bentonite Mixtures

ANDREA CANCELLI,[a] RAFFAELLO COSSU,[b] FRANCESCA MALPEI[c]
& ALESSANDRA OFFREDI[a]*

[a] *Department of Earth Sciences, University of Milan, Via Mangiagalli 34, 20133 Milano, Italy*
[b] *Department of Land Engineering, University of Cagliari, Piazza d'Armi, 09123 Cagliari, Italy*
[c] *Institute of Sanitary Engineering, Politecnico di Milano, Piazza Leonardo da Vinci, 32-20133 Milano, Italy*

INTRODUCTION

Beside other mineral liner materials, bentonite and bentonite–soil mixtures play an important role as a lining material for landfills. This material is, in general, much more uniform and predictable in its behavior than, for example, excavated clay without any pretreatment. Moreover, in many areas adequate native soils are not available, so in those cases bentonite purchased from anywhere in the world might be used.

In most cases bentonite is mixed with soil using different recipes and techniques (Lundgren, 1981; Alther, 1982; Hoeks *et al.*, 1987).

In the Northern River Po Plain, for example, natural clays are very rare and many sanitary landfills are placed into old quarries for coarse-grained aggregates, to be reclaimed. There, conspicuous quantities of fine-grained soils, ranging from silts to fine silty sands, are available at the bottom of old settling ponds, related to the washing of coarse aggregates. In order to reach a sufficiently low hydraulic conductivity, such by-product soils have to be mixed with some

---

* Present address: Kora s.r.l., Via degli Orombelli, 7 A-20131 Milano, Italy.

sealing agents, e.g. commercial bentonite or equivalent swelling clays (Cancelli & Francani, 1984). Adding bentonite to the soil, in proportions of 4–8% by weight, attains values of the coefficient of permeability k as low as 10^{-9}–10^{-10} m/s (Lundgren, 1981; Fayoux, 1983).

Dry-mixing bentonite powder with soil is recommended for putting-in-place, in order to obtain a satisfactory soil–bentonite blanket (Ingles & Metcalf, 1972; Lundgren, 1981); a minimum thickness of 0·3 m is required, to reduce the risk of interruptions of the sand–bentonite layer. The application of a surcharge (about 0·5 m of sandy gravel) before adding water, together with the downward seepage force, is considered to be sufficient to inhibit untimely swelling of bentonite particles (Cancelli, 1983; Fayoux, 1983).

In engineering practice, clayey soils are generally assumed to be 'impervious'. On the contrary, when water and pollutant movements are expected to occur over periods of hundreds of years, fine-grained soils cannot be simply assumed as 'impervious'; the hydraulic conductivity of a sand–bentonite layer generally tends to decrease with time (Hoeks & Ryhiner, 1989); on the other hand, the presence of chemical substances in percolating water can largely increase the permeability of compacted clay liners (Quigley *et al.*, 1984). Therefore, it becomes necessary to measure the hydraulic conductivity of compacted clays and sand–bentonite mixtures, also in presence of pollutant fluids, and to investigate how this, and other soil characteristics, are affected by a long-term exposition to leachates.

EFFECTS OF CHEMICALS ON BENTONITE PROPERTIES

Characteristics of Bentonite

Bentonite is a natural clay, whose peculiar characteristics (high swelling, tixotropy, gelling ability and cohesion) offer several different uses, both in engineering applications (sealing, slurry trenching, etc.) and in industrial activities (foundry, etc.). It consists mainly of the clay mineral montmorillonite, followed by beidellite. Both these minerals belong to the smectites group.

The structure of the minerals of the smectite group consists of two

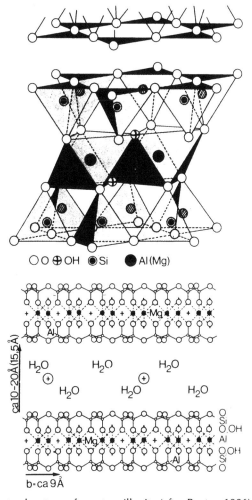

Figure 1. Structural pattern of montmorillonite (after Reuter, 1986).

layers of silica tetrahedon, with an octahedral layer interposed (Fig. 1). The arrangement of the three layers is such that the oxygen atom which is at the tethraedon vertex is shared with one vertex of an octahedron. The vertices of the octahedron not bound with the silica tetrahedrons are taken up by hydroxy groups.

The combination of these three layers forms the lattice: the lattices are arranged one on top of another and form the platey-shape clay micelle. The upper base of a lattice is bonded to the lower base of the contiguous lattice by van der Waals' forces and by the cations adsorbed between the lattices: these bonds are weak and can be easily divided as a result of water adsorption.

The lattice of montmorillonite is negatively charged, because the silicon and aluminium cations placed in the centre of tetrahedrons and octahedrons are in part substituted by cations with a lower valence (magnesium or iron can substitute aluminium and aluminium can substitute silicon). This results in adsorption of cation-like sodium (Na^+), potassium (K^+), calcium (Ca^{+2}) and magnesium (Mg^{+2}) on the surface of the micelle and between the lattices.

The electric charge is not uniformly distributed on the micelle: the planar bases of the micelles are negatively charged while the tips are positively charged, because here the lattices are broken off. For this reason, in natural deposits the micelles are not parallelly arranged, but edge-to-face assembled (see Fig. 2), forming the so-called wedge bond.

The resulting frame is very loose and expanded, and allows a lot of water to be adsorbed and held between the micelles.

One of the most important properties of bentonite is its ability to adsorb water. The main difference between bentonite and other clay soils is that bentonite is able to adsorb water not only on the surface of the micelle but also between the lattices. Depending on the quantity of water adsorbed between the lattices, there is a progressive separation of the lattices up to the complete separation.

Water molecules are electrically bound as dipoles in the diffuse double layer surrounding the clay minerals: they are rigidly bound to the clay minerals and are not able to flow freely.

The thickness and the electric potential of the double layer are usually described mathematically by the Gouy–Chapman model and depend both on soil and fluid characteristics. The thickness of the double layer is expressed by:

$$t^2 = 8 \frac{\varepsilon KT}{\pi n e^2 v^2} \tag{1}$$

where:

t = double layer thickness,
ε = dielectric constant,

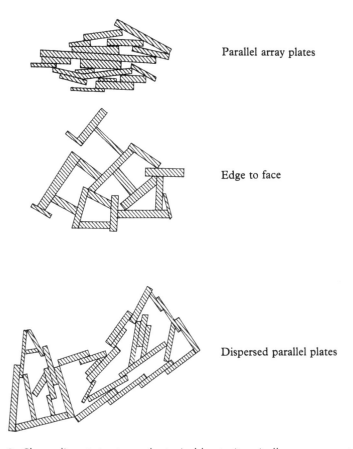

Parallel array plates

Edge to face

Dispersed parallel plates

Figure 2. Clay sediment structures: the typical bentonite micelles arrangement is the edge-to-face type (after Lambe, 1958).

$K =$ Bolzman's constant,
$T =$ temperature,
$n =$ electrolyte concentration,
$e =$ elementary charge,
$v =$ valence of cations in pore fluid.

However, the Gouy-Chapman theory does not take into account some factors likely to influence behavior: among them, the effects of pH in the pore fluid and the size of the hydrated exchangeable cations.

The Gouy-Chapman theory, in fact, deals with point charges on the surface of the clay lactices, whereas this charge (CEC) is associated with ions (cations) of finite size. The smaller an ion plus its 'shell' of hydration water, the closer it can approach the colloidal surface, the smaller the hydration radius and the smaller the adsorbed water.

Two sodium ions have a larger hydration radius than one calcium ion and this is why Ca-bentonite has a relatively low water-adsorbing capacity compared to Na-bentonite. Many commercially-available bentonites are processed with sodium carbonate to exchange the dominant cation from calcium to sodium and to improve the swelling characteristics of bentonite.

The capacity of bentonite to adsorb water ranges from about 200% (Ca-bentonites) up to more than 600% (Na-activated bentonites and natural Na-bentonites). The ratio of double layer thickness to the thickness of smectite micelles is up to 20 times.

Sand–Bentonite Mixtures

In sand–bentonite mixtures, the bentonite is present in pores between the sand granules and coats the granules with a uniform, very thin layer. When the mixture is moistened with water, the bentonite swells and seals the pores. At the same time the bonding forces due to the double layer make the mixture cohesive.

The reduction in hydraulic conductivity of a sand–bentonite mixture depends on the bentonite type used, as well as on the percentage of bentonite introduced in the mixture. For a medium-fine sand (50% finer than 0·25–0·50 mm), the addition of 3–4% of natural Na-bentonite allows a decrease in the permeability of two orders of magnitude (Sima & Harsolescu, 1979); an addition of 7–8% of natural Na-bentonite can be required to get a decrease of five orders of magnitude (Hoeks *et al.*, 1987).

Theoretical Discussion of Factors of Influence on Bentonite Soil Structure

Changes in the hydraulic conductivity in the presence of different permeants can be attributed to differences in the characteristics of the fluid or to modifications in the structure of soils, due to interaction between the permeating fluids and the soil.

It is known that the hydraulic conductivity k (m/s) is correlated with the intrinsic permeability (i.e. the typical permeability of a given soil structure) K (m^2), by the Kozeny-Karman Law:

$$k = K * \frac{\tau}{\mu} \tag{2}$$

where μ is the viscosity of the permeant fluids and τ is the specific weight.

This theoretical relationship takes into account the influence of mechanical characteristics of the fluid, but does not take into account the modification induced by the nature of the permeant to the structure of the soils.

Since in clayey soils, it is the modification in the soil structure that gives the largest contribution to changes in hydraulic conductivity, it is necessary to study the modification in the bentonite structure caused by the nature of permeant, in order to understand, and possibly predict, the changes in hydraulic conductivity of a given soil.

The first and probably most important element that can modify the structure of bentonite is a modification in the double layer thickness and electric potential. A reduction in the double layer thickness causes the clay particle to flocculate, the soil skeleton to shrink and cracks to form.

Equation (1) shows very clearly which are the parameters that influence the double layer thickness:

- *Dielectric constant*: the dielectric constant is a measure of the ease with which molecules can be polarized and oriented in an electric field. It measures the ability of a material to perform as an insulator, i.e. it describes a material's capacity to reduce the strength of an electric field. The higher the dielectric constant of a material, the more the material behaves as an insulator. Equation (1) results in a direct proportionality between the thickness of the double layer and the dielectric constant of the pore fluid: the higher the dielectric constant, the thicker the double layer. Pure water has a very high dielectric constant ($\varepsilon = 80\cdot4$), but the dielectric constant of salt water is probably much smaller.
- *Electrolyte concentration*: the concentration of ions (ionic strength) in the pore fluid is in inverse proportion to the double layer thickness, though high concentrations of solutes in water

result in a smaller double layer. Note that a concentration change from 10^{-4} M NaCl to 10^{-2} M NaCl (two orders of magnitude) reduces the thickness of the double layer by one order of magnitude (Mitchell, 1976).

• *Valence of cations in pore fluids*: for solutions of the same molarity, a change in cation valence affects the thickness of the double layer: the higher the valence (Ca^{+2} versus Na^{+}), the smaller the thickness of the double layer.

• *Temperature*: according to eqn (1), an increase in temperature causes an increase in double layer thickness.

In the previous paragraph, it was explained that the double layer thickness depends also on the size of hydrated ions. Another element that can cause a decrease in the double layer thickness is the exchange of the dominant cation with another of smaller hydrated radius.

The sodium ions at the surface of the bentonite lattices are easily and quite rapidly exchanged with multivalent ions such as magnesium or calcium, which have a smaller hydrated radius.

Na-bentonites, whose swelling characteristics and sealing capability are much better than Ca-bentonite, are nevertheless more exposed to cation exchange and to partial loss of their distinguishing properties. Therefore, a permeant fluid can increase the permeability of a bentonite by two distinct mechanisms, due to:

• higher ionic concentration and ionic valence of the reference fluid (water);
• exchange of the Na cation with a multivalent cation.

Apart from the reduction of the double layer thickness, which is to be regarded as the main cause for the increase in hydraulic conductivity, other phenomena can destroy the barrier-like characteristics of the bentonite.

Acids and bases can react with the bentonite, destroying and dissolving the soil mineral (Yariv & Cross 1979; Peterson & Gee, 1985): acids can dissolve alluminium and alkaline metals, while bases can dissolve silica (Grim, 1953). It appears that strong inorganic acids (e.g. HCl) cause a higher dissolution than organic acids (e.g. acetic acid) (Carroll & Starkey, 1971). Anyhow, only if the attack is long and severe, the mineral structure is destroyed and completely dissolved.

As permeability changes depend on structural changes, there must be a relationship between modifications induced by a given permeant to the permeability of bentonite and modifications of other properties and characteristics of the bentonite that depend on, or influence, its structure, such as Atterberg's limit, free swelling, exchangeable cations, etc.

Several authors (Daniel & Liljestrand, 1984, Bowders *et al.*, 1986) have studied the correlation between Atterberg's limits and hydraulic conductivity of clays, and generally agree that an increase in the permeability of a clay permeated with a fluid different from water corresponds to a decrease in the plasticity index of the clay moulded with the permeant.

Atterberg's limits (liquidity and plasticity limits) reflect the consistency and the structure of cohesive soils and therefore are a good parameter with which to detect modifications in the soil structure. As their determination is quite easy and fast, they can be a useful tool for a first and qualitative evaluation of clay–permeant compatibility.

Similarly, modifications in the bentonite structure can be reflected by changes in equilibrium sediment volume and free swelling characteristics. Hettiaratchi *et al.* (1988) studied the volume shrinkage in soils with a high bentonite percentage (higher than 65%) exposed to contaminants and suggested that a high decrease in the free swelling volume (as obtained in a laboratory test with contaminant liquids) can result in a high shrinkage of the clay-liner and crack formation, with an increase in the hydraulic conductivity.

State of Knowledge on Hydraulic Conductivity of Clayey Soils (Literature Review)

Laboratory investigations on the hydraulic conductivity of clayey soils have been widely conducted, using different types of soil–liquid mixtures and testing techniques, by many researchers (Carroll & Starkey, 1971; Gordon & Forrest, 1981; Nasiatka *et al.*, 1981; Anderson *et al.*, 1982; Brown & Anderson, 1983; Daniel & Liljestrand, 1984; Quigley *et al.*, 1984, 1987, 1988; Fernandez & Quigley, 1985; Peterson & Gee, 1985; Simons & Reuter, 1985; Peirce *et al.*, 1987; Cancelli *et al.*, 1987; Uppot & Stephenson, 1989; Broderick & Daniel, 1990.

Specific data on bentonite clay are quite limited; however, data from a wide survey on the literature specifically refer to bentonites, smectitic and montmorillonitic clays, and to sand–bentonite mixtures, and are summarized in Table 1.

TABLE 1. Literature Data on the Effects of Different Permeants on Hydraulic Conductivity of Bentonites, Sand–Bentonite Mixtures, Smectitic and Montorillonitic Clays, with Respect to Hydraulic Conductivity with Water

Permeants	Effect	Type of soil	Reference	Comments
Acids				
Acetic acid	+	Pure Mg-bentonite	Uppot & Stephenson (1989)	Flex. p., direct, a.s.w.
		Noncalcareous smectite		
Acetic acid (pure)	0		Brown & Anderson (1983)	Fix. p., direct, a.s.w.
5% (acetic acid + propionic acid) in water	0/+	Natural Na-bentoite	Simons & Reuter (1985)	Flex. p., direct
5% (acetic acid + propionic acid) in water	0	Sand + 33% Na-bentonite	Simons & Reuter (1985)	Flex. p., direct
H_2SO_4	−	Sand + Na-bentonite	Nasiatka et al. (1981)	Direct
H_2SO_4	++	Sand + Ca-bentonite	Nasiatka et al. (1981)	Direct
H_2SO_4 (1%)	0/+	Silty sand + bentonite	D'Apollonia (1980)	
HCl (5%)	0/+	Silty sand + bentonite	D'Apollonia (1980)	
Leachates				
MSW	+	Sand + 8%Na-bentonite	Alther (1982)	Direct
MSW	+	Sand + 5% Na-bentonite	Hoeks et al. (1987)	Fix. p., direct, a.s.w.
MSW	++	Sand + 5% Na-bentonite	Hoeks et al. (1987)	Fix. p., direct, a.s.w.
MSW	+	Sand + 5% Na-bentonite	Cancelli et al. (1987)	Flex., p., direct, a.s.w.
MSW	−	Smectite	Griffin & Shrimp (1978)	Lab., direct
MSW	−	Smectite	Daniel & Liljestrand (1984)	Flex. p., direct, w.s.w
Bases				
Aniline	no flow	Pure Mg-bentonite	Uppot & Stephenson (1989)	Flex. p., direct, a.s.w.
		Noncalcareous smectite		
Aniline	++		Brown & Anderson (1983)	Fix, p., direct, a.s.w.
NaOH (5%)	+	Silty sand + bentonite	D'Apollonia (1980)	
CaOH (1%)	+	Silty sand + bentonite	D'Apollonia (1980)	

Neutral organic compounds

Ethyl alcohol	+++	Pure Na-bentonite	Mersi & Olson (1971)	Indirect, w.s.w.
Benzene	>+++	Pure Na-bentonite	Mersi & Olson (1971)	Indirect, w.s.w.
Heptane	>+++	Sand + 5% Na-bentonite	Broderick & Daniel (1990)	Fix. p., direct, a.s.w.
	++	Noncalcareous smectite	Brown & Anderson (1983)	Fix. p., direct, a.s.w
Methanol	>+++	Sand + 5% Na-bentonite	Broderick & Daniel (1990)	Fix. p., direct, a.s.w.
	+	Pure Mg-bentonite	Uppot & Stephenson (1989)	Flex. p., direct, a.s.w
Methanol (5% dilute in water)	>+++	Noncalcareous smectite	Brown & Anderson (1983)	Fix. p., direct, a.s.w.
	–	Smectite	Daniel & Liljestrand (1984)	Flex. p., direct, w.s.w.
Ethylene glycol	+++	Noncalcareous smectite	Brown & Anderson (1983)	Fix. p., direct, a.s.w.
Xylene	no flow	Pure Mg-bentonite	Uppot & Stephenson (1989)	Flex. p., direct, a.s.w
	+++	Noncalcareous smectite	Brown & Anderson (1983)	Fix. p., direct, a.s.w
Xylene (196 ppm in water)	–	Smectite	Daniel & Liljestrand (1984)	Flex. p., direct, w.s.w.

Neutral inorganic compounds

$FeCl_3$ (0.5 g/l in water)	0/+	Montmorillonite	Pierce et al. (1987)	Fix. and Flex. p., a.s.w.
$Ni(NO_3)_2$ (0.3 g/l in water)	0/+	Montmorillonite	Pierce et al. (1987)	Fix. and Flex. p., a.s.w.
10% ($NaCl + Na_2SO_4$)	0	Natural Na-bentonite	Simons & Reuter (1985)	Flex. p., direct
10% ($NaCl + Na_2SO_4$)	0/+	Sand + 33% Na-bentonite	Simons & Reuter (1985)	

+ = increase of 1 to 0 times
++ = increase of 10 to 100 times
+++ = increase of 100 to 1000 times
>+++ = increase more than 1000 times
0 = no effect
– = decrease of 1 to 10 times
– – = decrease of 10 to 100 times
– – – = decrease of 100 to 1000 times

<– – – = decrease more than 100 times
Fix. p. = fixed wall permeameter
Flex. p. = flexible wall permeameter
a.s.w. = after saturation with water
w.s.w = without saturation with water
direct = direct permeability measures
indirect = not direct measures (drives from oedometric tests)

When comparing these data and trying to explain the increase in the hydraulic conductivity according to the double layer theory described earlier, it is necessary to respect the different experimental procedures used by different researchers. These differences can be very significant. In fixed-wall permeameters, any physical–chemical reaction that causes even a slight volumetric reduction in the soil specimen will decrease the boundary stress condition and increase the potential for permeant to pass between the specimen and the rigid wall. The amount of this short-circuiting is difficult, if not impossible, to determine and relate it to the increase in the hydraulic conductivity. Moreover, in fixed-wall permeameter it is not common to apply a back pressure to facilitate the saturation of the specimen and it is therefore not possible to verify if the full saturation is realized or not. Flexible-wall permeameters seem much more adequate to investigate and compare the hydraulic conductivity of clayey soils in the presence of different fluids. This could be the reason of the great differences between results obtained with xylene and aniline as the permeant fluid by Uppot & Stephenson (1989) and Brown & Anderson (1983) (see Table 1): in the former case a flexible-wall permeameter was employed (no xylene flow was detected), while in the latter case a fixed-wall permeameter was employed and a three orders of magnitude increase in hydraulic conductivity was found.

Different testing procedures can also cause significative differences in the results: the main differences derive from saturating the specimen with water or directly with the permeant fluid. If the specimen has been previously saturated and permeated with standard water, the permeant fluid must displace and substitute all the pore water to induce the changes in structure and hydraulic conductivity. If the permeant fluid is not or poorly miscibile with water (e.g. xylene, aniline, benzene) and/or is highly viscous and far less polar than water, it is difficult for the permeant to displace the water and to adsorb to the clay particles, giving rise to the structural modification described in the previous paragraph. If the permeant is directly put in contact with the specimen, the bentonite swelling will be conditioned by the physico-chemical characteristics of the fluid (e.g. dielectric constant, electrolyte concentration) and cation exchange may eventually take place. This is clearly shown in Table 1, where the hydraulic conductivity to municipal solid waste (MSW) leachate is ten times greater when the specimen has not been previously saturated with water (Hoeks *et al.*, 1987). Finally, it is useful to note that if the specimen is not completely saturated with water (as may be the case in fixed-wall permeameters), immiscibile fluids can permeate through the air voids in the specimen and there generate the physico-chemical reactions.

The use of pure or dilute fluids as permeants can make a large difference: pure fluids may cause an increase in the hydraulic conductivity of bentonite of orders of magnitude. If dilute test fluids are used, they do not cause any significant effect, or even a decrease, e.g. the case of methanol and xylene, as shown in Table 1. This is a very relevant aspect because it is not very likely that in a sanitary landfill sand–bentonite liners will get in contact with pure contaminant fluids.

Data in Table 1 show that most of the tested liquids affect in some way the structure of bentonite or sand–bentonite mixtures, causing an increase in hydraulic conductivity. The largest increase—three and more orders of magnitude—is caused by pure organic fluids, mainly alcohols and solvents: however, as already mentioned above, it appears that the same fluids, when diluted and permeated on a bentonite liner saturated with water in advance, would have a far less detrimental effect in terms of permeability. Similar conclusions are confirmed also by Daniel and Liljestrand (1984) and by Uppot and Stephenson (1989).

Other fluids, like acids, diluted inorganic compounds and leachates from MSW-landfills, do not show a clear tendency towards increasing the permeability of the clayey soils, since the same fluid can cause a decrease or an increase in the permeability: this can be due to differences in the testing conditions or to differences in the composition of the fluid, as for municipal landfill leachates.

It is clear that in a sand–bentonite mixture the sealing properties are a function of the bentonite content: the higher this content, the more severe the increase in the hydraulic conductivity of the mixture may be, due to the partial loss of bentonite unique properties.

EXPERIMENTAL PROGRAMME

Scope and Methods

The scope of the experimental programme in the laboratory was to study the effects that leachates and chemical fluids that may be present in a sanitary landfill have on the hydraulic conductivity of a sand–bentonite mixture.

Since the effects on the hydraulic conductivity are due to modifications in the bentonite properties, geotechnical characteristics of bentonite (Atterberg's limits, porosimetry) were also investigated. Therefore, the research program was characterized by:

- comparison of the results from permeability tests using water, leachates from MSW sanitary landfills and leachates from some industrial waste landfills;
- use of the same sand–bentonite mixture, in constant proportion and with the same sample preparation procedure, i.e. Proctor compaction according to Standard AASHO (American Association of State Highways Officers) method;
- use of the same apparatus and testing procedures (triaxial apparatus as a flexible-wall permeameter; saturation of samples with water, followed by addition of the specific permeant to be tested; use of back-pressure to facilitate preliminary saturation; use of the same values of cell pressure, back-pressure and hydraulic gradient for all tests; long term tests, until a constant value of hydraulic conductivity is reached);
- chemical, mineralogical and geotechnical analyses on soils, before and after permeability tests;
- chemical and physical analyses of leachates, before and after permeability tests.

Tested Materials

Sand–bentonite mixtures were prepared, by admixing different percentages of bentonite to an inert siliceous sand; a commercially-available sodium-activated bentonite was used.

With these materials, the following physico-chemical and geotechnical characterization tests were run, in accordance with ASTM and CNR-UNI (Italian) standard methods of testing materials:

- physico-chemical and mineralogical analyses, on all basic components (sand and bentonite) and on mixtures (the mineralogical analyses were carried out by means of a PW1800 X-Ray Philips Diffractometer);
- particle size analyses (by sieving and sedimentation), on both basic components (the results are reported in Fig. 3);

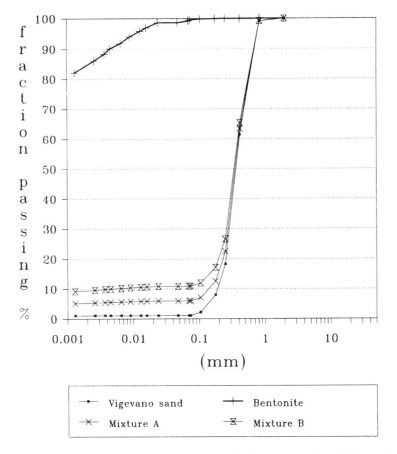

Figure 3. Granulometric distribution of sand, bentonite and sand–bentonite mixtures.

- Atterberg's limits, on bentonite;
- compaction tests, on mixtures;
- specific surface, on mixtures.

The most important physico-chemical and geotechnical properties (Atterberg's limits and plasticity index, CEC, swelling) of the bentonite are reported in Tables 2 and 3. The high values of plasticity (IP = 527%) and free swelling are outstanding.

TABLE 2. Geotechnical Characteristics of the Sand and Bentonite used in the Experiments

Parameter	Units	Sand	Bentonite
Sand fraction (SF)	%	99	0·5
Silt fraction (MF)	%	1	14·5
Clay fraction (CF)	%	—	85
Effective diameter D_{10}	mm	0·19	—
Significant diameter D_{60}	mm	0·41	—
Uniformity coefficient C_u		2·16	—
Specific gravity G	%	2·68	2·74
Liquidity limit[a]	%	—	615
Plasticity limit[a]	%	—	35
Plasticity index[a]			580
Activity index		—	6·82
CEC	meq/100 g		90–100

[a] Values determined with distilled water; for tap water see Fig. 9.

Figure 4 shows the diffractograms of bentonite. Bentonite was analysed in different hydrated–dehydrated conditions, in order to assess the swelling–shrinkage behaviour. In the natural state, the basal reflection is well marked and pointed at 12·32 Å; this confirms that sodium (Na^+) is the exchangeable cation present between the lattices. The mineralogical analysis of sand states that it is prevalently formed of quartz and feldspar grains, with a minor amount of mica, amphibole, chlorite and kaolinite.

Two mixtures were prepared, with different percentages of bentonite (on a dry weight basis), respectively:

- *Mixture A*: sand 95%–bentonite 5%
- *Mixture B*: sand 90%–bentonite 10%

The grain size composition of the so-obtained mixtures is reported in Fig. 3.

The mixtures were compacted according to the Proctor Standard AASHO Method. According to the results reported in Fig. 5, increasing the bentonite content in the mixture results in an increase of the dry density and a decrease of the optimum water content.

The specific surface, determined by means of the porosimeter, was 3·42 m^2/g for mixture A and 6·63 m^2/g for mixture B (in accordance with a higher percentage of flake-shaped bentonite particles).

TABLE 3. Mineralogical and Physico-chemical Characteristics of Bentonite

Mineralogical composition		
Smectite	%	80
Plagioclase	%	10
K-feldspar	%	4–5
Quartz	%	2–3
Calcite	%	2–4
Dolomite		traces
Amphiboles		traces
Chemical composition		
SiO_2	%	59·39
Al_2O_3	%	18·15
TiO_2	%	0·25
Fe_2O_3	%	4·39
MnO	%	0·03
P_2O_5	%	0·05
MgO	%	4·22
CaO	%	2·82
K_2O	%	0·93
Na_2O	%	2·64
Fire loss	%	7·13
Physical characteristics		
Swelling (2 g in 100 ml water)	cm^3	15–20
Eslin's index:		
after 2 h	%	300
after 8 h	%	500

Permeants Used for Tests

The permeants selected for the hydraulic conductivity tests were:

- water (tap and distilled);
- leachate I, from a MSW sanitary landfill;
- leachate II, derived from an industrial solid waste sanitary landfill;
- dilute acetic acid (20% in water);
- dilute propanol (20% in water).

Table 4 contains the chemical composition of leachates. Before the permeability tests, leachates were filtered on a GF/C filter (45 μm), to avoid clogging and any modifications of the permeability that may be caused by mechanical factors. The residual suspended solid content was about 1 g/litre. Specific weight and viscosity of leachates were also

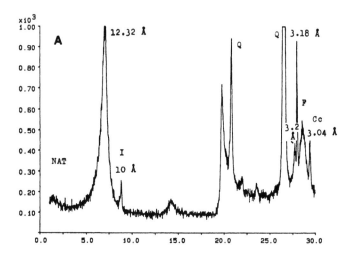

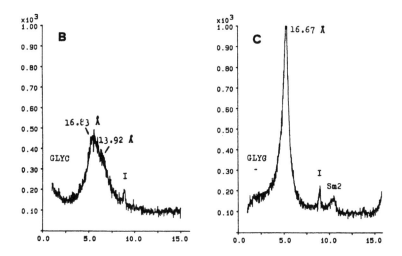

Figure 4. X-ray diffractograms of bentonite P. 120. (A) natural preparation; (B) glycolated preparation; (C) glycolated preparation heated up to 60°C for 24 h; (D) heated preparation up to 550°C; (E) natural preparation heated up to 60°C for 24 h.

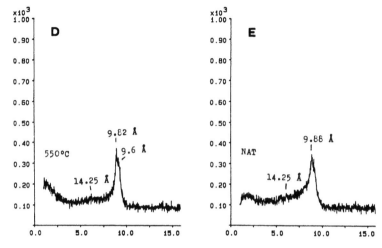

Figure 4. (*Continued*)

determined. The ratio of specific weight to viscosity for all the leachates was very close to that of water, therefore the effect of a different specific weight or viscosity is negligible.

The choice of the pure dilute fluids to be tested was based on the following demands:

- possible presence in a sanitary landfill;
- complete miscibility with water;
- low dielectric constant (water: $\varepsilon = 80$; acetic acid: $\varepsilon = 6 \cdot 2$; propanol: $\varepsilon = 20$).

Permeability Tests

A flexible wall permeameter, adapted from a triaxial apparatus such as used in soil mechanics laboratories was selected for tests (see Fig. 6). The main advantages of this solution are:

- for low permeability materials, a flexible wall avoids to a high degree undesired flow along irregularities of the specimen lateral surface; anyway, a silicon grease film was interposed between the specimen surface and the latex membrane;

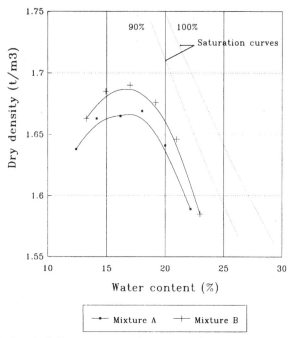

Figure 5. Standard Proctor compaction curves for mixtures A (sand 95%, bentonite 5%) and B (sand 90%, bentonite 10%).

- the application of a back-pressure to the hydraulic circuit allowed an easier, and more controllable, preliminary saturation of the testing specimens.

Constant hydraulic head tests were carried out; within the specimen, upward flow was imposed, in order to furtherly facilitate saturation.

The testing conditions were characterized by the following values (see also Fig. 6):

- specimen diameter: 70 mm
- specimen height: 150 mm
- all-around cell pressure: 397 kPa
- hydraulic head (applied at the specimen bottom):   265 kPa
- back-pressure (applied at the specimen top):   245 kPa
- net hydraulic load: 20 kPa
- constant hydraulic gradient: 13·3

TABLE 4. Chemical Analyses of the Tested Leachates (I = MSW Leachate; II = Industrial Solid Wastes Landfill Leachate) Before and After Permeability Tests

Parameter[a]	I		II	
	Before	*After*	*Before*	*After*
pH	6	7·5	7·3	7·65
COD	38·520	39·600	1·924	3 010
BOD	3·000	15·000	1·230	850
Volatile fatty acids (C)	1·574	n.a.	n.a.	n.a.
Organic nitrogen (N—NH_4)	60	n.a.	173	n.a.
Ammonia (N—NH_4)	1·293	1·421 6	222	61·90
Total phosphorus (P—PO_4)	n.a.	n.a.	2·50	n.a.
Alkalinity ($CaCO_3$)	5·125		n.a.	n.a.
Chlorine (Cl)	2·231		n.a.	n.a.
Sulphate (SO_4)	1·600		n.a.	n.a.
Calcium (Ca)	175	329·8	53·20	74·50
Sodium (Na)	1·400	1·700	n.a.	n.a.
Potassium (K)	1·200	1·800	n.a.	n.a.
Magnesium (Mg)	1·469	562·3	193·90	297·30
Iron (Fe)	47	5·9	0·30	n.a.
Manganese (Mn)	42		n.a.	n.a.
Zinc (Zn)	7	4·4	0·10	n.a.
Lead (Pb)	1	0·8	0·23	n.a.
Nickel (Ni)	1·6	1·6	0·58	n.a.
Copper (Cu)	0·2	0·2	0·07	n.a.
Cadmium (Cd)	0·1	0·1	0·04	n.a.
Aluminium (Al)	n.a.	n.a.	0·02	n.a.
Surface-active agents—ABS	n.a.	n.a.	6·30	3·20
Phenols	n.a.	n.a.	6·30	n.a.

[a] All values in milligrammes per litre except pH.
n.a. = Not analysed.

Fluids were always introduced into the hydraulic circuit after a preliminary complete saturation of the specimen with de-aerated water (the condition of full saturation was assessed by values of the Skempton's pore pressure parameter B close to the unit) and after some measurements of permeability to water on the same specimen.

The outflow was measured with a special burette, inserted in the back-pressure line, and the hydraulic conductivity values were computed with the following formula, resulting from Darcy's Law applied

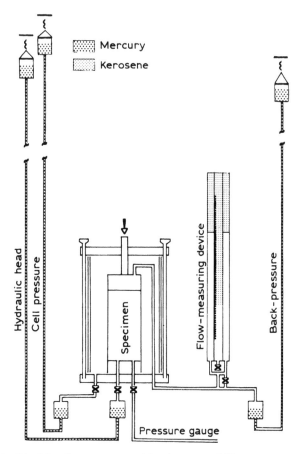

Figure 6. Triaxial cell apparatus used for the permeability.

to constant hydraulic head tests:

$$k = \frac{V}{A \cdot t \cdot i} \quad [LT^{-1}]$$

where:

V is the volume outflowing during time t (litres^3) [L^3],

A is the section area of the soil specimen [L^2] and

t is the elapsed time between subsequent measures [T].

Two test-series were conducted during the research programme:

• *1st Series*: mixture A, permeated with leachates I and II (tap water as reference fluid);
• *2nd Series*: mixture B, permeated with dilute acetic acid and propanol (distilled water as reference fluid).

Tap water was employed in the first series of tests to simulate the actual condition of the liner in a sanitary landfill (Malpei, 1987), while distilled water was preferred in the second series as a standard fluid, allowing comparison and reproduction of the testing conditions (Offredi, 1990).

Depending on the permeability of the mixtures, every test lasted for a minimum of 20 days to a maximum of 139 days.

RESULTS OF PERMEABILITY TESTS

First Series of Tests

The test results from the first phase of the research are reported in Fig. 7. For each day, a minimum of three measurements were made: the average daily value of computed hydraulic conductivity is reported versus the elapsed time.

Permeability to tap water test. The trend of the curve shows a decrease of the hydraulic conductivity with time: the final value, reached after 23 days, is about $2 \cdot 3 \times 10^{-7}$ m/s. It is possible that the tendency of hydraulic conductivity to decrease with time can be ascribed to a progressive swelling of bentonite particles into the largest pores of the aggregate.

Permeability to leachate I test. Filtered leachate I was used for the second test. The hydraulic conductivity, being about $4 \cdot 2 \times 10^{-7}$ m/s after the preliminary saturation with water, gradually increased to about $6 \cdot 3 \times 10^{-7}$ m/s in about two weeks and remained approximately constant during the following two weeks. The whole period of the test

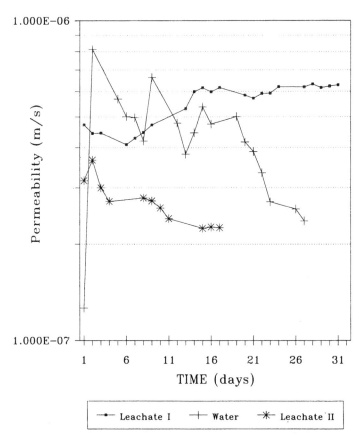

Figure 7. Results of the first series of permeability tests with mixture A permeated with tap water, leachate I and leachate II.

was 33 days. The final value of the hydraulic conductivity was nearly three times higher than the value for water.

Permeability to leachate II test. The general trend of the test performed by using a leachate taken from a particular industrial waste landfill (leachate II) is very similar to that obtained from the permeability test to water. The hydraulic conductivity shows a tendency to decrease with time, reaching a final value of about $2{\cdot}1 \times 10^{-7}$ m/s.

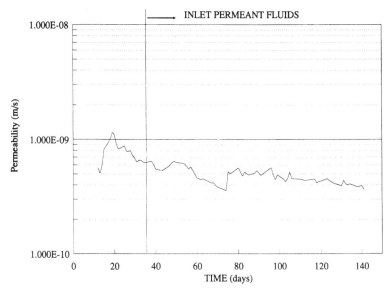

Figure 8. Results of the second series of permeability tests with mixture B permeated with distilled water.

Second Series of Tests

Figure 8 shows the results of the tests that were performed during the second phase of the research. Only the curve referring to distilled water is reported, because with acetic acid and propanol it was impossible to measure a reliable value of hydraulic conductivity, as will be explained below.

Permeability to distilled water test. The permeability test on mixture B using distilled water lasted 139 days; during this period of time, a volume equivalent to the pore void volume migrated through the specimen (equal to $232\,cm^3$). The mean daily flow decreased from $3\,cm^3$/day to $2\,cm^3$/day at the end of the test and the hydraulic conductivity coefficient decreased from $6\cdot4 \times 10^{-10}\,m/s$ to $3\cdot6 \times 10^{-10}\,m/s$. This was probably due to the progressive swelling of bentonite in the specimen. The trend to decrease with time was also observed in the test with tap water of the first series. It can be noted that the increase in the percentage of bentonite from 5 to 10%

decreased the permeability to water of the sand–bentonite mixture by nearly three orders of magnitude.

Permeability to acetic acid test. Permeability test to dilute acetic acid lasted 78 days. No flow was detected during this period, while the formation of a large amount of gas could be observed. The gas flowed in the measuring device (burette) and tended to accumulate at the interface with the volume measurement fluid. So, it was impossible to measure experimentally a reliable value of the hydraulic conductivity coefficient.

Permeability to propanol test. Permeability test with dilute propanol lasted 80 days. As for dilute acetic acid, no flow was detected during all the testing period and gas formation was observed.

EFFECTS OF PERMEANTS ON TESTED SPECIMENS

Laboratory tests were performed to find out the effects of the permeation of the fluids on bentonite and sand–bentonite mixtures and on the fluids themselves.

First Series of Tests

Atterberg's limits. The effects of different leachates on bentonite were investigated by remoulding dry bentonite powder with:

(A) tap water;
(B) filtered leachate I;
(C) leachate II.

It was preliminarly observed that the amount of liquid necessary to achieve a complete swelling was relatively low in the second case, slightly higher in the third one and largely higher when tap water was used. The visual aspect of the slurry was also quite different.

Atterberg's limits were successively determined on the same slurries, with the following results:

• slurry A: LL = 569%, PL = 42%; PI = 527%;

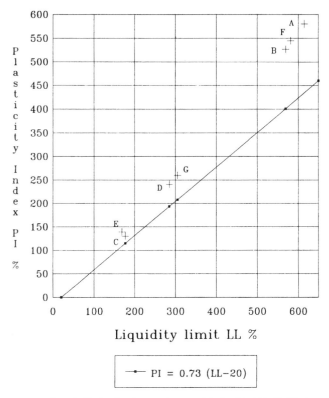

Figure 9. Atterberg's limits for bentonite moulded with (A) distilled water, (B) tap water, (C) leachate I, (D) leachate II, (E) acetic acid and (F) propanol. G represents the limits for bentonite moulded with tap water and then put in contact with leachate I.

- slurry B: LL = 177%; PL = 47%; PI = 130%;
- slurry C: LL = 285%; PL = 44%; PI = 241%.

The same values are reported in Casagrande's plasticity chart (Fig. 9), where the effect of leachates on the plasticity of bentonite becomes particularly evident. As a consequence of reduced plasticity, a reduced quality as a sealing material has to be expected.

On the other hand, if a dry bentonite powder is formerly remoulded with tap water and the so-obtained slurry gets into contact with leachate, the decrease of plasticity should be comparatively less dramatic. This kind of test should be more applicable to the field situation and to the permeability test procedures that are followed in this research work.

From tests carried out on slurry A, subsequently put in contact with filtered leachate I for 20 days, the following values were determined (see Fig. 9): LL = 300%; PL = 40%; PI = 260%.

It is evident that leachate–bentonite interactions still have negative effects on sealing properties of bentonite itself, though the decrease of LL and PI were comparatively smaller (see Fig. 9).

Chemical analyses on leachates. The results of chemical analyses on leachate I after seepage through the compacted sand–bentonite mixture, as reported in Table 4, can be summarized as follows:

- increase of pH;
- outstanding increase of BOD (without any significant variation of COD);
- increase in Ca, Na and K content;
- decrease of Mg and Fe content;
- no significant variation in the heavy metals content.

The increase in Ca content, probably due to leaching of carbonates, is responsible for the increase of pH (from 6·0 to 7·5).

The increase in pH might be the reason for the increase in BOD: biological activity is strongly influenced by pH and a pH value of 6 can already be a limiting factor for biological growth.

The decrease of Fe (and also of Zn) content could be ascribed to precipitation processes, also related to the increase of pH.

The variation in Na, K, and Mg content may be associated with cation exchange processes between bentonite and leachate.

Second Series of Tests

Atterberg's limits. The effect of acetic acid and propanol on bentonite swelling was determined by remoudling bentonite powder with distilled water and successively adding the fluids (20% acetic acid and 20% propanol, in solution). The mixtures were left hydrating for 20 days and finally, Atterberg's limits were determined, with the following results:

- distilled water only: LL = 615%; PL = 35%; PI = 580%;
- acetic acid solution: LL = 169%; PL = 30%; PI = 139%;
- propanol solution: LL = 581%; PL = 36%; PI = 545%.

Therefore, acetic acid had a drastic effect on the plasticity characteristics; on the contrary, propanol caused only a slight decrease of the liquidity limit.

Gas chromatography. The protracted contact between the sand–bentonite mixture and the propanol solution resulted in a gas yield (not allowing a reliable determination of the hydraulic conductivity). Chromatographic analysis of gas produced during the test gave a very high content of methane (82·0%) and a smaller percentage of carbon dioxide (0·5%).

Anaerobic digestion of propanol seems the most likely explanation of the phenomenon of methane generation, but the specimen was not analysed for microbiological identification. This phenomenon deserves to be studied in more detail: it is known that there can be biological activity in a specimen subjected to permeability tests, but such a phenomenon of methane generation is not reported in any of the similar tests described in literature.

The acetic acid solution, too, produced some gas but it was not analysed.

X-ray diffractometer analyses. For each tested specimen, X-ray diffractometer analyses were performed on only the finest fractions (passing through a 0·074 mm sieve), due to the fact that the observed phenomena are controlled only by the bentonite particles. Moreover, it has to be remembered that all permeability tests were conducted with upward flow, therefore different parts at different heights within each tested specimen were subjected to mineralogical analyses.

All the diffractograms of the specimens permeated with distilled water, acetic acid and propanol show similar behaviour. In the lower part of the specimens (see for example Fig. 10), an apparent increase of Ca-bentonite (peak at about 14 Å) content could be observed, while Na-bentonite (peak at about 12·6 Å) was still prevailing in the upper part (see Fig. 11).

This behaviour is probably due to the leaching of Na^+ ions by distilled water flow (this also applies to the specimens permeated with

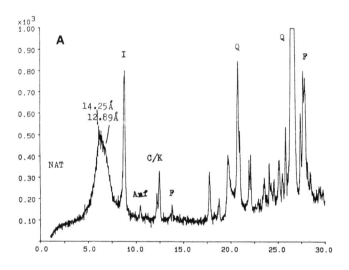

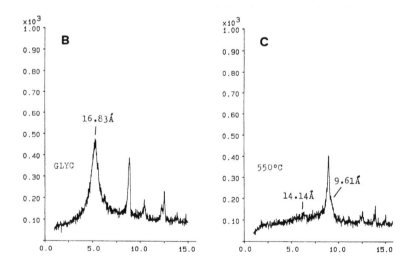

Figure 10. X-ray diffractograms of mixture B with propanol (bottom part of the specimen, fraction <74 μm): (A) natural preparation; (B) glycolated preparation; (C) preparation heated up to 550°C.

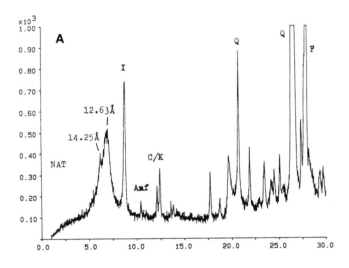

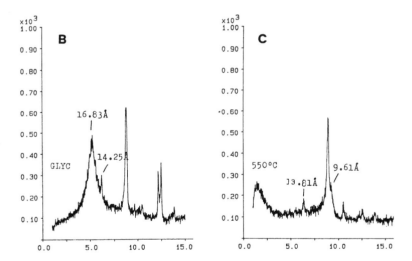

Figure 11. X-ray diffractograms of mixture B with propanol (top part of the specimen, fraction <74 μm): (A) natural preparation; (B) glycolated preparation; (C) preparation heated up to 550°C.

acetic acid and propanol which were first permeated with distilled water to achieve saturation) and does not seem to correlate with the presence of acetic acid and propanol as permeants.

CONCLUSIONS AND RECOMMENDATIONS

Hydraulic conductivity and other properties of bentonite (like Atterberg's limits) can be affected by the presence of fluids other than water.

This can be explained by a partial loss of colloidal and swelling properties of the bentonite, due to cation exchange processes and to variation of the electrical potential within the double layer, in relation with the physico-chemical characteristics of the fluid. These phenomena are well known from literature and were confirmed by the authors' permeability tests described in this chapter.

However, while some of the literature data showed a dramatic increase of permeability with some organic fluids, the authors' results do not show such a marked difference.

The permeability of MSW-leachate through a compacted sand–bentonite liner was about three times higher if compared with the permeability to water, while permeability to a leachate coming from an industrial waste disposal plant was similar to that determined for water.

Subsequent tests, conducted with diluted pure fluids (acetic acid and propanol) did not allow the measurement of the hydraulic conductivity, because no flow was detected after 70 days or more, but gas formation was detected. It is noticeable that no similar behaviour was reported by other authors. Literature data regarding hydraulic conductivity with acetic acid, pure and diluted, show that there is significant difference compared to hydraulic conductivity with water.

Anyhow, laboratory and in-situ conditions differ in stress and confinement status, in hydraulic head and flow direction, in degree of saturation, etc. Therefore, laboratory tests may not be fully representative of in-situ behaviour.

For the following reasons the authors have reached the conclusion that operating conditions are far less severe than laboratory conditions:

- leachates contain a noticeable amount of suspended solids, which can cause a physical clogging of the liner and can reduce the permeability by more than one order of magnitude (Cancelli *et al.*, 1987);

• no pure and concentrated fluids are likely to be found in a sanitary landfill and the literature data show that the effects of some pure organic fluids are much more dramatic compared with MSW-leachates or other diluted organic fluids.

Nevertheless, it is necessary not to rely on permeability to water values and to measure experimentally the possible increase in permeability in the presence of a permeant other than water. On the basis of literature data, this procedure is particularly important when a leachate with a high amount of organics such as aniline, benzene, methanol, etc., is expected.

The determination of the Atterberg's limits using the permeants also gives a good indication of the loss of swelling characteristic of bentonite caused by the permeants.

The choice of the optimum amount of bentonite to be added to the sand, in order to get the required permeability value should be based on the results of tests using the actual permeant.

REFERENCES

Alther, G. R. (1982). The role of bentonite in soil sealing applications. *Engineering Geology,* **19**(4), 401–9.

Anderson, D., Brown, K. W. & Green, J. (1982). Effects of organic fluids on the permeability of clay soil liners. *US EPA, Rep. no. 600/9-82-002,* Cincinnati, Ohio.

Bowders, J. J., Daniel, D. E., Broderick, G. P. & Liljestrand, H. M. (1986). Methods for testing the compatibility of clay liners with landfill leachate. *ASTM, Spec. Tech. Publ. 886,* Philadelphia, PA.

Broderick, G. P. & Daniel, D. E. (1990). Stabilizing compacted clay against chemical attack. *ASCE, J. Geotech. Engng,* **117**(10), 1549–67.

Brown, K. W. & Anderson, D. C. (1983). Effects of organic solvents on the permeability of clay soils. *US EPA, Rep. no. 600/2-83-016,* Cincinnati, Ohio.

Cancelli, A. (1983). Caracterisation géotechnique d'un mélange imperméabilisant sol/bentonite. *Coll. Etanchéité Superficielle,* **I**, 250/1–6, CEMAGREF, Paris.

Cancelli, A. & Francani, V. (1984). Quarry reclamation in the Lombardy Plain. *Bull. Int. Ass. Engng Geology,* **29**, 237–40.

Cancelli, A., Cossu, R. & Malpei, F. (1987). Laboratory investigation on bentonite as sealing agent for waste disposal. In *Proc. First Intern. Symp. on Sanitary Landfill, Process Cagliari*.

Carroll, D. & Starkey, H. C. (1971). Reactivity of clay minerals with acids and alkalies. *Clays and Clay Minerals*, **19**, 321–33.

Daniel, D. E. & Liljestrand, H. M. (1984). Effects of landfill leachates on natural liner systems. Report to Chemical Manufacturers Association, University of Texas, Dept. of Civil Eng., Austin, Texas.

D'Appolonia, D. (1980). Slurry trench cut-off walls for hazardous waste isolation. In-house report, Engineered Construction International, Inc., Pittsburgh, PA.

Fayoux, D. (1983). Etanchéité superficielle par traitement de sol. Rapp. Gén., *Coll. Etanchéité Superficielle*, **I**, 113–28, CEMAGREF, Paris.

Fernandez, F. & Quigley, R. M. (1985). Hydraulic conductivity of natural clays permeated with simple liquid hydrocarbons. *Can. Geotech. J.*, **22**, 205–14.

Gordon, B. B. & Forrest, M. (1981). Permeability of soil using contaminated permeant. *ASTM, Spec. Tech. Publ. 746*, pp. 101–20.

Griffin, R. A. & Shimp, N. F. (1978). Attenuation of pollutants in municipal landfill leachate by clay minerals. *US EPA, Rep. no. 600/2-78-157*, Washington, D.C.

Grim, R. E. (1953). *Clay Mineralogy*, 2nd edn. McGraw-Hill, New York.

Hettiaratchi, J. P. A., Hrudey, S. E., Smith, D. W. & Sego, D. C. C. (1988). A procedure for evaluating municipal solid waste leachate components capable of causing volume shrinkage in compacted clay soils. *Environmental Technology Letters*, **9**(2), 23–34.

Hoeks, J. & Ryhiner, A. H. (1989). Surface capping with natural liner materials. In *Sanitary Landfilling: Process, Technology and Environmental Impact*, ed. T. H. Christensen, R. Cossu & R. Stegmann. Academic Press, London, pp. 311–22.

Hoeks, J., Glas, H., Hofkamp, J. & Ryhiner, A. H. (1987). Bentonite liners for isolation of waste disposal sites. *Waste Managem. & Research*, **5**, 93–105.

Ingles, O. G. & Metcalf, J. B. (1972). *Soil Stabilization*. Butterworths, Sydney.

Lambe, T. W. (1958). The structure of compacted clays. *J. of the Soil Mechanics and Foundations Division, Proc. of the ASCE*, **84**, SM2.

Lundgren, T. A. (1981). Some bentonite sealants in soil mixed blankets. In *Proc., 10th Int. Conf. Soil Mechanics & Foundation Engineering*, vol. 3. Balkema, Rotterdam, pp. 349–54.

Malpei, F. (1987). Bentonite as sealing agent in sanitary landfill. Master Thesis, Politecnico di Milano (in Italian).

Mesri, G. & Olson, R. E. (1971). Mechanism controlling the permeability of clays. *Clays and Clay Minerals*, **19**, 151–8.

Mitchell, J. K. (1976). *Fundamentals of Soil Behaviour*. J. Wiley & Sons, Toronto.

Nasiatka, D. M., Shepherd, T. A. & Nelson, J. D. (1981). Clay liner permeability in low pH environments. In *Proc., Symposium on Uranium Mill*

Tailings Management, Colorado State University, Fort Collins, Colorado, pp. 627–45.

Offredi, A. (1990). Geotechnical and mineralogical characterization of sand–bentonite mixtures used as sealing agent in sanitary landfills. Master Thesis, University of Milan (in Italian).

Peirce, J. J., Sallfors, G., Peel, T. A. & Witter, K. (1987). Effects of selected inorganic leachates on clay permeability. *ASCE, J. Geotech. Engng.,* 113(8), 915–19.

Peterson, S. R. & Gee, G. W. (1985). Interactions between acidic solutions and clay liners: permeability and neutralization. *ASTM, Spec. Tech. Publ.* 874, pp. 229–45.

Quigley, R. M., Crooks, V. E. & Fernandez, F. (1984). Engineered clay liners. In *Proc., Semin. on Design and Construction of Municipal and Industrial Waste Disposal Facilities.* Can. Geotech. Soc., Toronto, pp. 115–34.

Quigley, R. M., Fernandez, F., Helgason, T., Margaritas, A. & Whitby, J. L. (1987). Hydraulic conductivity of contaminated natural clay directly below a domestic landfill. *Can. Geotech. J.,* 24, 377–83.

Quigley, R. M., Fernandez, F. & Rowe, R. (1988). Clayey barrier assessment for impoundment of domestic leachate (Southern Ontario) including clay-leachate compatibility by hydraulic conductivity testing. *Can. Geotech. J.,* 25, 574–81.

Sima, N. & Harsulescu, A. (1979). The use of bentonite for sealing earth dams. *Bull. Int. Ass. Eng. Geology,* 20, 222–6.

Simons, H. & Reuter, E. (1985). Entwicklung von Prüfverfahre und Regeln zur Herstellung von Deponieabdichtungen aus Tonzum Schutz des Grundwassers. *Mitteilung des Instituts für Grundbau und Bodenmechanik Technische Universität Braunschweig,* Heft N. 18, Braunschweig.

Uppot, J. O . & Stephenson, R. W. (1989). Permeability of clays under organic permeants. *ASCE, J. Geotech. Engng.,* 115(1), 115–31.

Yariv, S. & Cross, H. (1979). *Geochemistry of Colloid Systems for Earth Scientists.* Spinger-Verlag, New York.

3.5 Investigation and Assessment of the Deformability of Mineral Liners

RAINER SCHERBECK[a] & HANS LUDWIG JESSBERGER[b]

[a] Jessberger + Partner Ltd, Am Umweltpark 5, D-4630 Bochum 1, Germany

[b] Institute for Soil Mechanics and Foundation Engineering, Ruhr-University Bochum, Universitätsstrasse 150, D-4630 Bochum 1, Germany

INTRODUCTION

Typical lining materials preferentially used in geotechnical practice are clayey minerals, asphaltics and geomembranes with different geotechnical properties concerning permeation behaviour, sensitivity on settlements, workability, stability etc. In landfill construction practice compacted mineral liners have long predominated; modern liner systems combine the mineral component with flexible synthetics to form an almost impermeable hydraulic barrier. However, the synthetic components have to be assessed for long term behaviour for which there is a lack of information, since present experience with these materials does not exceed a few decades. It is obvious, therefore, that the mineral component is of continued relevance for the case of long term behaviour. The mineral liner thus forms a key element of the whole lining system and is the main object of this investigation.

Mineral liners consist of compacted, fine grained clayey soils which provide hydraulic barriers due to their low permeability and mineralogical structure. Various factors can affect the liner performance, for example chemical interaction between liner and fluids (Gordon & Forest, 1979), climatic conditions such as desiccation and frost action (Kleppe & Olson 1985, Andersland & Al-Moussawi, 1987) and—last but not least—deformation (Gilbert & Murphy, 1987, Larson & Keshian, 1988). Such deformations may occur as the result of the

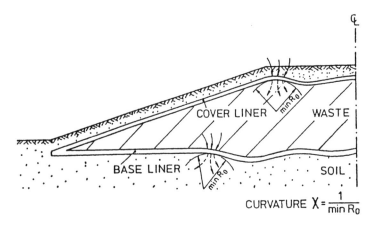

Figure 1. Liner response to differential settlement.

compressive nature of the underlying soil and waste. They may lead to curvature loading of the liner, as shown in Fig. 1, when appearing as uneven, differential settlements.

It is of major interest to study the effect of such deformations on the performance of compacted mineral liners as effective hydraulic barriers, and furthermore to provide recommendations for design of mineral liners to cater for uneven settlements. In this paper a description of centrifuge model tests on this topic carried out in Bochum Geotechnical Centrifuge is given, followed by the presentation of laboratory test results from investigations on the tensile behavior of cohesive soils. Finally, the investigated features of deformability of mineral liners were condensed into a technical recommendation of the ETC 8 (European Technical Committee), which is presented here in a draft form.

CENTRIFUGE MODEL TESTS

Application of Centrifuge Modelling

The principal reason for using the centrifuge model technique here is the possibility of simulating prototype stresses in a small scale model. For the case of base liners, prototype overburden depths of about 30 m waste or more are common, so that correct modelling of self weight stresses is essential. Previous studies by Stone and Wood (1987) showed from comparison between model tests under earth gravity (1 g) and increased gravity (n . g) that cracking is a condition heavily influenced by the presence of prototype stress levels. Since cracking is expected to be partly significant in this investigation, the centrifuge modelling principle is applicable. The centrifuge modelling technique also provides excellent opportunities to change and vary boundary conditions during the model investigation.

The centrifuge model tests reported here were performed on the Bochum 10 m balanced beam Geotechnical Centrifuge under an acceleration field of 50 gravities (= scaling factor n). Details of the operating characteristics and capabilities of the Bochum Geotechnical Centrifuge can be found in Jessberger and Güttler (1988).

Centrifuge scaling factors—the links between model and prototype behaviour—have been described extensively elsewhere (see Jessberger & Güttler, 1988). The geometry of a prototype lining situation, e.g. liner thickness, depth of overburden and water table etc., is scaled down in the centrifuge model by the factor of increased gravity level n. Other scaling laws between prototype and model do not match like this and have to be taken into account when analyzing test results. For example, the time t for consolidation processes is accelerated in centrifuge by a factor of n^2 relative to the corresponding prototype. On the other side, the coefficient of permeability k is scaled by n (Tan & Scott, 1985), meaning that the k-value in the model situation is n times the value of the prototype. When presenting model tests results later, diagram values are standardized to prototype conditions.

Model Test Package, Material and Test Procedure

The centrifuge model tests were performed in a rectangular strong box representing a liner area of 650 m^2 in protoype scale. The test package involves a trap-door mechanism to initiate the differential deformation of the mineral liner under controlled conditions. Since chemical interaction between the liner material and the permeating fluid was not desired in this investigation, water was chosen to permeate the liner. A detailed description of the model package and experimental technique can be found in Stone *et al.* (1989) and Jessberger *et al.* (1989) respectively. Figure 2 gives a sketch of the basic function elements used in the centrifuge model.

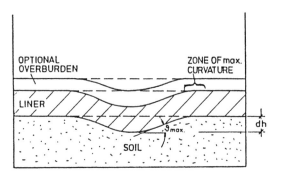

MODEL IDEA FOR DEFORMATION TESTS

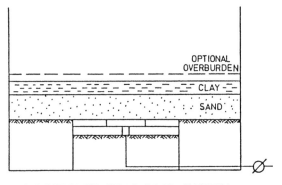

UNDEFORMED TRAP DOOR SYSTEM

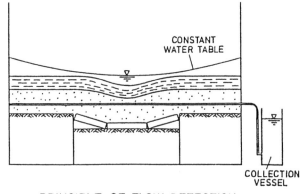

PRINCIPLE OF FLOW DETECTION

Figure 2. Schematic function of centrifuge model (cross section).

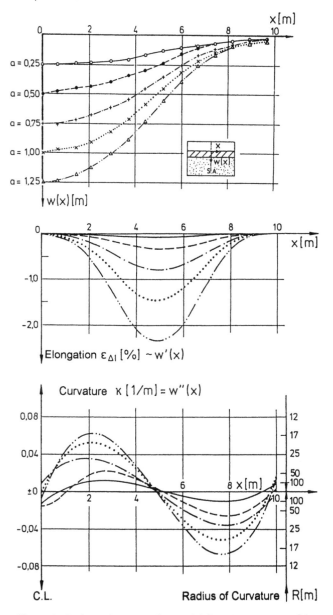

Figure 3. Deformation input for model liner (prototype scale).

The data obtained from the model tests consists of displacement and flow rate data recorded by LVDT measurements and optical records. Two CCD-cameras observe the liner surface in-flight to record the onset of liner deterioration. Photographs showing the position of discrete markers placed in the mineral liner enable post-test analysis via digitizing techniques, to allow displacement fields to be found. Derived from several post-test analysis, Fig. 3 shows the quantity of differential deformation attainable with the centrifuge model package depending on the central piston settlement, *a*. Mathematically derived from the shape of the deformed liner, curvature κ and elongation $\varepsilon_{\Delta l}$ are the two major components to describe the deformation input.

The liner material preferentially tested in this study was a commercially available kaolin clay of low plasticity. The major geotechnical properties for this liner material are shown in Table 1. The kaolin liners were consolidated inside the strong box from a slurry to a maximum consolidation pressure of 630 kPa following the normal-consolidation curve to produce an equivalent 95% proctor density. The kaolin for use in centrifuge is highly overconsolidated, equivalent to an overburden prestress of about 40 m landfill material. After the liner material was placed into the model container, centrifuge is accelerated to 50 gravities with fixed trap-door mechanism and the surface water level is raised to its designated value. Due to the circular water table in centrifuge the hydraulic gradient differs slightly along the longitudinal section of the liner with a mean hydraulic gradient *i* between 1 and 2(−) in all tests. Consolidation of the liner to its equilibrium effective stress is—depending on liner thickness—achieved

TABLE 1. Material Data

Material property	Kaolin
Liquid limit (%)	44·4
Plasticity index (%)	16·3
Water content (%)	37·5
Density (kN/m^3)	15·0
Permeability (m/s)	$1·0 \times 10^{-9}$
Particle diameter at . . .	
. . . 10% weight in mm	<0·001
. . . 60% weight in mm	0·002

approximately after one hour and dropping of the trap-door starts at a predetermined rate of 0·1 mm/min until detection of failure or full trap-door settlement.

Test Results

A test series with more than 30 single centrifuge model tests was carried out, varying mainly the overburden load, the liners' consistency and the type of the liner material. A typical result from the tests using kaolin clay without overburden is shown in Fig. 4. The

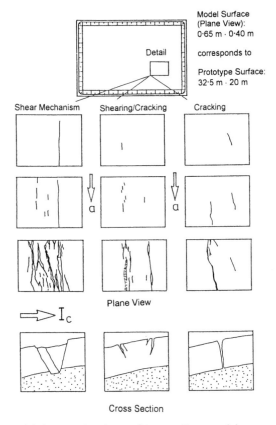

Figure 4. Typical failure modes observed in centrifuge model tests.

observed response of the cohesive liners to induced curvature is characterized by two features, shear mechanism and cracking, depending on the degree of deformation, a, and the material consistency, indicated by the consistency index I_c.

Shear mechanism was observed as the predominant failure in tests with an overburden load where shear bands were seen to extend the full depth of the liner with increasing curvature load. Tension cracking was the most significant failure mode in tests without overburden. The onset of tension cracks seemed to be influenced by material plasticity, consistency—quantified by the consistency index I_c—and thickness. The hydraulic performance was considerably different and dependent on the mode of liner failure. Whereas shearing mode did not significantly influence the flow rate through the liner, tension cracking could lead to total failure of the liner indicated by uncontrolled permeation (see Fig. 5).

When shearing mechanism is indicated to be the dominating failure mode, a maximum increase in flow rate and mean permeability k_m, respectively, by the factor of two was observed. Since this increase reflects the behaviour of the total liner surface which is in most parts unaffected by any deformation action, the results have to be transferred

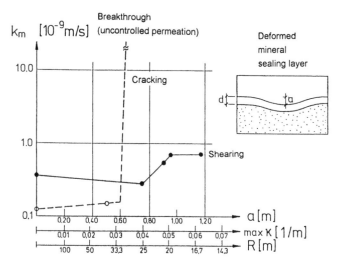

Figure 5. Typical data record showing the relation between deformation and permeation.

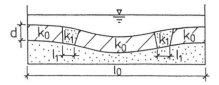

$$R_a = \frac{\text{ruptured}}{\text{non – ruptured}} \text{ Length}$$

$$= \frac{\Sigma l_1}{l_0} = 10\%$$

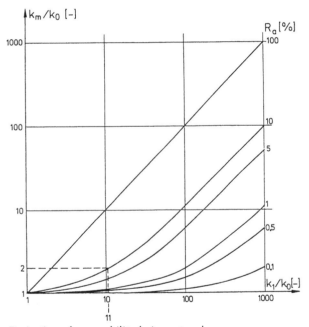

Figure 6. Derivation of permeability k_1 in ruptured areas.

geometrically to the deformed areas (see Fig. 6). In case of shearing mode a maximum increase in ruptured zone permeability k_1 by the factor 11 could be derived from the test results.

During variation of liner materials, the most significant changes in behaviour were observed, when using a coarse granular material with a pore filling mixture of fine sand and bentonite. This material was developed in collaboration between Dywidag AG, Essen, and the geomechanics group of Ruhr-University, Bochum. It has so far successfully been tested in prototype scale such as mineral lining for sanitary landfills, cut off walls and embedding around leaking pipes. Major advantages in construction practice result from the ability of pre-mixing and from the low water content, allowing the use of vibro-compaction to attain the desired compaction. This material with its granular behaviour seemed to be unaffected by the induced deformations and neither shear bands nor cracking mode could be observed. Mineral liner materials of a non-cohesive type are obviously not sensitive towards differential settlements.

INVESTIGATIONS OF TENSILE PROPERTIES

The results from centrifuge model tests illustrate that curvature of compacted clay liners can lead to failure conditions resulting in an uncontrolled breakthrough of leachate. It was observed that this failure condition appears predominantly when the liner is able to form tension cracks, i.e. for uncovered kaolin liners. The ability of cohesive material to fail in tension will be further analyzed. The tensile behavior of soil is not normally of significance in most geotechnical applications and therefore not very well investigated. There are several different test methods described in the literature to measure the tensile strength and strain of cohesive soils, for example the direct tension test, modified triaxial tests or beam element tests (see Ajaz & Parry, 1975 and Henne, 1989). To perform tensile tests in this study, a direct tension test device was developed to determine the ultimate tensile strain properties of some cohesive soils, especially those used in the centrifuge model tests. Figure 7 shows a sketch of the laboratory device and Fig. 8 gives an impression about a typical stress–strain relationship of a cohesive soil in tension.

Laboratory testing of tensile properties is significantly influenced by randam effects, resulting in deviations of about 25% concerning the ultimate tensile strain ε_{zq} and about 10% concerning tensile strength q_z. Since the tensile strain is more important when interpreting deformability of mineral liners, an dequate number of single element

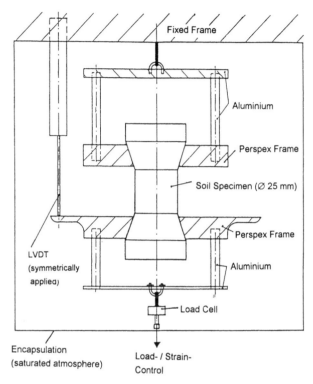

Figure 7. Laboratory device for determinating stress–strain relationships in tensile zone.

tests have to be carried out. Table 2 gives an overview of typical results from tensile tests based on investigations using the laboratory device shown in Fig. 7 for different types of clays.

The influence of compactive work on the tensile properties—additionally to the influence of the consistency—is demonstrated by varying between two kinds of compaction energy (Proctor: $0.6\,\mathrm{MNm/m^3}$, modified Proctor: $2.7\,\mathrm{MN\,m/m^3}$). The results from Table 2 show, that there is no obvious tendency in strain behaviour regarding the influence of plasticity, consistency and compactive effort. Maximum values for ultimate tensile strains are reached in tests using specimens compacted to 95% proctor density as well as with specimens compacted to 100% modified proctor density. The most

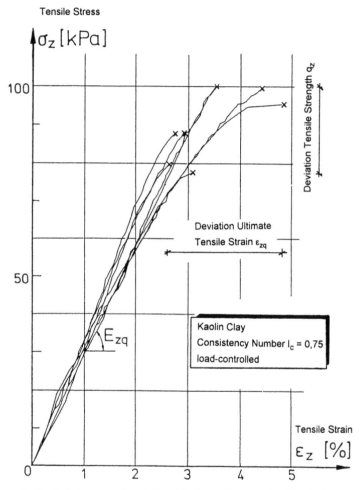

Figure 8. Typical stress–strain relationship in tensile zone (kaolin clay).

significant message deduced from these elementry tests concerns the quantity of ultimate tensile strain. Typical tensile strain values ε_{zq} for clays when compacted with proctor energy can be found between 1·5 and 4·4%. This range of values is identical with those observed in the centrifuge model tests, where liners withstand curvature loadings with radii of curvature down to 200 m without cracking.

TABLE 2. Classification and Tensile Properties of some Cohesive Soils

Subject	*Clay A*	*Clay B*	*Clay C*	*Kaolin clay*
Particle weight (%)				
Clay	58	70	56	61
Silt	37	25	41	39
Sand	5	5	3	0
Atterberg limits				
Liquid limit (%)	59	76	63	44
Plasticity index (%)	31	49	32	16
Proctor density (t/m^3)				
Simple: ρ_{pr}	1·55	1·50	1·67	1·55
Modified: ρ_{mpr}	1·79	1·71	1·82	1·68
Water content w (%), Consistency index I_c (−)				
100% ρ_{pr}	22, 1·19	24, 1·06	20, 1·34	26, 1·13
95% ρ_{pr}	29, 0·97	33, 0·88	26, 1·16	30, 0·88
100% ρ_{mpr}	16, 1·39	19, 1·16	18, 1·41	21, 1·44
95% ρ_{mpr}	21, 1·23	24, 1·06	24, 1·22	25, 1·19
Tensile strain[a] ε_{zq} (%)				
100% ρ_{pr}	2·4	2·4	3·3	2·9
95% ρ_{pr}	4·1	2·1	1·7	1·7
100% ρ_{mpr}	1·5	1·6	4·4	4·2
95% ρ_{mpr}	2·5	3·3	4·2	3·0

[a] The ultimate tensile strain ε_{zq} was computed as the mean value from minimum 4 single tests.

RECOMMENDATIONS FOR ASSESSING DEFORMED MINERAL LINERS

The ETC 8 publishes technical recommendations to standardize the works in the field of geotechnics of landfills and contaminated land (see GLC 1991). A new recommendation, R 2-11, is based on the research work described above and deals with the assessment of mineral sealing layer deformation. It therefore gives a condensed

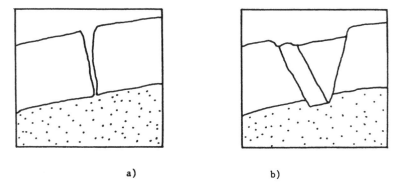

a)   b)

Figure 9. Settlement induced deformation in mineral sealing caused by a) cracking and b) shearing mechanism.

interpretation of the centrifuge model tests and the investigations on tensile properties, and will be introduced here. For more details concerning interpretation of the centrifuge model test and the determination of tensile properties see Scherbeck (1992).

General

Mineral sealing layers should not be subject to settlement to such an extent that their barrier function is compromised by deformation. Such deformations can cause cracking or shearing mechanisms in mineral sealing layers, as shown in Fig. 9. An unacceptable influence on the barrier function caused by settlement-induced deformation can be expected when the tensile strength of the mineral sealing material is exceeded and open cracks develop. If deformation action leads to shear mechanisms without open cracks, no significant increase in leakage rates through the liner will occur. In particular, this situation may occur when overburden pressure prevents cracking and under these circumstances sealing performance will not be affected.

Assessment Procedure

Input data for the following assessment approach are: the extent of deformation (e.g. uneven settlement of the landfill base or cap) and the stress–strain behaviour of the sealing material, including its

TABLE 3. Assessment Procedure

Content	Formulae
STEP 1	
Check deformation input and material properties	$\dfrac{\varepsilon_{zq}}{\varepsilon_{RF}} > \eta_v \quad (>2\cdot0)$
STEP 2	
a) Check vertical stresses (overburden pressure) σ_0	$\sigma_0 > 2 \times c \times \text{Tan}\,(45° + \phi'/2)$ (see Fig. 10)
b) Check the deformation mechanism (shearing is acceptable)	*req. c' > ex. c'* (see Fig. 11)

properties under tension. Firstly, the induced deformation is compared with the deformation capacity of the mineral sealing material. Secondly, the stability of an open crack and the influence of bending stresses is investigated. The concept of the validation procedure is shown in Table 3.

Deformability. The first step in assessing cohesive mineral sealing material is to compare the induced deformation and the deformation capacity of the mineral material in terms of tensile strains

$$\varepsilon_{zq}/\varepsilon_{RF} > \eta_v \qquad (1)$$

The maximum deformation capacity of the mineral sealing material can be defined by the maximum tensile strain at the onset of cracking. This value is equivalent to the ultimate strain ε_{zq} measured in direct tension tests—as described above—alternatively the ultimate strain can be determined from bending tests (see Ajaz & Parry, 1975).

The shape of the deformed mineral liner is given by the settlement curve $w(x)$, from which the maximum edge fibre strain, ε_{RF}, can be specified by the geometrical values of elongation $(\varepsilon_{\Delta l})$ and curvature (κ) based on simple beam analysis. Shear stresses acting at the interface of mineral liner and landfill base lead to displacement of the neutral axis in the cross section. This effect is considered by a factor $2/3$ in eqn (4). The following mathematical relations exist between the settlement curve $w(x)$ and elongation and curvature, respectively

$$\varepsilon_{\Delta l}(x) = w'(x)/(\text{Sin}\,(\text{Arctan}\,w'(x)) - 1) \tag{2}$$

$$\kappa(x) = w''(x) \tag{3}$$

$$\varepsilon_{RF} = 2/3 \times d \times \max \kappa + \varepsilon_{\Delta l} \tag{4}$$

where $\varepsilon_{\Delta l}$ is maximum elongation $(-)$, $w(x)$ is the shape of settlement curve (m), $w'w''$ is the first and second derivation from $w(x)$, κ is the curvature [1/m], ε_{RF} is the maximum strain in edge fibre $(-)$, d is the thickness of mineral sealing layer (m), $\max \kappa$ is the maximum curvature $(= \max w''(x) = 1/R)$ (1/m) and $R =$ radius of curvature (m). The safety factor of deformability, η_v is defined by the quotient of strains in eqn (1) and should exceed a recommended value of $2 \cdot 0$. If this is proved, the sealing material is considered capable of withstanding induced deformations without cracking and the barrier function of the mineral sealing is unaffected.

Overburden loads can be considered in this validation step by comparing the deformation induced edge fibre strain, ε_{RF}, with the overburden induced transverse strain. If the transverse strain exceeds the edge fibre strain, cracking of the mineral sealing layer is prevented.

Stress. If strain conditions satisfying eqn (1) are not fulfilled, stress conditions should be taken into consideration for the second validation stop to determine the significant deformation mechanism. Local shear mechanisms can be accepted; but on the other hand, cracking will lead to a loss of liner integrity.

(i) Required overburden load (excluding bending stresses). Overburden loads reduce the danger of cracking by suppressing the development of deformation-induced tensile stresses. The required overburden load to avoid cracking can be deduced from the situation shown in Fig. 10. In this figure the equation for calculation of the height of a stable crack depth, z, is analyzed in a diagram which shows that cohesive soils with friction angles of about 20° resist cracking when the overburden load is a minimum three times the cohesion value c'. The overburden load acting on the mineral sealing layer has to exceed the required load from Fig. 10 by a safety factor as used in slope stability analysis (e.g. by $1 \cdot 4$).

(ii) Influence of bending stresses. If the overburden stress described under (i) is not sufficient to suppress cracking, it should be determined if bending stresses will lead to shearing. In this case the existing

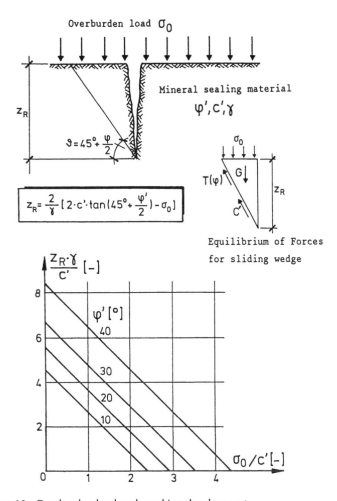

Figure 10. Overburden load and cracking development.

material cohesion, *ex. c'*, is compared with a computed cohesion, *req. c'*; required for equilibrium at the sliding wedge at the moment before cracking, (see Fig. 11). If the required cohesion is less than the existing cohesion, there is no stable equilibrium in the system and further deformation would lead to cracking. On the other hand, if the relation *ex. c' < req. c'* is fulfilled, the shear resistance is actually too

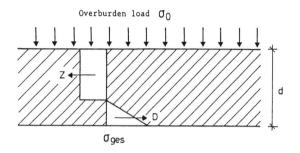

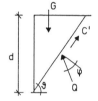

Equilibrium of Forces
(tensile zone not cracked)

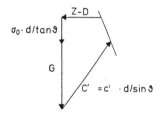

| req. c' < ex. c' : Cracking |
| req. c' > ex. c' : Shearing |

a) Overburden pressure $\sigma_0 = 0$:

$$req.\ c' = 1/2 \cdot \gamma \cdot d \cdot (F_1 - 2/3 \cdot F_2)$$
$$+ q_z \cdot (2/3 - 1/12 \cdot R_g) \cdot F_2$$

b) Overburden pressure $\sigma_0 > 0$:

$$req.\ c' = (1/2 \cdot \gamma \cdot d + \sigma_0) \cdot F_1$$
$$- \sigma_0 \cdot (1/3 + 1/12 \cdot R_g) \cdot F_2$$
$$+ q_z \cdot (2/3 - 1/12 \cdot R_g) \cdot F_2$$

where: γ is the specific weight of mineral sealing material (kN/m³), d is the mineral layer thickness (m), σ_0 is the overburden pressure (kPa), q_z is the uniaxial tensile strength of mineral sealing (kPa), R_g is the stiffness ratio: tangential stiffness from uniaxial compression tests/tangential stiffness from direct tension tests (−), $F_{1,2}$ is the parameter (−), see right.

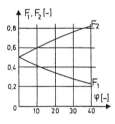

Figure 11. Bending tensile stresses in the deformed sealing layer, tensile zone not cracked (Scherbeck, 1992.)

low to maintain equilibrium. At this stage shear mechanism have already occurred and hence the validation in the second step is fulfilled.

SUMMARY

From centrifuge model tests it could be derived that deformed compacted mineral liners only stop to perform as effective hydraulic barriers when cracks occur. Depending on the plastic behaviour of the liner material, tension cracks could appear, leading to an uncontrolled permeation through the liner; presence of overburden loads will suppress this tendency. Instead, shearing mechanisms may occur and will lead to inner deformations, but this response is unlikely to evoke a loss of integrity of the liner. Additional laboratory element tests quantified the influence of consistency, plasticity and compaction on the tensile strain behaviour of some cohesive soils, showing that they can typically attain ultimate strains between 1·5 and 4·4%. For design purposes the ETC 8 recently published a draft version about assessment of mineral sealing layer based upon the results published in this paper. This recommendation is refered to in this paper when interpreting test results.

REFERENCES

Ajaz, A. & Parry, R. H. G. (1975). Stress–strain behaviour of two compacted clays in tension and compression. *Geotechnique*, 5(3), 495–512.

Andersland, O. B. & Al-Moussawi, H. M. (1987). Crack formation in soil landfill covers due to thermal contraction. In *Waste Management and Research*; No. 5; 445–52.

Gilbert, P. A. & Murphy, W. L. (1987). Settlement and Cover Subsidence of Hazardous Waste Landfills. EPA Report No. EPA/600/2-85/035. Environmental Protection Agency, Cincinnati, Ohio, USA.

GLC (Geotechnics of Landfills and Contaminated Land), (1991). Technical Recommendations. Ed. German Geotechnical Soc. International Soc. of Soil Mechanics and Foundation Engng. Ernst & Sohn Verlag, Berlin, 76 pp.

Gordon, B. B. & Forrest, M. (1979). Permeability of soils using contaminated permeant. *Permeability and Groundwater Contaminant Transport*, ASTM STP 746. pp. 101–20.

Henne, J. (1989). Versuchsgerät zur Ermittlung der Biegezugfestigkeiten von bindigen Böden. *Geotechnik,* 12(2), 96–9.

Jessberger, H. L. & Güttler, U. (1988). Bochum geotechnical centrifuge. In *Proc. Inter. Conf. on Geotechnical Modelling—Centrifuge '88,* ed. J. F. Corté. Balkema Publishers, Rotterdam, pp. 37–44.

Jessberger, H. L., Güttler, U. & Stone, K. J. L. (1989). Centrifuge modelling of subsidence effects on clay barriers. In *Proc. 2nd Int. Landfill Symposium, Sardinia '89,* Vol. I. Porto Conte, Sardinia: XIII–XIIII.

Kleppe, J. H. & Olson, R. E. (1985). Dessication cracking of soil barriers. *Hydraulic Barriers in Soil and Rock,* ASTM STP 874: pp. 263–75.

Larson, N. B. & Keshian, B. (1988). Prediction of strains in earthen covers. In *Hydraulic Fill Structures.* Geotechnical Special Publication No. 21, ASCE, Fort Collins. Ed. D. J. A. Van Zyl & S. G. Vick. ASCE Geotechnical Special Publication No. 21, New York, pp. 367–88.

Scherbeck, R. (1992). Geotechnisches Verhalten mineralischer Deponieabdichtungsschichten bei ungleichförmiger Verformungseinwirkung. Schriftenreihe des Institutes für Grundbau, Ruhr-Universität Bochum, Heft 16, 191 pp.

Stone, K. J. L. & Wood, D. M. (1987). Model Studies of Soil Deformation over a Moving Basement: Rep. Engineering Departement, Cambridge University, U.K.

Stone, K. J. L., Güttler, U. & Jessberger, H. L. (1989). Some observations of the influence of deformation on a clay liner. In *Proc. 10th National Conference Superfund '89.* Washington, DC, 537–42.

Tan, T. S. & Scott, R. F. (1985). Centrifuge scaling considerations for fluid-particles systems. *Geotechnique,* 37(4), 131–3.

4. PROPERTIES AND QUALITY CONTROL OF SYNTHETIC MATERIALS

4.1 Basic Composition and Properties of Synthetic Materials in Lining Systems

HENRY E. HAXO & PAUL D. HAXO

Matrecon Inc., 815 Atlantic Avenue, Alameda, California 94501, USA

INTRODUCTION

In order to function as needed in the environment existing in landfills over extended time periods, a synthetic lining system must meet performance requirements for a wide range of properties, including permeability, chemical compatibility, mechanical compatibility, and durability.

Low Permeability

The installed liners must have sufficiently low permeability to *all* constituents of the waste so that the level of constituent transmission through the lining system does not pose a threat to human health or the environment. An excessive transmission level for a particular constituent is specific to the site, the constituent's toxicity, and the mobility and biodegradability of the constituent.

Chemical Compatibility

All liner system components that may contact the waste leachate must be chemically compatible over an extended time period with the leachate to be contained. The geosynthetics should maintain their permeability and mechanical properties, as required, after exposure to

317

the waste so that they can continue to function as part of the lining system. Chemical incompatibility of a geosynthetic with a waste leachate is manifest in the loss in values of important properties resulting principally from: (1) absorption of large amounts of waste constituents, or (2) extraction of components of the original polymeric compound. In many countries, the chemical compatibility of the components of a lining system for a hazardous waste landfill must be demonstrated during the permitting stage.

Mechanical Compatibility of the Components

The components of a lining system must maintain their integrity on exposure to mechanical stresses. Short-term mechanical stresses can include stresses encountered during installation such as those caused by placement of a granular drainage layer and the traffic of heavy equipment, stresses caused by thermal shrinkage, and stresses related to the weight of the materials placed on top of the system. Long-term mechanical stresses are more often the result of the materials on top of the liner system or differential settlement of the subgrade. In addition, there must be adequate friction between the membranes on the slope and the soil and components of the leachate drainage and collection systems to ensure that no slippage or sloughing occurs. In response to this need various textured synthetic membranes that increase the friction between a membrane and soil have been developed and are in use.

Capability of Being Installed

A synthetic membrane used to line a land disposal unit must be capable of being installed in such a way that it can form a continuous durable membrane. The ability to form a continuous durable membrane is dependent on the ability of the material to be seamed. The difficulty encountered in preparing good durable seams of crosslinked membranes has led to discontinuance of their use in many applications. Membranes must also be sufficiently durable to withstand the rigors of construction and installation.

Durability

The components of a lining system need to be able to maintain their integrity and performance characteristics over the operational life of the unit and the post-closure care period. Ultimately, the service life of the components of the system will depend on the intrinsic durability of the material and on the conditions under which they are exposed. In particular, these components must be able to resist the combined effects of chemical, mechanical, and biological stresses.

SYNTHETIC MATERIALS USED IN THE CONSTRUCTION OF LINING SYSTEMS

A wide range of synthetic materials is often used in constructing a lining system for a landfill. These materials, which fulfill a variety of functions, include membranes, geotextiles, geogrids, geonets, geocomposites and pipes, and are based on a wide range of polymers, i.e. elastomers, plastics, fibers, and resins. Table 1 lists some of the major

TABLE 1. Principal Polymers Currently Used in the Manufacture of Major Products for the Construction of Lining Systems (Matrecon Inc., 1988)

Polymer	*Type*	*Product*			
		Mem-branes	*Geo-textiles*	*Geogrids and geonets*	*Plastic pipe*
Chlorinated polyethylene (CPE)	Rubber	X	—	—	—
Chlorosulfonated polyethylene (CSPE)	Rubber	X	—	—	—
Polyester terephthalate (PET)	Fiber/ resin	X[a]	X	X	—
Polyethylene (PE):					
Linear low-density (LLDPE)	Resin	X	—	—	—
High-density (HDPE)	Resin	X	X	X	X
Polypropylene (PP)	Resin	X	X	X	—
Polyvinyl chloride (PVC):					
Plasticized	Resin	X	—	—	—
Unplasticized	Resin	—	—	—	X

[a] Used as a reinforcing fabric in membranes.

polymers that are being used in the manufacture of geosynthetics and pipe used in landfills.

Synthetic Membranes

A synthetic membrane is a material of low permeability and serves as the barrier in the liner system between mobile polluting substances and the groundwater; synthetic membranes are also used in final cover systems as barrier layers to minimize the amount of rainwater entering closed units.

The first polymeric membranes were based on butyl rubber; since then, a wide variety of synthetic membranes based on different polymers have become available. Sheetings are produced by calendering, extrusion, or spread-coating processes, and are produced in widths ranging from 1·5 to 10 m. The narrower sheetings are seamed in the factory to make large panels which are transported to a construction site where they are seamed to form the liner. Some of the wider membranes made by extrusion (e.g. polyethylene) are brought to the field in rolls for seaming and installation.

The physical and chemical properties of polymeric membranes vary considerably, as do methods of installation and seaming, costs, and interaction with different wastes. The composition and properties of membranes even of a given generic polymer type can differ considerably, depending on the compound formulation. Polymers are rarely used alone in a membrane; whether used singly or in blends, they are usually compounded with a variety of ingredients (e.g. fillers, plasticizers or oils, antidegradants, and curatives) to improve either selected properties or the balance of properties depending on end-use. Table 2 presents generalized formulations for synthetic polymeric membranes. Properties of a polymeric membrane also depend on its construction, e.g. thickness, whether or not it is fabric reinforced, the type of fabric reinforcement used, and the number of plies. Because the grade and source of polymers of a given generic type vary, differences between membranes also arise from the polymer.

At present, most polymeric membranes are not crosslinked and, therefore, are thermoplastic. This is true even for membranes that use crosslinkable polymers such as chlorinated polyethylene (CPE) and chlorosulfonated polyethylene (CSPE), which are more chemically

TABLE 2. Basic Compositions of Synthetic Membranes

Component	Composition of compound type, parts by weight		
	Cross-linked	Thermo-plastic	Semicrys-talline
Polymer or alloy	100	100	100
Oil or plasticizer	5–40	5–55	0–10
Fillers:			
Carbon black	5–40	5–40	2–5
Inorganic	5–40	5–40	—
Antidegradants	1–2	1–2	1
Crosslinking system:			
Inorganic system	5–9	*a*	—
Sulfur system	5–9	—	—

a An inorganic curing system that crosslinks over time is incorporated in chlorosulfonated polyethylene (CSPE) membrane compounds.

resistant when crosslinked. Thermoplastic membranes, which can be seamed by various solvent and thermal methods, are now preferred because they are easier to seam and repair effectively during installation.

In addition to polymeric synthetic membranes, a wide variety of synthetic membranes based on asphalt, coal-tar, and other bitumens have been successfully used to contain water in reservoirs and canals, and to some extent to contain wastewaters and line municipal solid waste (MSW) landfills. These membranes are of two general types:

- sprayed-on liners prepared in the field from either hot-blown or emulsified asphalt;
- prefabricated laminates supplied in rolls or panels.

Sprayed-on liners are formed in the field by spraying the asphaltic material onto a prepared soil surface on which a geotextile may or may not have been placed. The sprayed-on liquid solidifies in place to form a continuous seam-free membrane. Such liners have been used in canals, small reservoirs and ponds for water control and for storing brine solutions. Water storage applications have used air-blown asphalt; however, membranes from asphalt blends containing additives of elastomeric polymers (e.g. urethanes, styrene-butadiene thermoplastics, neoprene) and fillers are now being used in solar ponds for containment of brines, and are being promoted by manufacturers as suitable materials for waste storage applications in the mining industry.

Though sprayed-on membranes are seam-free, bubbles and pinholes, which are extremely difficult to detect, may form during field installation causing serious difficulties at a later date. The proper preparation of the surface to be sprayed is important. The asphaltic materials are thermoplastic and of low molecular weight, and will react adversely with many wastes. However, in carefully controlled conditions, and when protected from mechanical damage and ultraviolet degradation, they can be used to form a serviceable liner for brines and many inorganic solutions. Coal tar, which has greater solvent resistance than asphalt, is sometimes used as a sprayed-on membrane in conjunction with additives such as urethanes and epoxies.

The prefabricated asphaltic membrane described by Hoekstra (1984) consists of a laminate of bitumen, polyester fabric reinforcement, polyester film, and sand. This membrane is fabricated in wide rolls which can be placed in the field and seamed with heat guns. A similar type of asphalt membrane was described by Le Coroller et al. (1984). In the United States, a membrane was marketed based on an emulsified asphalt sprayed onto a fabric. A variety of these sprayed-on membranes is described in Matrecon Inc. (1988).

Geotextiles

Geotextiles are used in liner systems to provide separation between solid wastes and the leachate collection system or between the membrane and cover or embankment soils, to reinforce the membrane against puncture from the subgrade or the waste that is placed above it, and to provide filtration around collection pipes. A number of polyester and PP fiber types (monofilament, multifilament, slit film) are used to make a variety of woven and nonwoven fabrics. These fabrics are manufactured using a large variety of weaving patterns (plain, modified, etc.) and nonwoven manufacturing techniques (heat set, needle punched, resin bonded, etc.).

Geonets

Geonets are grid-like products based on polyethylene (PE) or polypropylene (PP) and are used exclusively as in-plane drainage systems; they are always used with geotextiles, membranes, or other materials in the planes above and below them. For example, a geonet can be placed between two membranes to function as a leak-detection system, or between a geotextile filter and a membrane, as in a leachate-collection system.

Geogrids

Geogrids are used in the construction of waste disposal units to reinforce soils in the dikes; they are also used within landfills to steepen earth slopes or to create embankments between cells. Geo-grids should not be confused with geonets which are used exclusively for drainage. The geogrids that are available differ in the directionality of their strength, the size and shape of their apertures, and in their node construction.

Geocomposites

'Geocomposites' is a term loosely used to identify a wide range of composite materials that consist of two or more geosynthetics. Drainage geocomposites are sometimes used as leachate-collection subsystems with a geotextile filter attached, but they appear to be particularly useful as surface water collectors in a landfill closure system where normal stresses are relatively low. A great variety of drainage geocomposites is available. The drainage cores themselves take the shape of columns, piers, cuspations, dimples, etc. Drainage geocomposites vary widely from product to product in both mechanical and hydraulic properties.

Pipes and Fittings

Plastic pipes based primarily on PVC or high-density polyethylene (HDPE) are used in constructing leachate-collection and leak-detection systems and in gas venting applications. The pipes used in these applications are either perforated or slotted; they are also used to convey wastes into and out of the system and in monitoring systems. Thermoplastic pipes are preferred over nonplastic pipes for leachate collection and drainage above a liner because of the chemical resistance of the thermoplastics, particularly to inorganic chemicals.

IMPORTANT PROPERTIES OF POLYMERIC GEOSYNTHETIC MATERIALS

All of the geosynthetic materials, as well as plastic pipe, discussed in this chapter are based on synthetic polymers (Table 1), which are products of the chemical, plastics, rubber, and fiber industries. From the viewpoint of composition, an almost infinite range of polymeric materials can be produced. Polymers vary in properties from polymer to polymer; furthermore, polymers within a type can vary according to grade and manufacturing process. In addition, considerable variation among compositions based on the same polymer is introduced by the product manufacturer through compounding with ingredients designed to enhance or develop specific characteristics (Table 2). Knowledge of the composition of each geosynthetic and pipe can be important when dealing with landfill leachates containing organics.

Four general types of polymers are used in the manufacture of geosynthetics:

- thermoplastics and resins, such as chlorosulfonated polyethylene (CSPE) and polyvinyl chloride (PVC);

- elastomers that are crosslinked during manufacture, such as ethylene propylene rubber (EPDM) and butyl rubber;
- semicrystalline thermoplastics polymers, such as the polyethylenes (PE);
- highly crystalline, oriented polymers, such as polypropylene (PP) and polyester fibers used to manufacture geotextiles.

The following paragraphs briefly discuss some of the properties that are important for geosynthetic materials used in the construction of lining systems, particularly double-liner systems.

Permeability of Synthetic Membranes to Leachate Constituents

The principal component of a synthetic liner system is the membrane which is the nonporous barrier to the escape of mobile waste constituents from a waste disposal unit. Synthetic membranes are being used as barriers to many different permeating species in many different environments. This wide range of uses requires different test procedures to assess the effectiveness of a given membrane for a given application. The permeating species range from a single gas to highly complex mixtures of constituents, such as those found in waste liquids and leachates. For actual service it is important to know the actual transmission or migration of a species that would take place under the specific conditions and environments. Specific tests for measuring the permeability of membranes to different permeants for various applications have been suggested (Haxo, 1990).

In discussing the permeability of synthetic membranes, it is important to recognize the difference between barriers that are nonporous homogeneous materials and those that are porous, such as soils and concretes. The transmission of permeating species through hole-free synthetic membranes occurs when the species is absorbed by the membrane and diffuses through the membrane on a molecular basis. The driving force for transport of a specific species is its chemical potential across the membrane. A liquid permeates porous materials in a condensed state that can carry dissolved constituents, and, in such a case, the driving force for permeation is hydraulic pressure. Due to the 'perm-selective' nature of synthetic membranes, the permeation of the dissolved constituents in liquids can vary greatly, i.e. some components of a mixture can permeate more readily than others. Synthetic membranes are semipermeable to inorganic aqueous salt solutions because the water (but not the ions) can be transmitted; the water that is transmitted through a hole-free membrane does not carry inorganics. The direction of permeation is determined by the chemical potential across the membrane. For example, the water or leachate on the contained side is at a lower potential than the purer water on the opposite side. Thus, the purer water can permeate into the leachate by osmosis, unless the head of leachate is sufficient to overcome osmotic pressure.

Although inorganic salts do not permeate the membranes, organic species can permeate at a rate which depends on the solubility of the organic in the membrane and its diffusibility through the membrane as driven by the chemical potential. The diffusion within the membrane is controlled by such factors as the microstructure of the polymer, the flexibility of the polymer chains, and the size of the permeating molecules. The rate is also determined by the gradient within the membrane.

The extent to which mobile constituents will move through a hole-free synthetic membrane in service is related to factors such as:

- the composition of the membrane with respect to the permeating species with particular reference to the solubility of the permeating species in the membrane;
- the thickness of the membrane;
- the service temperature;
- the temperature gradient across the membrane in service;
- the chemical potential across the membrane, which includes pressure and concentration gradient;
- the composition of the contained fluid and the mobile constituents;
- the solubility of constituents of the particular membrane in the waste liquid being contained;
- the diffusibility of the permeant in the membrane;
- the ion concentration of the contained liquid;
- the ability of the permeating species to move away from the surface of the membrane on the downstream side.

Because of the great number of variables, it is important to test the permeability of a membrane under conditions that simulate as closely as possible the actual environmental conditions in which the membrane will be in service.

Examples of the permeability of different membranes to different

TABLE 3. Permeability of Synthetic Membranes to Selected Gases at 23°C, Determined in Accordance with ASTM D1434, Procedure V

Synthetic membrane description				Gas transmission rate $(ml(STP^{[b]})\,m^{-2}\,d^{-1}\,atm^{-1})$			Gas permeability coefficient (P), $(barrier^{[c]})$		
Base polymer	Density $(g\,cm^{-3})$	Thickness[a] (mm)	Compound type[d]	CO_2	CH_4	N_2	CO_2	CH_4	N_2
Butyl rubber	—	1·60	XL	512	120	19·7	12·5	2·92	0·480
Chlorosulfonated polyethylene	—	0·82R	TP	122	21·6	26·2	1·52	0·27	0·33
Elasticized polyolefin	—	0·86	TP	418	124	27·1	5·47	1·62	0·36
	—	0·58	CX	1450	280	125	12·8	2·47	1·10
Ethylene propylene rubber	—	0·89R	TP	2720[e]	—	—	36·8[e]	—	—
	—	0·90	XL	5260	1400	314	72·0	19·2	4·30
Neoprene	—	0·90	XL	716	80·9	31·1	9·81	1·11	0·43
Polybutylene	—	0·71	CX	818	248	62·3	8·84	2·68	·0·67
Polyethylene[f] (low density)	0·921	0·25	CX	6180[e]	1340[e]	—	23·5[e]	5·10[e]	—
Polyethylene (linear low density)	0·923	0·46	CX	1370	322	—	9·59	2·25	—
Polyethylene (high density)	0·945	0·61	CX	729	138	—	6·77	1·28	—
	0·945	0·86	CX	467	104	—	6·11	1·36	—
Polyvinyl chloride (plasticized)	—	0·25	TP	7730[e]	1150[e]	—	29·4[e]	4·38[e]	0·81
	—	0·49	TP	3010	446	108	22·4	3·32	—
	—	0·81	TP	2840[e]	285[e]	—	35·0[e]	3·51	—
Polyethylene terephthalate[g]	—	0·022	CX	357	—	—	0·119	—	—

[a] R = fabric-reinforced.
[b] STP = standard temperature and pressure.
[c] One barrier = 10^{-10} ml(STP) cm cm^{-2} s^{-1} cmHg^{-1}.
[d] XL = crosslinked; TP = thermoplastic; CX = semicrystalline.
[e] Measured at 30°C.
[f] Natural resin (no carbon black).
[g] NBS Standard Material 1470. The determination was made at 15·0 psi, under which condition the NBS Certified CO_2 transmission rate was calculated to be 338 ml(STP) m^{-2} day^{-1} atm.

TABLE 4. Permeability of Selected Synthetic Membranes to Various Solvents, Measured in Accordance with ASTM E96, Procedure B (Modified to Test Solvents)

Polymer[a]	ELPO	HDPE	HDPE	HDPE-A	LDPE	PB	Teflon
Average thickness (mm)	0·57	0·80	2·62	0·87	0·75	0·69	0·10
Type of compound				Semicrystalline			
SVT (g m^{-2} day^{-1})							
Methyl alcohol	2·10	0·16	—	0·50	0·74	0·35	0·34
Acetone	8·62	0·56	—	2·19	2·83	1·23	1·27
Cyclohexane	7·60	11·7	—	151	161	616	0·026
Xylene	359	21·6	6·86	212	116	178	0·16
Chloroform	3230	54·8	15·8	506	570	2 120	0·16
Solvent vapor permeability[b] (10^{-2} metric perms cm^{-1})							
Methyl alcohol	0·11	0·01	—	0·04	0·05	0·02	0·003
Acetone	0·23	0·02	—	0·09	0·10	0·04	0·006
Cyclohexane	0·49	1·05	—	14·7	13·6	47·8	0·0003
Xylene[c]	292	24·6	25·6	262	124	175	0·002
Chloroform	103	2·46	2·32	24·6	24·0	82·2	0·12

[a] ELPO = elasticized polyolefin; HDPE = high-density polyethylene; HDPE-A = high-density polyethylene alloy; LDPE = low-density polyethylene; PB = polybutylene.
[b] The median thickness value of the specimens was used to calculate the permeability.

chemical species that are present in MSW or in leachates of the waste are presented in Tables 3 and 4. Table 3 presents the permeability of synthetic membranes to gases that are encountered in sanitary landfills. The data are reported in gas transmission rates and in gas permeability coefficients for carbon dioxide, methane, and nitrogen for a variety of synthetic membranes. Table 4 presents the permeability of synthetic membranes to various solvents that have been encountered at low concentrations in sanitary landfill leachates. These data were obtained on individual solvents following a test procedure often used for measuring the moisture transmission rates.

Some general observations that can be made on the permeability of membranes based on the data in Tables 3 and 4 and other available data (Haxo, 1988, 1990, Matrecon Inc., 1988) are:

• The transmission of an individual species varies from polymer to polymer.

- The transmission of different species through a single nonporous membrane can vary over several orders of magnitude.
- The presence of other permeating species can affect the transmission of a species through a membrane.
- Because of the selective permeation due to high partitioning of the organic to the membrane, permeation of an organic in a dilute aqueous solution through a membrane can be substantially higher than what would be expected from its low concentration.

Chemical Compatibility of Synthetic Membranes with Waste Liquids

In the early 1970s, the United States Environmental Protection Agency (USEPA) became concerned with effects on the environment of the landfilling of wastes, particularly with respect to possible pollution of the groundwater. Because of these concerns, the agency became interested in the use of various man-made materials as possible linings for waste storage and disposal facilities. However, the USEPA was concerned about the ability of these materials to function adequately on extended service as liners for waste storage and disposal facilities, because there was essentially no experience of membranes as liners in these applications. As a consequence, the USEPA undertook a series of laboratory and pilot experimental research programs to assess the compatibility and durability of various potential lining materials, including synthetic membranes, on exposure to a range of waste liquids for durations of several years and to assess the various factors that affect the compatibility of membranes and waste liquids. These wastes included municipal solid waste (MSW) leachate and a range of hazardous wastes. The projects also studied the effects of exposing membrane seams to various chemicals and attempted to develop acceptability criteria. A program was initiated to determine the relationship between laboratory studies and the field performance of membranes. Data from many of these projects have been summarized in USEPA technical resource documents (Matrecon Inc., 1983, 1988).

In the initial project, 12 different potential lining materials, which had been used in water containment, were exposed in landfill simulators to MSW leachate generated within the simulators. These liner materials included six synthetic membranes, five asphaltic compositions, and a soil–cement composition (Haxo *et al.*, 1985;

TABLE 5. Properties of Synthetic Membrane Liners After 12 and 56 Months of Exposure to Leachate in MSW Landfill Simulators Compared with Original Properties (Matrecon Inc., 1983, 1988)

Parameter	Test method[b]	Exposure time (mths)	Base polymer[a]					
			Butyl rubber	CPE	Potable CSPE-R	EPDM	LDPE	PVC
Type of compound[c]	—	—	XL	TP	TP	XL	CX	TP
Analytical properties								
Volatiles, 2 h at 105°C (%)	—	0	—	0·10	0·29	0·50	0·00	0·09
		12	2·02	6·84	12·78	5·54	0·02	3·55
		56	2·37	7·61	13·90	5·74	1·95	2·08
Extractables after removal of volatiles (%)	ASTM D3421, modified	0	11·0	7·5	3·8	31·8	—	37·3
		56	9·8	5·1	3·4	28·3	3·37	34·4
Solvent[d]	—	—	MEK	n-Heptane	Acetone	MEK	MEK	CCl$_4$ + methanol
Physical properties								
Thickness (mm)	—	0	1·60	0·81	0·91	1·30	0·30	0·53
		12	1·63	0·89	0·96	1·30	0·28	0·53
		56	1·63	0·94	0·94	1·24	0·25	0·56
Tensile strength (MPa)	ASTM D412	0	10·0	15·9	12·3	10·3	15·0	18·0
		12	9·75	12·6	11·5	10·2	17·2	16·4
		56	10·2	13·7	14·8	10·2	18·1	19·2

Property	Test method							
Elongation at break (%)	ASTM D412	0	400	410	250	415	505	280
		12	410	400	300	435	505	330
		56	405	385	235	375	540	340
Stress at 200% elongation (MPa)	ASTM D412	0	4·86	9·31	10·7	5·28	8·82	13·8
		12	4·80	7·63	8·72	5·18	8·44	10·8
		56	5·25	7·98	12·8	5·60	9·28	12·7
Tear strength, Die C (kN/m)	ASTM D624	0	30·6	44·6	_a_	31·5	68·3	58·6
		12	35·0	56·0	_a_	34·1	86·7	78·8
		56	32·4	29·8	_a_	22·8	70·9	49·9
Hardness (Duro A points)	ASTM D2240	0	51	77	79	54	—	76
		12	50·5	65·5	64	51·1	—	64
		56	51	70	70	51	—	70
Puncture resistance Force (kg)	FTMS 101B, Method 2065	0	19·5	20·5	14·3	17·2	6·05	11·2
		12	21·6	21·7	24·8	17·5	6·45	13·1
		56	21·8	22·6	25·4	18·1	7·45	13·6
Deformation (mm)		0	31	26	15	37	19	18
		12	30	25	22	30	20	18
		56	32	25	22	30	32	21

[a] CPE = chlorinated polyethylene; CSPE-R = chlorosulfonated polyethylene (fabric-reinforced); EPDM = ethylene propylene rubber; LDPE = low-density polyethylene; PVC = polyvinyl chloride.
[b] ASTM = American Society for Testing and Materials; FTMS = Federal Test Method Standard (USGSA, 1980).

Matrecon Inc., 1988). Samples of the liner materials were sealed as liners into the bases of the simulators; thus, one side of the samples was exposed continuously to the leachate, which was ponded on the liners at a constant depth of 30 cm. Provisions were also made to collect any leachate that might have seeped through the liners. The effects of exposing the six synthetic membranes to leachate for 12 and 56 months are reported in Table 5.

The results show significant absorption by some of the membranes during the course of the exposure; the magnitude of absorption ranged between a low for the low-density polyethylene (LDPE) to a high for the potable-grade CSPE. After 56 months of exposure, the extractable contents of the membranes had decreased, indicating extraction by the leachate that had flowed over the membrane. Depending on the property, the physical properties appeared to reach a minimum or maximum value at 12 months; this effect may have reflected the maximum concentration of organic acids in the leachate at 12 months, as is shown in Table 6. After the end of the first year of exposure, the concentration of all leachate constituents decreased due to the continuous leaching of the constant amount ($\sim 0 \cdot 8 \, m^3$) of shredded municipal solid waste placed initially in the simulators.

The results of testing strip samples that were buried in the sand above the liners were similar to those of testing the liners which had

TABLE 6. Analysis of MSW Leachate[a] Produced in Simulators and Used for Membrane Exposure Tests (Matrecon Inc., 1983, 1988)

Test	Value
Total solids (%)	3·31
Volatile solids (%)	1·95
Nonvolatile solids (%)	1·36
Chemical oxygen demand (COD) (g liter^{-1})	45·9
pH	5·05
Total volatile acids (TVA) (g liter^{-1})	24·33
Organic acids (g liter^{-1})	
Acetic	11·25
Propionic	2·87
Isobutyric	0·81
Butyric	6·93

[a] At the end of the first year of operation when the first set of liner specimens was recovered.

been exposed on one side, indicating a correlation between immersion and one-sided exposure. The results of testing the seam incorporated in the strip varied greatly depending on the method of seaming.

To achieve a broader representation of synthetic membranes, 28 additional samples were immersed in a series of tanks through which a constant flow of leachate from the simulators was pumped. This group included several samples of CPE, potable-grade CSPE, ethylene propylene rubber (EPDM), neoprene, and PVC so that additional polymers and different sources and compositions of a given polymeric type were represented. The results of testing these samples after 8 and 31 months of immersion are summarized in Table 7; for the types of membranes that were represented by more than one sample, a range of values is reported. The data vary from one polymeric type of membrane to another, and also show that, even for a given polymeric

TABLE 7. Summary of the Effects on Membranes of Immersion in MSW Leachate for 8 (8M) and 31 (31M) Months[a]

Membrane		Weight increase (%)		Original property			
				Tensile		Elongation (%)	
Base polymer[b]	Number in test	8M	31M	8M	31M[c]	8M	31M
Butyl rubber	1	1–8	25	90–97	92	104–106	90–92
CPE	3	8–10	25–28	80–115	78–106	64–135	71–103
CSPE-P	3	13–19	19–32	82–124	103–138	97–107	69–86
ELPO	1	0·1	8	86–94	98–106	91–92	96–98
EPDM	5	1–21	8–24	64–107	94–113	76–138	88–138
Neoprene	4	1–19	5–88	69–100	68–105	82–103	78–146
PB	1	0·1	—	96–99	84–97	96–97	86–89
PEL	1	2·0	16	99–115	81–90	101–108	80–96
LDPE	1	0·6	3	110–180	118–161	96–181	100–168
PVC	7	1–3	4–24	91–110	87–117	98–139	79–120
PVC + pitch	1	6	14	92	101–104	109–133	80–103

[a] Ranges of retention values for tensile strength and elongation are lowest and highest averaged values obtained for either machine or transverse directions of all tensile specimens within the group of slab specimens of a given polymer type.

[b] CPE = chlorinated polyethylene; CSPE-P = chlorosulfonated polyethylene (potable-grade); ELPO = elasticized polyolefin; EPDM = ethylene propylene rubber; PB = polybutylene; PEL = polyester elastomer; LDPE = low-density polyethylene; PVC = polyvinyl chloride.

[c] Some samples were inadequately cleaned, so some values are high.

type of membrane, there can be a substantial range of effects on properties.

Information on HDPE has not been included in the data presented as this material was not available at the time the research project was initiated. However, studies with hazardous wastes and limited studies with MSW show that the response of HDPE in exposure is similar to or better than that of LDPE (Haxo, 1989, 1991).

In addition to the polymeric membranes that were exposed to MSW leachate in leachate simulators, two asphaltic membrane liners were exposed. These were a 7·6-mm-thick catalytically-blown asphalt used in canal linings and an emulsified asphalt sprayed on a nonwoven fabric. Each was exposed for 12 and 56 months. Neither of these membranes allowed any seepage in the 56 months of exposure, during which a constant depth of 30 cm of leachate was allowed to pool.

When the catalytically-blown asphalt membrane was recovered after 12 months of exposure, it had a moisture content of 2·9%, and a slightly higher viscosity at 25°C. Otherwise, it appeared to be unchanged from the original material. However, after 56 months of exposure, the membrane had a nonhomogeneous appearance in which some areas were 'cheesy' and cracked easily on deformation, whereas other areas maintained their original appearance and properties. Those areas that were 'cheesy' had a moisture content of 4·55%, whereas the normal-appearing areas had a moisture content of 1·5%.

As with the catalytically-blown asphalt membrane, the emulsified asphalt membrane showed no deterioration after 12 months of exposure to the MSW leachate, although analysis showed a moisture content of 4·8%. After 56 months of exposure the emulsified asphalt membrane continued to show no visible deterioration, although analysis showed a moisture content of 8%.

It appears that over the long exposure there can be a considerable absorption of water by asphaltic membrane liners.

In this test, hydraulic and paving grades of asphalt concrete and a soil cement were tested under the same conditions as the membranes. None of these liners showed any seepage; however, all showed severe losses in compressive strength (Haxo et al., 1982).

Overall, the results of testing indicate that the chemical resistance of polymeric products in waste environments can vary considerably and that it is largely dependent on the tendency of these products to swell in service. The major characteristics of a polymeric product that affect the magnitude of its swelling in a liquid include:

- the solubility parameters of the polymer with respect to those of constituents of the liquid;
- the degree of crosslinking and the crystallinity of the polymer;
- the filler content, plasticizer content, and the soluble constituents of the polymeric compound.

Due to differences in polymers and in compounding, the importance and applicability of these properties vary from one polymeric product to another. As part of the permit process the USEPA requires that all materials to be used in the construction of lining systems for hazardous waste disposal units be tested for compatibility using samples of the waste liquids to be contained.

Miscellaneous Physical Characteristics

Viscoelasticity. All polymeric materials are viscoelastic; that is, when undergoing a deformation they show in varying degrees both viscous and elastic behaviour. The elastic component behaves like a spring and is independent of rate of deformation. The viscous component behaves like a dashpot and is highly dependent on the rate of deformation and on temperature.

Most of the polymers used in geosynthetics and pipe vary greatly in properties with temperature, even within the temperature range in which waste disposal facilities operate. At low temperatures some polymers become glassy and brittle, and at high temperatures they become soft and plastic. These characteristics will greatly affect the applications in which a polymeric material can be used.

Thermal expansion. Polymeric materials have linear thermal expansion coefficients approximately 5 to 10 times greater than those of metals and concrete. Such coefficients can be important in the performance of materials that are exposed to temperature changes. For the lower modulus membranes, changes in dimension with temperature are not a major problem; however, for stiffer membranes such as the PEs, changes in temperature can cause considerable deformation, buckling, and flexing of a liner when exposed to normal weather, and high stress when exposed to cold weather if the liner was placed without sufficient slack in the sun during warm weather.

Memory or frozen strain. Characteristic of many polymeric materials is their tendency to have 'memory'; for example, deformation of the polymer during processing and forming into sheets leaves 'frozen' residual strains in the sheeting. This results in a 'grain' effect which generally leads to different property values in different directions. For some polymers, when the sheeting is warmed residual strain can cause shrinkage in the extrusion direction and expansion in the transverse direction.

Multiaxial stress. Most of the tensile and tear testing for specification purposes is performed uniaxially, which can yield high elongations. Performing the tests biaxially, i.e. deforming the materials simultaneously in machine and transverse directions, yields considerably lower elongation values. This can show up in the testing of puncture and hydrostatic resistances and in actual service.

Creep. Other properties of polymers can affect the durability of polymeric products. Compared with more traditional materials of construction, such as steel, concrete and wood, polymeric materials have a relatively high tendency to creep, that is, to increase in length or change dimensions under constant load or to relax in stress when placed in constant strain. During creep the molecules slip to new positions from which they do not recover, resulting in a permanent 'set'. This characteristic of creep is important to long-term exposure such as would be encountered by all components in a liner system. For example, a synthetic membrane on an uneven surface will tend to deform and be strained in accommodating the irregularities of the surface, resulting in polyaxial stresses (Koch *et al.*, 1988). In-place drainage nets and pipes are under constant load and a synthetic membrane placed over a protrusion is under constant stress. The absorption of organics can soften the polymer and aggravate these tendencies.

Stress cracking. In addition to their tendency to creep, as with all materials, polymeric materials are subject to loss of strength and to fracture when under mechanical stress for extended periods of time. Some semicrystalline polymeric compositions, e.g. some varieties of polyethylenes and polyester elastomers, when placed under stress in chemical environments, can be affected by one of the chemical species present and crack or craze in moderately short times. Such a phenomenon has been termed, 'environmental stress-cracking'.

DURABILITY AND QUALITY ASSURANCE

The basic conditions to which polymeric membranes and other components of a liner system are exposed in both MSW and hazardous waste landfills include comparatively low-ambient temperatures, lack of light, moisture, aerobic and anaerobic atmospheres depending on the component of the liner system and the location within the fill, and low concentrations of dissolved constituents. Thus, polymeric materials placed in service in liner systems do not generally encounter the critical conditions that are normally considered to cause degradation of the base polymeric resins, particularly when polyethylene, modified olefinic polymers and some polyesters are involved. Nevertheless, when these polymers or compounds are used in products such as membranes, drainage nets, geotextiles, and pipe, they are subject to mechanical and combined mechanical and chemical stresses which may cause deterioration of some of the important properties of these polymeric products.

Some of the principal areas of concern that may affect the service lives of components of liner systems and the functioning of the liner system as originally designed include (Haxo & Haxo, 1988, 1989):

• The combined mechanical and chemical stresses under which the liner system functions may cause cracking and breaking of the components due to environmental stress-cracking or possibly to mechanical fatigue under long service.
• Swelling and softening of polymeric components of a landfill liner system due to the absorption of waste constituents combined with the overburden load may cause creep-induced damage to the drainage, collection, and detection systems during the active life and post-closure care period.
• Plasticized polymeric materials under long exposures in landfill environments may lose plasticizer by extraction, by volatilization, or by biodegradation of the plasticizer, resulting in adverse changes in properties.
• Seams of synthetic membranes continue to be an area of concern,

as none of the test methods truly assess the effects of long-term exposure in landfills.

- Clogging of drainage and detection systems continues to present a problem. The clogging can be caused by biological growth or sedimentation or through precipitation of dissolved constituents.

Durability

Polymeric compositions used in the manufacture of components of a lining system can be subject to the following degradation mechanisms resulting from chemical stresses:

- breakdown of the polymer resulting in depolymerization or a reduction in molecular weight. Mechanisms by which polymers can break down include oxidation, photodegradation (particularly by ultraviolet light), ozone degradation, thermal degradation (e.g. after prolonged exposure to high service temperatures), hydrolysis, and biodegradation;
- crosslinking and gelling with subsequent embrittlement of the polymer;
- swelling and possible dissolution of the polymer by constituents of the waste leachate to which the polymeric material is exposed;
- volatilization of ingredients of the polymeric compound (e.g. plasticizers) when the polymeric component is exposed directly to the weather for prolonged periods of time;
- extraction of ingredients of the polymeric compound (e.g. plasticizers or antidegradants) when the polymeric component of the lining system is exposed to the waste leachate, especially if the leachate is flowing and contains organics.

The extent to which any of these degradation mechanisms need be of concern depends on the individual component, its composition, and all of the conditions to which it is exposed.

Quality Assurance

Generally, synthetic materials in lining systems, when combined with proper design and construction, can function satisfactorily in waste disposal environments for moderately long periods of time.

Short-term tests of a lined unit are not available for predicting the long-term performance, such as can be done with manufactured products, e.g. pneumatic tires. At the present time, owners and designers of disposal facilities must depend empirically on methods that have developed (as conventions) for setting specifications, quality assurance programs to cover both materials and construction, and materials that have demonstrated satisfactory performance.

By assuring that the design of the unit is being met, strict conformance to a well-planned quality assurance plan for the construction of a landfill has been found to be one of the most important factors contributing to the success of a disposal unit (Matrecon Inc., 1988). Rigorous quality assurance may make the difference between a unit that functions with a minimum number of problems throughout its service life and one that falls short of its maximum performance goals.

Construction quality assurance in the context of constructing a landfill is a planned system of activities that provides assurance that the unit is constructed as specified in the design. A system of activities has been developed and after public review has been published as policy by the USEPA (Northeim & Truesdale, 1986). Such a system should be initiated by the facility owner to ensure that the construction of the disposal facility, including manufacture, fabrication, and installation of the various components of the lining system, meets design specifications and performance requirements. Construction quality assurance activities include inspections, verifications, audits, and evaluations of materials and workmanship necessary to determine and document the quality of the constructed facility. These activities are often performed by a third-party quality assurance team that is independent of the designer, manufacturer, fabricator, installer, and owner/operator to ensure impartiality. Various aspects of quality assurance of membranes are reviewed in a special issue of *Geotextiles and Geomembranes* (Ingold, 1986).

CONCLUDING REMARKS

Current technology and experience with synthetic liner systems and with individual polymers and components indicate that with proper design of the unit, proper selection of materials, quality construction, and adequate control of the wastes placed in the unit, geosynthetics and plastic pipe are available that will be chemically compatible and serviceable in such service environments. Failures though, have occurred, and some of these failures appear to be associated with design and construction. In addition, some failures may have been the result of inadequate recognition of basic characteristics of the materials used. In spite of the recognized nondegradability of the polymeric materials used in constructing waste disposal units, uncertainty remains about the length of the service lives of lining systems, particularly after closure of the unit.

At present, there is a lack of quantitative data on the mechanical, chemical, biological, and environmental stresses to which lining systems are exposed in actual service. In addition, few of these stresses are well defined because they almost always operate in combinations. At this time, neither the magnitude nor the relative importance of a given stress or factor can be stated in a given situation. More actual field studies need to be conducted to quantify these stresses.

The performance of waste disposal units is being monitored principally by measuring the amount and composition of leachate that is collected above the liner in a leachate collection and removal system and below the top liner in a leak detection system. Monitoring wells placed outside a unit are also used to assess any possible escape of waste constituents from the unit. The information derived from such monitoring indicates the overall performance of the unit at the time; it does not supply information on the materials themselves, though it might indicate whether components have failed, e.g. whether there is a breach in the liner system. However, even though the unit may not have failed, the materials used in constructing the lining system may be deteriorating. Direct information on the in-service condition of these materials is needed in order to develop correlations with laboratory exposure tests and models for predicting service lives of materials in the different environments within containment units.

Limited efforts have been made to monitor the condition of in-service liner systems through the use of coupons placed within containment units (Matrecon Inc., 1988). Coupons have been placed in landfill sumps and withdrawn at various time intervals and their properties measured. This type of exposure is useful for assessing the chemical compatibility of the material with the leachate or waste liquid over extended periods during which the composition varies. However, coupons are not being exposed in the intended service environment of the material, nor are they being exposed to mechanical stresses that would normally be encountered in service.

Another source of information on the condition of in-service lining systems and the individual materials is field studies on liner materials taken from units that are being dismantled (Matrecon Inc., 1988). The limited information that has been obtained in such studies has severe limitations. Overall, the amount of information in the open literature regarding the service conditions and the performance of geosynthetics under service conditions is limited. Both the engineernig profession and the general public depend on such information so that neither major disasters occur nor are containment units left to become superfund sites for future generations. This lack of information leaves a significant gap in our knowledge regarding waste disposal on land and one which should be filled.

REFERENCES

ASTM. *Annual Book of ASTM Standards*. Issued annually in several parts, American Society for Testing and Materials, Philadelphia, PA:
D412-83.   Test Methods for Rubber Properties in Tension, Sections 08.01, 09.01. 09.02.
D624-86.   Test Method for Rubber Property—Tear Resistance, Section 09.01.
D1434-84. Test Method for Determining Gas Permeability Characteristics of Plastic Film and Sheeting to Gases, Section 08.01.
D2240-86. Test Method for Rubber Property—Durometer Hardness, Sections 08.02 and 09.01.
D3421-75. Recommended Practice for Extraction and Analysis of Plasticizer Mixtures from Vinyl Chloride Plastics, Section 08.03.
E96-80.   Test Methods for Water Vapor Transmission of Materials, Sections 04.06, 08.03 and 15.09.
Haxo, H. E. (1988). Transport of dissolved organics from dilute aqueous solutions through flexible membrane liners. In *Proceedings of the Fourteenth Annual Solid Waste Research Symposium: Land Disposal, Remedial Action, Incineration and Treatment of Hazardous Waste, May 9–11, 1988,* EPA/600/9-88/021. US Environmental Protection Agency, Cincinnati, OH, pp. 145–66.
Haxo, H. E. (1989). Compatibility of flexible membrane liners and municipal solid waste leachate. In *Proceedings of the 15th Annual Research Symposium Remedial Action, Treatment and Disposal of Hazardous Waste, April 10–12, 1989.* EPA/600/9-90/006. US Environmental Protection Agency, Cincinnati, OH, pp. 350–68.

Haxo, H. E. (1990). Determining the transport through geomembranes of various permeants in different applications. In *Geosynthetic Testing for Waste Containment Applications, ASTM STP 1081,* ed. Robert M. Koerner. American Society for Testing and Materials, Philadelphia, PA, pp. 75–94.

Haxo, H. E. (1991). Compatibility of flexible membrane liners and municipal solid waste leachates. In *Geosynthetic Testing for Waste Containment Applications,* NTIS PB 91-231 522, US Environmental Protection Agency, Cincinnati, OH.

Haxo, H. E. & Haxo, P. D. (1988). Consensus report on the *ad hoc* meeting on the service life in landfill environments of flexible membrane liners and other synthetic polymeric materials of construction. EPA 600/x-88-252, US Environmental Protection Agency, Cincinnati, OH.

Haxo, H. E. & Haxo, P. D. (1989). Synthetic lining systems for land waste disposal facilities. Proceedings Sardinia 89, Second International Landfill Symposium, Alghero (Italy), October 9–13, 1989.

Haxo, H. E., White, R. M., Haxo, P. D. & Fong, M. A. (1982). Final report: evaluation of liner materials exposed to municipal solid waste leachate, NTIS No. PB 83-147-801. US Environmental Protection Agency, Cincinnati, OH.

Haxo, H. E., White, R. M., Haxo, P. D. & Fong, M. A. (1985). Liner materials exposed to municipal solid waste leachate. *Waste Management and Research,* **3,** 41–54.

Hoekstra, S. E. (1984). Reliable groundwater protection with fabric-reinforced bitumen membranes. In *Proceedings of the International Conference on Geomembranes, Denver, CO,* vol. II. Sponsored by Industrial Fabrics Association International, St Paul, MN, pp. 415–19.

Ingold, T. S. (ed.) (1986). Special issue on geomembrane quality assurance. *Geotextiles and Geomembranes,* **3**(4).

Koch, P., Gaube, E., Hessel, J., Gondro, Chr. & Heil, H. (1988). Long-term strength of dump lining sheets made of polyethylene. Translation of 'Langzeitfestigkeit von Deponiedichtungsbahnen aus Polyethylen, *Muell und Abfall,* **8/88,** 3–12.

Koerner, R. M. (1986). *Designing with Geosynthetics.* Prentice Hall, Englewood Cliffs, New Jersey.

Le Coroller, A., Roland, S. & Bernard, I. (1984). Use of asphaltic geomembrane for protecting railway subgrade. In *Proceedings of the International Conference on Geomembranes, Denver, CO,* vol. II. Sponsored by Industrial Fabrics Association International, St Paul, MN, pp. 415–19.

Matrecon Inc. (1983). *Lining of Waste Impoundment and Disposal Facilities, SW-870.* Revised. US Environmental Protection Agency, Washington, DC, Government Printing Office No. 055-00000231-2.

Matrecon Inc. (1988). *Lining of Waste Containment and Other Impoundment Facilities,* 2nd edn. US Environmental Protection Agency, EPA/600/2-88/052, Cincinnati, OH.

Northeim, C. M. & Truesdale, R. S. (1986). Technical guidance document: construction quality assurance for hazardous waste land disposal facilities. US Environmental Protection Agency, EPA 530-SW-86-031. OSWER Policy Directive No. 9472.003, Washington, DC.

US General Services Administration (1980). Method 2065: Puncture resistance and elongation test (1/8-inch probe method), and Method 2031: Tetrahedral-tip probe method. In *Federal Test Method Standard 101C*. US Services Administration, Washington, DC.

4.2 Quality Control of Flexible HDPE Liners

MARK W. CADWALLADER & PAUL W. BARKER

Gundle Lining Systems, 19103 Gundle Road, Houston, Texas 77073, USA

INTRODUCTION

The approach to the quality control of HDPE liners must revolve around three major aspects: liner raw material, liner product, and liner installation. These aspects are discussed in this chapter.

LINER RAW MATERIAL

Many polymer resins carry the tradename 'polyethylene'. However, different methods of joining the ethylene molecules together result in different characteristics of the final product. The different polyethylene materials have traditionally been classified by their densities. Low-density polyethylene (LDPE) categorizes polyethylenes in the range of about $0.915-0.935 \, g/cm^3$, while high-density polyethylene (HDPE) covers polyethylenes roughly from $0.935-0.970 \, g/cm^3$ in density.

Proper selection of polyethylene resins is very important, even when the material is confined to the HDPE classification. Different catalyst processes and the use of different co-monomers alongside ethylene are ways of manufacturing different HDPE resins.

Some of these resins are used for injection molding, others for wire and cable applications, and still other resins are for manufacturing pipe. The cost of these materials can vary by as much as a factor or two, and only the higher quality and thus more expensive resins are suitable for rigorous geomembrane applications.

With proper resin quality, HDPE geomembrane liner technology is tough to beat in terms of strength, toughness, durability, and chemical resistance.

Polyethylene resin selection is important because improper resin can result in poor product performance. This is most easily understood when it is realized that the molecular structure forms literally the 'backbone' of the product, and that molecular structure is, to a very great extent, a function of resin quality. Improper resin has been linked with, among other things, poor environmental stress crack results for the product.

Since different polyethylene resins vary in molecular structure, some of the consequences of these differences will influence product performance.

Correlations between basic molecular structure and material properties of polyethylenes have been difficult to establish with certainty but they do exist. Polyethylene has essentially three structural variations which affect its properties (Cassady et al., 1981):

- molecular weight and its distribution;
- chain branching (affected by comonomer type and concentration as well as by polymerization process);
- amount and type of crystallinity (density is a reflection of the amount of crystallinity).

Because of the interplay of these variables, commercially-available materials possess distinctly different properties. Hence, the various grades of HDPE resin mentioned above.

Polyethylene is classified as a crystalline polymer. This means that the polyethylene chains tend to pack into a regular crystal lattice. The molecules crystallize by folding of the polymer chains, forming what are called lamellae, or plate-like polymer crystals. Not only are the polymer chains arranged to form lamellae, but these lamellae are usually arranged in larger aggregates known as spherulites. Figure 1 illustrates the lameallar structure (verified by X-ray measurements) as well as the shape and appearance of spherulites (Rosen, 1982).

Spherulites grow out radially from the nucleus until other spherulites are encountered. The size, shape, arrangement and interaction of the spherulites impact upon the physical properties of polyethylene.

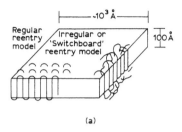

(a)

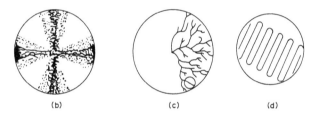

(b)   (c)   (d)

Figure 1. Lamellar structure and appearance of spherulites in HDPE liners (Rosen, 1982): (a) polymer crystal showing folded chain arrangement which forms lamellar crystal plates; (b) appearance of a spherulite between crossed polaroids of an optical micrograph; (c) branching of lamellae in a spherulite; (d) close up showing the orientation of lamellar chains.

For example, smaller typical diameters of spherulites have been correlated with greater resistance to stress cracking (Shanahan *et al.* 1980). Large spherulites contribute to brittleness in polymers. Developing cracks tend to avoid spherulites whenever possible and trace a path around spherulitic boundaries. Where there are smaller spherulites, they are more numerous, and the probability is that a growing crack front will meet a spherulite. (The analogy can be drawn here between a box filled with billiard balls and one filled with marbles.)

Molecular weight distribution is important with regard to this type of stress crack development. Resins with broad molecular weight distributions have many low molecular weight species which tend to be excluded from the spherulitic structure, thus occupying the amorphous zones. Resins with narrow molecular weight distributions have

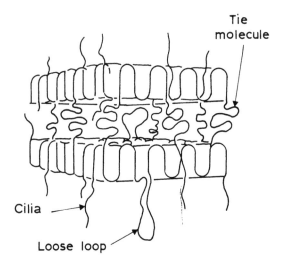

Figure 2. Basic structural elements between polyethylene lamellae.

less low molecular weight species to congregate in the spherulitic boundaries where crack propagation takes place.

The presence of crystallites and spherulites is one aspect of the structure and morphology of polyethylene which determines its physical properties. Another very important aspect is the type of molecular arrangement in the non-crystalline or amorphous regions of the polymer. This amorphous material is incorporated within and between spherulites, and between lamellae. It has already been mentioned that crack fronts tend to avoid crystalline regions and propagate through the amorphous zones. In these amorphous zones are polymer chains which are much more 'loosely' arranged than in the crystallites. As Fig. 2 illustrates, there are essentially three types of intercrystalline, amorphous material:

- cilia—chains suspended from the end of a crystalline chain;
- loose loops—chains which begin and end in the same lamellae;
- tie molecules—chains which begin and end in adjacent lamellae.

Since they are partially crystallized in two lamellae, 'tie molecules' form a bridge between lamellae, and if the lamellae are in separate spherulites they tie neighboring spherulites together. Tie molecules are

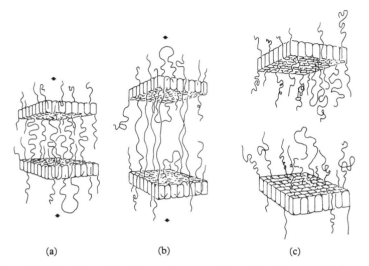

(a)   (b)   (c)

Figure 3. Polyethylene lamellae under stress (a), and leading to a brittle-type failure ((b) and (c)).

the 'cement' holding the lamellar 'bricks' of crystalline polyethylene together.

Poor mechanical properties and poor resistance to various types of stress crack and other brittle-type failues are associates with the lack of these tie molecule interconnections between crystallites. Figure 3 illustrates a sequence of events leading to failure by breaking of the tie molecules under stress. Long-term, low-level stresses can result in failure of the polymer if the number of tie molecules are not very many and not very well entangled. When such is the case, the molecules can begin to untangle and relax. Ultimately the load cannot be supported by the few tie molecules remaining and as a result the material fails in a brittle manner.

Several parameters are important in determining a sufficient level of tie molecules. These include: molecular weight, co-monomer content, and density.

- *Molecular weight*: high molecular average weights are indicated by low melt flow indices of the polymer melt. Therefore, low melt flow indices correspond to longer average polymer chains. Long average polymer chains result in more length outside the crystall-ine lamellar lattice and thus more tie molecules as well as more effective tie molecule entanglements. Molecular weight distribution is also important since lower molecular weight species tend to form poor and innumerous tie molecules.

* *Co-monomer content*: recent polyethylene technology has incorporated the use of co-monomers alongside ethylene in polyethylene manufacture. These co-monomers are longer chain olefins such as 1-butene or 1-hexene. The longer chain olefins inhibit crystallinity by providing short branches in the polyethylene chain. The short branches are not as able to enter into the lamellar lattice and hence add to the intercrystalline tie molecule material. The short chain branching also provides increased effectiveness of tie molecule entanglements thereby inhibiting their ability to relax and slip past one another.
* *Density*: density is an indirect measurement of crystallinity. The more crystalline the material or the thicker the crystalline lamellae, the fewer intercrystalline tie molecules that hold them together.

The problem of properly specifying HDPE for use in flexible membrane liners can be thought of as two-fold:

* maximizing tie molecules in the amorphous zones in order to supply high strength and resistance to brittle-type failures;
* optimizing the level and type of crystallinity in order to provide 'blocking agents' against crack propagation and the infiltration of foreign substances such as chemicals.

Actually, instead of 'optimizing' the level of crystallinity, the strategy should be one of maximizing the crystallinity while maintaining a maximum number of tie molecules. The problem is that increases in density are at the expense of a decrease in the number of tie molecules. Proper specification of the polyethylene resin thus becomes a situation where the following parameters need to be addressed:

* molecular weight and its distribution.
* comonomer type and concentration.
* amount and type of crystallinity.

The incorporation of co-monomers into HDPE technology is highly advantageous to resin users. The co-monomers essentially increase the amount and effectiveness of tie molecules without producing a significant drop in density. But the kind of co-monomer and its concentration are very important. The best way to exercise quality control in this area is to select a copolymer resin (made with ethylene and co-monomer) with proven properties in the area of stress crack failure categories.

The particular ethylene polymerization process is also very important in establishing final properties since it is the primary determinant of molecular weight distribution and type or texture of the crystallinity. These parameters are not so easily tested for. However, average molecular weight and amount of crystallinity can be easily and routinely monitored although indirectly, by testing for melt flow index and density.

Thus liner raw material quality control strategy should incorporate the following:

- Specification of a highly reputable copolymer HDPE resin and polymerization process, know to give good results in stress crack and brittle-type failure categories.
- Consistent and thorough monitoring of the melt flow index.
- Consistent and thorough monitoring of density.

ASTM tests for melt flow index and density are outlined below.

Melt Flow Index (MFI) Test Method

Description. At fixed temperature and pressure a molten sample is squeezed through a standard orifice. The extrusion rate is measured in g/10 min.

Official detailed method: ASTM D1238.

Test apparatus.
MFI plastometer, including: die, die plate holder, piston, piston weights, timer, governor, cleaning tool, die ejector tool, charging tool, thermometer
Analytical balance (0·1 mg)
Sample cutting knife
Fine cloth

Test procedure
1. Bring the plastometer to the operating temperature. A typical condition is 190°C.
2. Charge the cylinder with polymer until full. If the sample is not

in granular form, cut or chop it up to allow access to the plastometer cylinder.

3. Position the piston in the cylinder and place the weight on the piston. Two weights are typical, $2 \cdot 16$ kg and $5 \cdot 0$ kg.

4. Preheat for 6–8 min, then manually purge material down to the scribed mark on the piston.

5. Cut off the purged material and begin timing extrusion of the polymer under pressure of the weight.

6. At specific intervals, cut off the extruded material at the bottom of the die.

7. Weigh extrudate on an analytical balance.

8. Adjust numbers to report flow rate in g/10 min.

9. Clean the plastometer free of polymer using die ejector tool, cleaning tool, and fine cloth.

Density Gradient Column Test Method

Description. A sample of plastic is allowed to find its level in a liquid density gradient column and its position in relation to air-filled glass beads of known density is used to determine graphically the sample's density in g/cm^3 at 23°C.

Official detailed method: ASTM D1505.

Test apparatus.
Constant temperature bath
Density beads (floats) for appropriate ranges, e.g. $0 \cdot 930$– $0 \cdot 970$ g/cm^3
Graduated density column(s)
Tweezers
Distilled water
Isopropanol

Test procedure.
1. Prepare density column(s) by properly mixing the alcohol and water. This is done in a controlled way according to the column manufacturer's instructions. The alcohol and water are very slowly introduced in precise ratios into the graduated column(s).

2. Cautiously lift the graduated density column(s) into the constant temperature bath set to maintain 23°C ± 1°C.
3. Place air-filled color-coded glass beads of known density into column(s).
4. Prepare calibration graph by plotting each bead's height in the column(s) versus its density on standard graph paper.
5. Draw the best fit line through the plotted points on the graph.
6. If available, use a section of extrudate from the MFI plastometer as a sample. (This insures that the sample contains no air bubbles.)
7. Cut off a small piece of the extrudate section and using tweezers wet the sample in alcohol to clean it.
8. With the tweezers, drop the samples into the density column.
9. When the sample has reached a constant height in the column, read and record the sample position.
10. Find the sample density from the calibration graph, to the nearest 0·001 g/cm³.

LINER PRODUCT

The next area of quality control involves monitoring the quality of the liner product. This is important in order to verify proper operation of the manufacturing process as well as to confirm proper polymer resin selection. Since carbon black is compounded with the HDPE resin to protect against ultraviolet light degradation, the product must be analysed to see that this is done correctly.

The product tests chosen should provide a representation of the performance categories of strength, toughness and durability of the liner as these terms were defined above. In addition, the liner must demonstrate resistance to chemicals as well as to stress cracking.

A listing of tests which cover the basic categories of liner performance and which can be monitored easily on a routine basis is presented in Table 1.

Environmental stress cracking is actually only one form of a broader range of phenomena classified simply as stress cracking. Stress cracking is a term for most any kind of rupture at stress levels lower than should be expected to cause the rupture. Solvent cracking, thermal cracking, and static fatigue failure are also considered types of

TABLE 1. Product Performance Tests for Flexible HDPE Liners. Typical Property Values are Reported Unless the Minimum or Maximum is Specifically Indicated. The Typical Values Should Not be Specified as a Limit to Acceptable Performance

Test property	Test method	Quality result
Tensile properties	ASTM D638,	
Tensile strength at yield	Type IV at	$20 \, \text{N/mm}^2$
Tensile strength at break	50·8 mm/min	$33 \, \text{N/mm}^2$
Elongation at yield		15%
Elongation at break		800%
Initial tear resistance	ASTM D1004	223 N for 1·5-mm specimen
Puncture resistance	FTMS 101B, Method 2065	316 N for 1·5-mm specimen
Carbon black content	ASTM D1603	2·5%
Carbon black dispersion	ASTM D3015	A-1
Accelerated heat aging	ASTM D573, ASTM D1349	No drop in tensile strength or elongation
Environmental stress	ASTM D1693	2000 h
Crack resistance	C Condition	

stress cracking. Chemical resistance testing and accelerated heat aging, however, should help indicate potential problems in these other types of stress cracking.

Each of the test methods in Table 1 is briefly outlined in the following paragraphs.

Tensile Properties Test Method

Description. Values of tensile strength and elongation are measured at both yield point and break point. This gives a picture of the stress–strain characteristics which is very important in evaluating product performance.

Official detailed method: ASTM D638.

Test apparatus.
    Tensile test machine with 1-inch (1 inch = 2·54 cm) grips, load cell and recorder chart
    ASTM D638 dumb-bell specimens (type IV dumb-bell specimens

having only 40 mm of test length perform best for materials with
large elongations)
Micrometer

Test procedure.
1. Using a standard die, cut specimens parallel to the machine
   direction (direction in which die lines run) and also perpendicu-
   lar to machine direction (transverse direction).
2. Warm up the tensile test machine and be sure the tester is
   calibrated.
3. Choose the load cell appropriate for the material's strength, zero
   the chart pen and calibrate the load cell with the chart.
4. Set the gap between grips to 25 mm.
5. Measure the width and thickness of thin section of the dumb-bell
   specimen.
6. Center the specimen between grips, tightening the top grip first
   and the bottom grip second.
7. Begin pulling the specimen apart at a rate of strain of
   50·8 mm/min. If the chart speed is synchronous with the
   crosshead movement, the stress–strain curve forms a picture of
   the experience of the specimen in 'real-time'.
8. The tensile strength at yield and break points may be calculated
   in both N/mm width and N/mm^2.
9. Per cent elongations can be determined from the strain and
   calculated according to the ASTM procedure.

Initial Tear Resistance Test Method

Description. A tensometer is used to find the force required to tear a
standard test piece.

Official detailed method: ASTM D1004.

Test apparatus.
Tensile test machine with 1-inch grips, load cell and recorder chart
ASTM D1004 die specimens
Micrometer

Test procedure.
1. Set up the equipment as for tensile testing. Prepare sample
   specimens in both the machine direction (MD) and the transverse direction (TD). The MD is such that when the sample is

pulled apart, the line of tear will be in the direction of the die lines. Measure the thickness in the middle of the specimen using the micrometer.
2. Pull the sample specimen apart at 50 mm/min.
3. Read the maximum tear resistance stress in newtons.
4. Newtons per millimeter may be calculated if desired.

Puncture Resistance Test Method

Description. A probe is moved by means of a cage-type apparatus in a tensile testing machine in order to puncture the specimen.

Official detailed method. FTMS 101B, Method 2065.

Test apparatus.
Tensile test machine with load cell and recorder chart
Compression cage containing specimen holder and probe (3·2 mm radius on end)
Square (50 mm × 50 mm) specimens
Micrometer

Test procedure.
1. Cut the specimen to size and measure the thickness.
2. Set up the equipment as for tensile testing. Fasten compression cage into the tensile tester in place of the grips.
3. Center and tighten the specimen in the holder.
4. Activate the tensile tester, pulling the compression cage apart at a rate of strain of 50·8 mm/min, and bringing the probe through the specimen.
5. Note the stress level indicated on the chart and record it.

Carbon Black Content Test Method

Description. Since polyethylene has a vaporization point lower than carbon, samples of the product are placed in a furnace at 600°C to allow the polymer to vaporize. In order to prevent oxidation of the carbon to carbon dioxide, a nitrogen gas blanket is applied.

Official detailed method: ASTM D1603.

Test apparatus.
Combustion tube furnace (>600°C capacity)
Combustion tube (hard porcelain or boro-silicate glass)
Combustion boats (high-temperature porcelain)
Reagent grade nitrogen with gas flowmeter and U-shape drying tube
Dessicator (filled with active drierite-$CaSO_4$)
Analytical balance (0·1 mg)

Test procedure.
1. Allow the furnace to stabilize at 600°C.
2. Clean and weight the porcelain boat.
3. Put ~3 g of the product into the boat and weigh. Calculate the sample weight.
4. Place the boat in the furnace and apply nitrogen at 5–8 cm^3/min flowrate
5. Allow the polymer to vaporize for 15 min.
6. Carefully retrieve the boat and cool in a dessicator.
7. Reweigh the boat and calculate the residual carbon black. Determine the per cent carbon black.
8. If an inorganic filler is suspected place the boat again into furnace and withdraw nitrogen. Carbon black will oxidize into the atmosphere, leaving the inorganic filler.

Carbon Black Dispersion Test Method

Description. The degree and quality of mixing or dispersion of the carbon black particles in the polyethylene medium is observed under a microscope and compared against standards.

Official detailed method: ASTM D3015.

Test apparatus.
Microscope
Glass slides
Hot plate

Test procedure.
1. Clean two slides with alcohol.
2. Cut a very small plastic specimen (<10 mg) and place it on one slide.

3. Put the slide on the hot plate, bring to around 200°C, and place the other slide on top.
4. Put a metal plate on top of slides and add a 5-kg weight on top of the plate.
5. Let melt for approximately 2 min, remove and allow the slides to cool.
6. Place the slides under 100× magnification. Adjust the light and focus.
7. Compare the sample to the ASTM dispersion classification chart. Find the best match in the chart and note it.

Accelerated Heat Aging Test Method

Description.   Tensile test specimens are suspended in an oven and kept at an elevated temperature for a certain period of time. After exposure the samples are tested for tensile strength and elongation and results are compared with unexposed samples.

Official detailed methods: ASTM D573, ASTM D1349.

Test apparatus.
   Dry air ovens with thermometers and shelves
   Paper clips
   China markers
   Tensile test dumb-bell specimens
   Micrometer

Test procedure.
1. Set the oven to 110°C (alternative temperatures may be used).
2. Cut the dumb-bell specimens with die (both MD and TD), measure the thickness at the center using a micrometer, and write the thickness on each dumb-bell with the china marker.
3. Hang the specimens by paper clips from oven shelves, segregating machine direction (MD) from transverse direction (TD) samples.
4. Remove samples at periodic dates up to 6 months or more and perform tensile testing.
5. Record the tensile strength and elongation. Compare with the control.

Alternative Short-Duration Method

If a differential scanning calorimeter (DSC) is available, the oxidative stability of material can be determined according to ASTM D3895. This amounts to a very accelerated heat aging test. The advantages of this kind of procedure are that test times are reduced to less than 1 h, and the end of the test always leads to a failure, yielding a meaningful result. This compares to standard heat aging which because of time constraints cannot differentiate between average, good and excellent material. If the DSC comes equipped with a pressure cell, oxygen under high pressures can be used to accelerate the aging while bringing temperatures down to levels closer to what the liner actually experiences during use. This makes for even more meaningful results.

Environmental Stress Crack Test Method

Description. Notched specimens of sheeting are bent 180° and placed under this constant stress in the presence of a surface active reagent and heat. They are observed for the development of cracks over time.

Official detailed method: ASTM D1693, Condition C.

Test apparatus.
Square-edged rectangular die specimens
Jig etching tool (for making controlled imperfection in specimens)
Specimen holders (brass)
Test tubes
Constant-temperature bath
Test tube rack
Bending clamp

Test procedure.
1. Cut 10 specimens 2-mm-thick using an ASTM D1693 die.
2. Scratch each specimen using the etching tool.
3. Bend the samples in the bending clamp.
4. Place each of the 10 specimens in a brass holder.
5. Place the loaded specimen holder on its end and in a test tube.
6. Fill the test tube with 100% Igepal CO-630 and insert the tube in a temperature bath at 100°C. (*Note*: 100% Igepal and 100°C (Condition C of ASTM D1693) corresponds to a rigorous test condition.)

7. Observe the tube for cracks in the specimens, indicating failure.

8. Note the elapsed time from the start of specimen immersion and record the percent specimen failure, if any.

In addition to testing strength, toughness, durability and environmental stress crack as the methods outlined above do, chemical resistance testing of the liner material should be a matter of ongoing research to determine which chemical waste mixtures in what concentrations, might be damaging to liner integrity. One such chemical resistance test is the USEPA's Method 9090. This procedure calls for immersion of test material in the waste mixture which is to be contained. The immersion period runs for 120 days at both 23 and 50°C, with comparison of liner properties to controls at 30-day intervals. There should be no significant drop in particles.

Once again, if proper quality resin is selected, dependence on completely comprehensive liner testing is reduced because high quality resin forms a literal 'backbone' to a high quality liner.

LINER INSTALLATION

The next area in the quality control sequence involves liner installation. Installation of flexible membrane liners requires seaming the sheets together to form leak free layers.

In general, there are four basic methods of joining the membrane material in the field. These are:

- solvent/adhesive
- hot air
- hot wedge
- extrusion bonding

For chemically-resistant materials such as HDPE, solvent/adhesive systems are very unusual if not nonexistent. In addition, the adhesive layer formed in these systems is usually more susceptible to chemical attack, resulting in potential subsequent failure at the seam.

Hot air systems are also generally regarded with skepticism. Consistent results are difficult to achieve in the field since the techniques rely so heavily on temperature. Wind, clouds, etc., thus become a factor. And because the intense heating process often oxidizes the membrane surface, long-term durability is affected.

Hot wedge and extrusion welding are considered the most reliable seaming methods to date. Hot wedge welding passes a wedge of hot steel between the overlapped sections of adjacent membrane, melting the sheet. Pressure rollers fuse the sheet together, most often in two tracks separated by an air gap to enable air pressure testing of the seam for leaks. Extrusion welding involves placing a bead of molten parent material either between the adjacent overlapped sheets, or over the top of the adjoining sheets.

Regardless of the seaming method used, quality control of the seams should involve both nondestructive and destructive testing.

Nondestructive testing of the seams comprises three main techniques:

- visual examination
- vacuum testing
- air pressure testing

Visual examination of the welds is possible if an extrusion-type of weld is applied over the top of adjoining sheets. Any suspect areas, breaks, or holes in the weld can be seen and repaired. In addition, the seams should be vacuum tested. In this type of testing a plexiglass-faced suction box, typically 1 m long and wide enough to cover the weld, is placed over a section of the seam which has been wet with a soap solution. Suction is applied to the seam through a suction pump connection on the box, and any leaks are detected by the formation of bubbles. The seam is maintained under suction for 10 s. During this time, the next section of seam to be tested is soaped. The pressure is then relieved, and the box moved to the next section, allowing a 76-mm overlap onto the previous section tested. Two men can carry out testing of all the seams quickly and efficiently at a rate of approximately 30 m/h. Even pinhole-size leaks can be detected and repaired by this technique.

Air pressure testing of dual track hot wedge welds has become a preferred nondestructive test and seam leak detection method. The speeds and effectiveness of leak detection are unmatched, and air pressure testing capability is one reason why the dual track hot wedge welding methods have become the most widely practiced seaming method in the lining industry. In this procedure, a small air pump is used to pressurize the air gap between welded tracks to 2–3 atm pressure, maintaining the pressure for 5 min. Even small pinhole leaks are quickly detected by noting on the pressure gauge a sudden drop in air pressure. Leak locations are determined, through close visual observation, through use of soap solution, directional microphones, or through a process of elimination by cutting the test distance in half and locating the section in which the leak occurs. Typical test distances are 100–200 m at a time.

Destructive testing of the seams is very important because it provides the only direct evaluation of seam strength and bonding efficiency. Samples of destructive testing are cut out of the seams in approximately 25 cm × 40 cm rectangular sections along the seam. The samples are removed at agreed upon intervals anywhere from 60 m to 400 m depending on the size of the facility. Description, significance and reliability of destructive tests for the assessment of HDPE geomembrane liners are reported in Chapter 4.4.

Repairs of failed sample sections in destructive testing are accomplished by welding a patch over the suspected weak section of seam.

In addition to proper quality control of seams, liner installation should provide careful attention to the underlying soil material that comes in contact with the membrane. Loose rock, debris and roots should be removed.

Other aspects of the installation are important but they tend to fall under the category of liner syste design (see Chapter 2.2). Design must also be subject to quality control and careful overview.

CONCLUSION

A secure and properly-working waste disposal facility depends on effective quality control all across the board. That includes all facets of facility design, material manufacture, and construction of the composite liner system. The geomembrane layer, however, is the liner system layer with greatest responsibility in terms of the containment of waste. Comprehensive quality control of flexible membrane liners is thus very important.

REFERENCES

Cassady, M. J. *et al.* (1981). Development of improved plastic piping materials and systems for fuel gas distribution. Annual Report Battelle Laboratories, Contract No. 5014-352-0152 for Gas Research Institute, Chicago, Illinois.

Rosen, S. L. (1982). *Fundamental Principles of Polymeric Materials.* John Wiley & Sons, New York.

Shanahan, M. E. R., Chen-Fargheon, C. & Szhultz, J. (1980). The influence of spherulitic size on the environmental stress cracking of low density polyethylene. *Macromolecular Chemistry,* **1816**, 1121–6.

4.3 Standard Tensile Test for Geomembranes

JEAN PIERRE GOURC, HUBERT PERRIER, BERNARD LECLERCQ
& JEAN PAUL BENNETON

IRIGM, University Joseph Fourier, BP 53X, Grenoble Cedex, France

INTRODUCTION

The utilization of geomembranes for landfill lining requires thorough knowledge of the mechanical behaviour of this material. Many different test standards have been proposed to study the tensile behaviour of different materials. This chapter focuses on three types of geomembranes of highly differing characteristics: a PVC, a bituminous membrane and HDPE with and without a welded seam. By referring to extensive experimental studies (Gourc *et al.*, 1986, 1987), the various problems raised by the tensile behaviour of membranes in association with the type of test specimen are presented.

TENSILE TEST PROGRAMME AND CONDITIONS

Standard Test Specimens

Four types of test specimen are considered: two strips and two dumb-bells. Their main dimensions and shapes reported in Fig. 1. The significant dimensions of the specimens are desorbed in Fig. 2: b_0 corresponds to the sample width (the central part for the dumb-bell specimens), h_0 is the distance between tensile machine clamps and h_c is the specimen length for a constant width $b = b_0$ (for the dumb-bells).

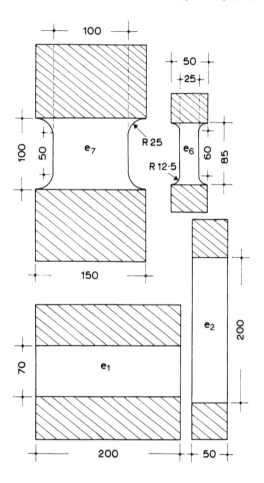

Figure 1. Geometry and main dimensions of the specimens used for tensile tests on geomembranes.

Specimen e_1 (Fig. 1) is a wide strip of dimensions close to those recommended for geotextile tensile tests. Specimen e_2 corresponds to the French standard NF G07001. Specimen e_6 is recommended in France by the LCPC and e_7 is used at the Ecole Polytechnique in Montreal (Rollin *et al.*, 1985) for tests on geomembranes.

In order to be able to disregard the 'strain rate' parameter, the

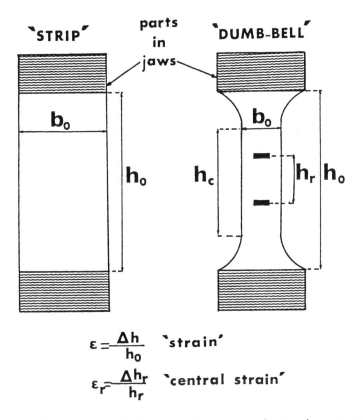

$$\varepsilon = \frac{\Delta h}{h_0} \quad \text{`strain'}$$

$$\varepsilon_r = \frac{\Delta h_r}{h_r} \quad \text{`central strain'}$$

Figure 2. Strip and dumb-bell types of specimen for tensile tests on geomembranes.

crosshead speed (rate of separation of the clamps) of the tensile machine is taken equal to $(h_0/2)$ per minute for all the specimens (e_1, e_2, e_6 and e_7). In addition, the test temperature is kept at 20°C.

The problem of weld quality between geomembrane sheets has also been studied. Various tests are available: a permeability test, a peeling test or the tensile test considered here. The geometry of the specimens was modified as follows: each sample had a central seam weld of constant length, $h_s = 40$ mm. This increased the distance between clamps which became $(h_0 + h_s)$. However, since welds are not generally very tensile, the same testing rate $(h_0/2)$ per minute was considered.

Measurements

The tensile force T was measured during elongation at a constant speed. In order to have valid comparisons for the different specimens e_1, e_2, e_6 and e_7, the force per unit width was used:

$$\alpha = T/b_0 \quad \text{(kN/m)}$$

The separation of the tensile machine clamps Δh was measured in order to determine the 'total elongation' ε:

$$\varepsilon = \Delta h/h_0$$

The parameter ε is not, however, very representative, as slipping or flowing of the geomembrane material between the clamps is always a possibility. Moreover, for dumb-bell specimens, which do not have constant width, the strain is obviously heterogeneous between the central part and the zones near the edge. It was for this reason that an optical system was used to measure the deformation between two marks made on the geomembrane with a spacing $h_r = 40$ mm (Fig. 2). The 'central strain', was then determined:

$$\varepsilon_r = \Delta h_0/h_r$$

For seam-welded specimens, the distance between these optical reference marks was taken to be $h_r + h_s = 80$ mm. Assuming the weld seam to be relatively untensile, the same method as above can be used to determine elongation:

$$\varepsilon = \Delta h/h_0 \quad \text{and} \quad \varepsilon_r = \frac{\Delta(h_r + h_s)}{h_r}$$

Types of Tested Geomembrane

Three types of geomembrane have been tested:

- a plastified, unreinforced PVC (Alkorplan-Solvay), 1 mm thick with a mass per unit area of 1250 g/m^2.
- a bituminous membrane (Teranap 431 TP-Siplast), 4 mm thick, with a mass per unit area of 4740 g/m^2.

- an HDPE (Gundle), 2·5 mm thick, with a mass per unit area of 2330 g/m².

Despite the fact that only the central strain was measured, it was still important to ensure that the geomembrane was held tightly in its clamps. For this purpose a special clamping system (designed by CER, Rouen) with constant hydraulic pressure ($p = 2$ MPa) was used.

MAIN RESULTS

The results of tests on different materials and different specimen shapes will be examined in order to elucidate a certain number of *idées-forces* (underlying factors) relative to the choice of a single specimen for all materials. The diagrams given hereafter represent

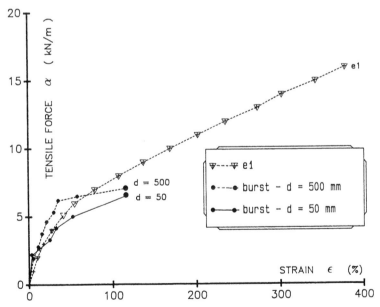

Figure 3. Three-dimensional and large-width strip tensile tests on PVC membrane. In the three-dimensional test (burst test) a circular sample of diameter *d* is subjected to hydraulic pressure until it bursts.

$\alpha(\varepsilon)$ or $\alpha(\varepsilon_r)$. Each graph represents the mean of five tests. Each test was continued until tensile failure of the specimen occurred.

PVC Geomembrane

Comparison with a three-dimensional tensile test. The result of a tensile test on a wide strip e_1 was compared with the result obtained from a burst test—where a circular sample of diameter d is subjected to hydraulic pressure until it bursts (Fig. 3). The results were interpreted assuming the strain in the burst test to be spherical and homogeneous (approximation). Results were obtained from laboratories using small diameter ($d = 50$ mm; Cemagref) and large diameter ($d = 500$ mm, Akzo) machines.

The burst test is of great interest, even though test conditions are complex. It is a severe test, as shown on Fig. 3. It simulates certain conditions encountered *in situ* and enables the creep phenomenon to be studied.

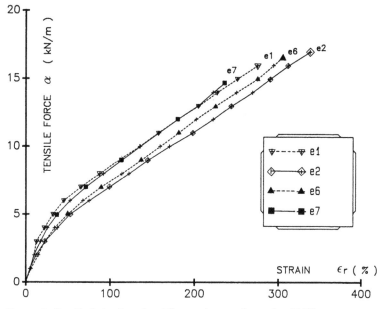

Figure 4. Tensile behaviour for different shapes of samples (PVC).

Influence of specimen shape. The results of the strip and dumb-bell tests could be compared only on the $\alpha(\varepsilon_r)$ diagrams as the strain ε for dumb-bells is hardly significant. The results obtained for the four specimens e_1, e_2, e_6 and e_7 are very similar (Fig. 4). The premature failure of specimen e_7 can be explained: the dumb-bell specimens should be cut using a die. A die for the standard specimen e_6 was available, but specimen e_7 had to be cut mechanically and it is clear that micro-tears in the area where the specimen widens out systematically act as failure initiation points, in spite of the great care taken in these tests.

Behaviour of specimens with seam welds. The $\alpha(\varepsilon_r)$ diagrams obtained for specimens with and without seam welds are similar before failure (Fig. 5). However, since failure of the welded specimen initiates at the weld boundary, the failure strain is smaller, as shown in Fig. 6.

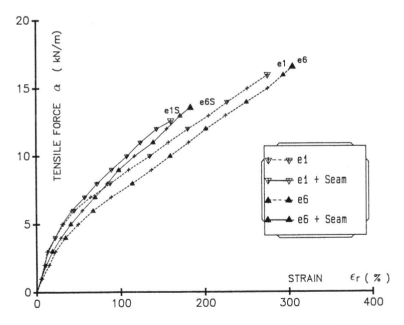

Figure 5. Tensile behaviour for samples of PVC with or without a welded seam.

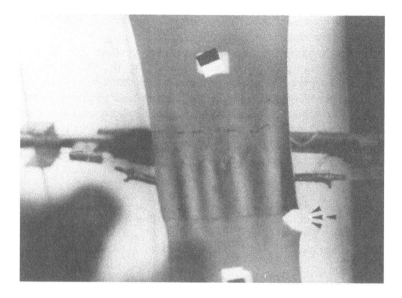

Figure 6. Initiation of the sample tearing near the seam, for a PVC e_7 sample.

Bituminous Geomembranes

Influence of specimen shape. It is interesting to compare the results obtained on the wide strip e_1 and the dumb-bell e_6 (Fig. 7). The difference between the $\alpha(\varepsilon)$ and $\alpha(\varepsilon_r)$ diagrams is indicative of bitumen flow and/or slipping of the material between the clamps. In addition, for a given value of α, the difference $(\varepsilon - \varepsilon_r)$ is found to be greater for e_1 than for e_6. This justified the use of the dumb-bell specimen for which less stress acts on the clamp jaws (tensile force distributed over a greater width).

Behaviour of specimens with seam welds. Here, consideration was given only to $\alpha(\varepsilon)$ diagrams (Fig. 8). The ε_r value for the seam weld specimen is no longer significant as failure is due to sliding of the weld seam (the length $h_s = 40$ mm is insufficient and smaller than lengths used on work sites). It is thus only normal that a much lower tensile strength is obtained for specimens with seam welds.

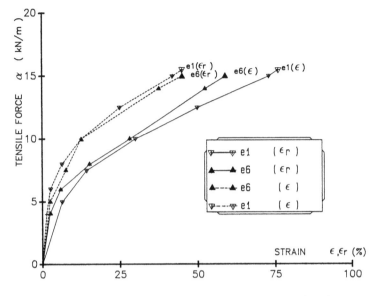

Figure 7. Tensile behaviour for different shapes of samples (bituminous geomembrane).

HDPE Geomembrane

Influence of specimen shape. HDPE geomembrane has a very special rheological behaviour:

- presence of a yield point for small strain levels. The yield point value would appear to be highly dependent on the type of specimen (see Table 1);
- material flows to give extremely high strains: this is illustrated in a series of photographs taken during a test on a type e_6 specimen (Figs 9 and 10).

The breaking strain is high (often greater than 50%). As a result, the initial specimen length h_0 must be limited in order to avoid exceeding the extension capacity of the machine (for this reason, the e_2 specimen will often not be allowed).

Moreover, the present study shows that only specimen e_6 enables the rise in force α to be obtained at the end of the test. This

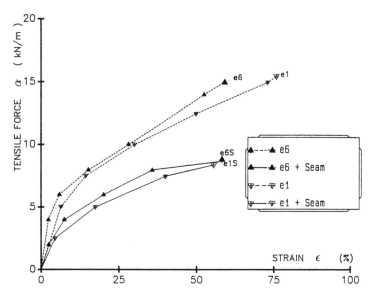

Figure 8. Tensile behaviour for samples of bituminous geomembrane with or without welded seams.

characteristic feature corresponds to the end of flow of material in the central zone (width h_c) and to the initiation of flow in the wider parts of the dumb-bell specimen, near the clamp jaws.

For the other dumb-bell specimen, e_7, the premature failure can be explained, as before, by the fact that a die was not available to cut it.

For strip specimens e_1 and e_2, failure occurs at the clamp jaws, at lower strain than for e_6.

TABLE 1. Yield Point Values of Different Parameters (α = Tensile Force, ε_r = Central strain, ε = Total Elongation) Related to the Type of Specimen (e_1, e_2, e_6 and e_7)

	e_1	e_2	e_6	e_7
α_{yp} (kN/m)	48	40	48	45
$(\varepsilon_r)_{yp}$ (%)	15	13·5	12	13
$(\varepsilon)_{yp}$ (%)	19	11	9·5	16

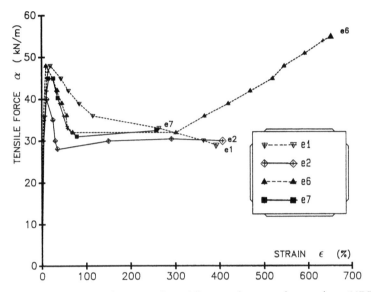

Figure 9. Tensile behaviour for different shapes of samples (HDPE geomembrane).

Behaviour of specimen with seam welds. Contrary to specimens e_1 and e_2 (wide strips), the dumb-bell specimens e_6 and e_7 do not fail in the vicinity of the seam weld but in the area where the specimens fan out (Fig. 11).

CONCLUSIONS

From the test results obtained, a dumb-bell shaped specimen is at present considered to be the best type of specimen. Owing to the absence of a cutting die for the e_7 specimen, a fair comparison between e_6 and e_7 was not possible. The dumb-bell has the advantage of exerting less stress on the clamps (less slipping) and restricting the strain to the central part (the HDPE case is exemplary in this respect).

By measuring strain ε_r in the middle of the specimen (necessary even for the strip specimens), equivalent results for both strip and

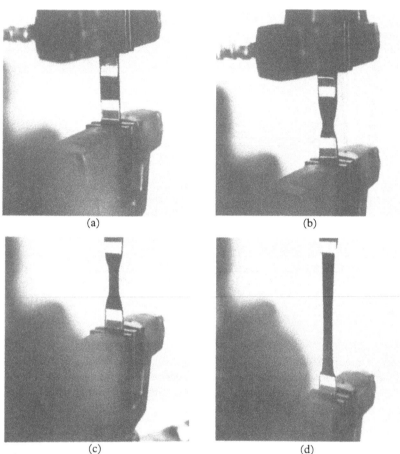

(a)   (b)

(c)   (d)

Figure 10. Tensile test on HDPE e_6 sample; flow stage by stage.

dumb-bell can be obtained for small strain levels. However, the optimum dimensions of the dumb-bell still have to be defined: h_0 sufficiently small in order to obtain long elongations, b_0 sufficiently large to allow the study of geomembranes reinforced with widely spaced wires, but sufficiently small to allow future studies in small temperature-controlled cells, for example.

With reference to the materials, the HDPE membrane was shown to have greater tensile strength than PVC and bituminous membranes).

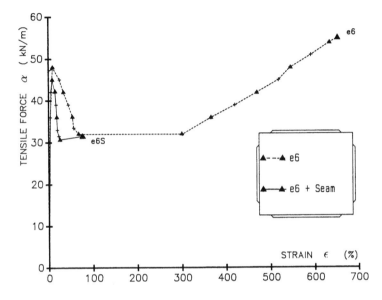

Figure 11. Tensile behaviour for samples of HDPE membrane with or without welded seams.

REFERENCES

Gourc, J. P., Leclercq, B., Benneton, J. P., Druon, M., Puig, J. & Delorme, F. (1986). Tensile strength behaviour of geomembranes. 3rd International Conference on Geotextiles, Vienna, Austria.

Gourc, J. P., Leclercq, B., Benneton, J. P., Druon, M. & Puig, J. (1987). Contribution à la détermination d'un essai de traction standardisé pour géomembranes. 1st International Congress RILEM, From Materials Science to Material Engineering, Paris, France.

Rollin, A. L. (1985). Testing of geomembranes. 2nd Canadian Symposium on Geotextiles and Geomembranes, Edmonton, Canada.

4.4   Assessment of HDPE Geomembrane Seams

ANDRÉ L. ROLLIN, ANA VIDOVIC & VICTOR CIUBOTARIU

Department of Chemical Engineering, Ecole Polytechnique de Montréal, CP 6079, Succursale A, Montréal, Canada H3C 3A7

INTRODUCTION

The quality of HDPE field seams depends on factors related to the geomembrane itself, welding parameters identified from the type of equipment and welding technique used and the field constraints such as climatic conditions (see Chapter 2.12) and project design. All of these factors may affect the seam strength and water tightness as well as the molecular structure of the plastic material.

Most of the welding equipment is designed to adequately seam the commercially-available geomembrane sheets under normal field conditions. The reliability of the seams' quality depends largely on the expertise of the trained operators, since proper equipment performance requires frequent calibration.

Significant progress in the improvement of field techniques is needed in order to achieve satisfactory quality assurance. Obviously, to achieve these goals, proper monitoring and control testing procedures on field seams have to be developed.

Various non-destructive test methods to assess geomembrane seams in the field are presently in use (see Chapter 4.2). They measure the continuity of a seam without measuring the relative strength of the bond.

Thus, to complement these non-destructive tests, destructive ones must be performed in the field and in the laboratory to allow proper calibration of welding machines or to quantify the bond strength of field seams. Usually these tests are done whenever a welding machine is restarted or a new set of conditions are encountered on a site.

Normally, tests will have been developed to evaluate the seam properties of HDPE geomembranes used in different applications. However, in order to assess the quality of work *in situ*, the bond strength of HDPE geomembranes is determined using two customary destructive test methods, the shear test and the peel test. Although results of these tests are used as acceptance criteria to evaluate the bond strength and consequently the tightness of seamed HDPE sheets, the data obtained from these tests should be used with care and, in many cases, must be supplemented by results obtained from other test methods like microscope analysis and/or impact tests.

OPTICAL MICROSCOPIC ANALYSIS

Using a microscopic technique, cross sections of seams can be analysed to observe molecular abnormalities, identify micro stress cracks within the bonded sheets, detect unbonded areas and observe slow crack growth phenomena. Recorded images can be captured using an optical microscope or/and a scanning electron microscope equipped with an appropriate camera.

Macro and micrographic analysis can be easily performed with an optical microscope. However, a detailed examination of fracture surfaces using optical micrography is practically impossible because of its limited resolution and depth of field. The analysis of rough surfaces can be accomplished only by a scanning electron microscope.

The examination of cross sections of welded HDPE geomembranes using an optical microscope can be performed on very thin slices like those used extensively in research laboratories (Peggs & Charron, 1989). To obviate the difficulties encountered with specimen preparation, optical fibres have been used to light up the surfaces of black polyethylene samples cut up using a sharp knife. The specimens, approximately 20 mm wide with a length equivalent to or greater than the weld itself, are directly installed under the microscope.

Using a minimum resolution of 100–200 nm and variable magnification up to 2000×, cross sections of seams can be observed and photographed to detect important defects within the seam that could not have been suspected following shear and peel tests.

In order to assess the quality of the HDPE geomembrane seams performed by an ultrasonic technique, the usual acceptance criteria consisting of shear and peel test results are used.

The information obtained from shear and peel test results are not sufficient to explain the specific behaviour of several welded geomembranes. The additional information required for these cases is acquired by optical microscopic analysis of the seam specimens and by the impact test carried out on 300 × 300 mm samples.

LABORATORY IMPACT TESTS

An impact test of a HDPE field seam can be performed with a drop-weight apparatus (Fig. 1), in order to produce proper immediate failure. The mean failure energy can be reported as the seam fail or pass upon application of an adequate drop-weight mean energy.

Apparatus developed at the Ecole Polytechnique of Montréal is a version of the one recommended for drainage pipe testing (Ami, 1987). It consists of a frame to which various pipes can be attached. Inside the pipe a test weight W is allowed to drop from various heights. Different test weights can be used for various sheet thicknesses, seam types or welding techniques.

The sample of a welded geomembrane is attached to a mandrel which is centred on the pipe axis. Special attention is given in aligning the welded strip equally spaced from the mandrel's borders. Before testing, the seamed samples are conditioned in a cool room at 0°C for a minimum of 24 h. The impact test itself is carried out 30 s (maximum) after taking samples outside the conditioning room.

The test weight used consists of a metal tup with a semi-spherical shape, the weight W varying between 2·8 and 5·5 kg. The tup is allowed to fall freely from the specified height and to impact the HDPE seam sample plane with its rounded extremity. The mandrel with a failed specimen of a HDPE seam obtained by the ultrasonic technique is shown in Fig. 2.

The tup should strike the specimen only once. The test result is determined by the energy necessary to produce failure of the welded geomembrane sample. The impact energy is varied until failure is obtained. With a constant height of 110 cm used in all impact tests,

Figure 1. Drop-weight impact test apparatus.

Figure 2. Mandrel and specimen fractured on impact test.

the impact energy varies between $30 \cdot 20$ J (for $W = 2 \cdot 8$ kg) and $59 \cdot 33$ J (for $W = 5 \cdot 5$ kg).

A test programme involving seamed HDPE geomembranes of various thicknesses has been performed. The programme, included impact tests and microscopic observations. The results were compared with shear and peel acceptance test results.

RELEVANT OPTICAL OBSERVATIONS AND IMPACT BEHAVIOUR OF HDPE SEAMS

All seams with abnormal behaviour with respect to usual acceptance criteria are in addition submitted to microscope analysis and to impact tests. The classification according to the test behaviour, as satisfactory or unsatisfactory, is made only on the basis of peel tests and visual examinations. Attempts to use the shear test results for classification purposes are unsuccessful because all results from this test are above the acceptance level.

The combined analysis of observations from the optical microscope and the results of the impact tests allows the improvement of the quality assessment of the seams. This has also made the identification of several behaviour types of seams obtained by ultrasonic technique possible.

Figure 3 shows three representative parts of a 1·0-mm geomembrane seam: the two edge parts and the middle part of the 25-mm width seam strip, as observed at 50× with the optical microscope. In order to identify the exact position of the photographs taken along the welded seam, a general location sketch is also shown on each of the following figures.

Careful examination of Fig. 3 indicates good adhesion of the extrudate to the geomembrane sheets on both edges of the seam which explains the high values of the shear (F_s) and peel (F_p) resistances $(F_s = 127\%$ and $F_p = 87\%)$.

The existence of some channels inside the seam could be the origin of the observed plastic deformations on both longitudinal borders of the welded strip following the impact test at a mean energy of 42·07 J.

A very comprehensive image of an unsatisfactory field seam is presented in Fig. 4. It shows a 0·75-mm geomembrane seam cross-section, seen with an optical microscope. Zones of melted polymer with punctual contacts and sharp angles, favouring tension concentrations, are observed at both edges of the seam strip. While the net section in the shear area is large enough to explain the shear factor of $F_s = 109\%$, the peel test has shown a reduced value of $F_p = 59\%$ and a complete fracture of the specimen was obtained on the impact test performed at an energy of 37·76 J.

A similar case is shown in Fig. 5 which shows both seam extremities of a 0·75-mm geomembrane. Very large unbonded areas and internal channels can be seen in the micrograph. The customary acceptance criteria for the shear test was satisfactory with a value $F_s = 109\%$; for the peel test it was unsatisfactory with a coefficient value $F_p = 56\%$. The specimen sustained plastic deformations on the borders of the welded strip at a low level of impact energy (30·2 J).

Figures 6 and 7 present the optical microscopic images obtained for a field seam of a 1·5-mm and a 1·0-mm thickness geomembrane respectively. The test results were large enough to obtain acceptable values using customary acceptance criteria: for the first one, F_s and F_p were 116% and 86% respectively and for the second one 123% and 81% respectively. No plastic deformation nor any kind of crack were discovered after impact tests performed with the 59·33 J and 48·54 J

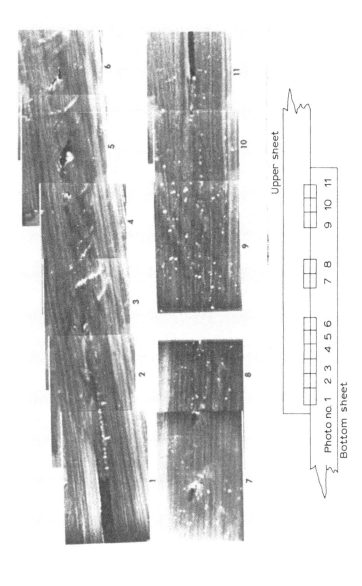

Figure 3. Photomicrograph of an HDPE welded seam (Specimen 26-F).

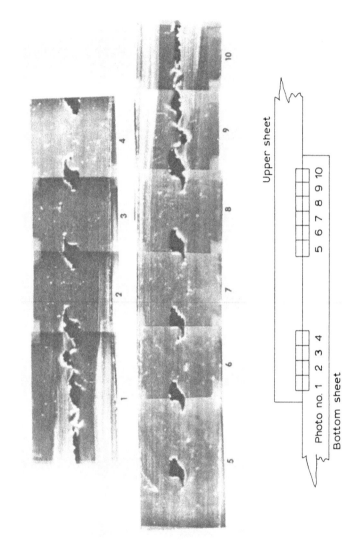

Figure 4. Photomicrograph of an HDPE welded seam (Specimen 24-D).

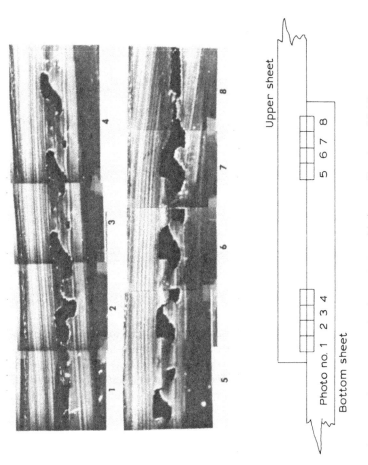

Figure 5. Photomicrograph of an HDPE welded seam (Specimen 33-M).

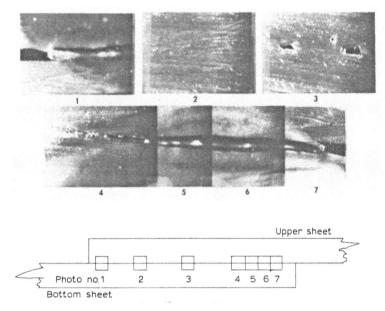

Figure 6. Photomicrograph of an HDPE welded beam (Specimen 15M).

respectively, although on the micrographs some minor flaws were observed.

Results from optical microscope analysis and impact tests provide answers to some practical aspects of the overall behaviour of geomembrane seams. The usual acceptance criteria tests must be used as standard comparative tests. However, several essential points presented below have been raised about the interpretation of these customary criteria.

SHEAR CRITERION

The shear and peel methods are modified versions of ASTM standards developed to measure the tensile and peel properties of very thin plastic-sheeting seams.

The shear test as recommended by the National Sanitation Foundation (NSF) as part of its standard test 54 on Flexible Membrane

Figure 7. Photomicrograph of an HDPE welded seam (Specimen 27-F).

Liners (revised 1985) is a modified version of the ASTM D-3083 test method covering Standard Specification for Flexible PVC Plastic Sheeting for Pond, Canal and Reservoir Lining of thicknesses between 0·2 and 0·76 mm. The shear tests simulate the stresses induced by mechanical and thermal contraction on a seam in service and the results are used to evaluate whether the seam fails before the liner itself. The NSF 54 defines minimum requirements for bonded seam strength as a coefficient value (F_s) between the tensile strength at yield of a welded specimen and the tensile strength at yield of the base material.

To evaluate the shear strength of HDPE seams performed by an ultrasonic technique, a strip specimen 1 inch (25·4 mm) wide and a length allowing a grip separation of 4 inches (101·6 mm) plus the width of the seam was used in accordance with NSF procedure. While the weld has to be centred between the clamps, the load is applied at a constant rate of elongation equal to 2 inches (50·8 mm) per minute. Concurrently with the test of a welded specimen, the tensile strength at yield of the base material has to be tested following the ASTM D-638 procedure (Standard Test Method for Tensile Properties of Plastics) on a standard dumb-bell shaped test specimen. Although the same rate of elongation is used by the two types of tests, the form and dimensions are different: the dumb-bell shaped specimen has an overall length of 115 mm, with an overall width of 19 mm, the length of the narrow section being 33 mm and the width of the narrow section being 6 mm.

The analysis of the stress conditions in the sheared area of the geomembrane shows that the failure by shear is usually possible when the thickness of the welded geomembrane is relatively important. This type of failure is obtained when the resistance in tension of the base material is larger than the total adhesion force that can be mobilized in the welded joint. It is important to point out that normal tension stresses which increase with the thickness of the geomembrane are mobilized at the same time as the shear stresses. The occurrence of normal stresses in a shear test suggests a failure mechanism closer to the one produced in the peel test.

Because the welded region is usually more rigid than the sheet itself, the large majority of failures occur outside the welded width of the specimen and its elongation is within the limits of the base material. Thus the NSF 54 shear test procedure yields in most cases not the shear strength of the bonded sheets but the tensile strength of the very base material measured on a invariable-width strip specimen. It is then very daring to compare tensile strength of specimens with different shapes and dimensions to assess the weld acceptability. It becomes evident that the acceptance requirement coefficient obtained by this procedure cannot be used as an adequate acceptance criterion for seam quality. This point of view has been clearly demonstrated by the experimental results obtained by Peggs and Rose (1987), Rollin (1987) and Rollin *et al.* (1989*a,b*).

PEEL CRITERION

To complement the shear resistance of a seam, a peel test is used to evaluate the adhesion strength between two welded geomembranes or between the extruded polymer and the sheets. The test is accomplished by submitting the specimen to a constant rate elongation of 2 inches (50·8 mm) per minute such that the interfaces are subjected to a peeling force that attempts to separate the adhered surfaces of the weld.

The testing procedure used has been recommended by the NSF 54 test which is a modified version of ASTM D-413 Standard Test Methods for Rubber Property—Adhesion to Flexible Substrate. The specimen strip type A, 1 inch (25·4 mm) wide, is subjected to 180° peel at a constant speed of deformation of 2 inches (50·8 mm) per min. The test results should include the behaviour of both edges of the seam.

The acceptance criterion proposed by the NSF is referred to as the Film Tearing Bond (FTB) criterion: the specimen must not fail before the welded interfaces separate. As pointed out by Peggs and Little (1985), the use of a peel strength acceptance requirement coefficient (F_p) for HDPE geomembrane seams, defined as the ratio of the peel resistance at break to the tensile strength at yield of the base material, is an improvement over the FTB criterion.

The required coefficient value usually recommended is such that the peel strength at break of a HDPE geomembrane seam exceeds 80% of the tensile strength of the base material. But it does define the necessity for a modified peel test to be performed as a sharp wedge penetration test.

Test results obtained (Peggs 1987; Rollin, 1988; Rollin *et al.*,

1989*a*) have indicated that the determination of seam resistance to peel forces by using the seam peel strength test evaluates seam quality more adequately than the shear test. Nevertheless, microscopic analysis of cross sections of welds that have passed both the shear and peel acceptance criteria have often been identified as having unbonded areas as well as changes in molecular structure of the base material.

CONCLUSIONS

The quality assessment of HDPE field seams cannot be satisfactorily made only on the basis of the usual criteria of shear and peel tests. Important improvements are necessary in order to predict the long-term behaviour of welded geomembranes.

Microscopic examination of the seams allows a better understanding of the shear and peel tests behaviour. However, the optical examination has to be complemented by mechanical tests which better simulate the field stress during installation and life service.

The impact test provides additional information on the behaviour of seams under dynamic stress conditions encountered mostly during the installation process.

REFERENCES

Ami, S. R. (1987). *Drainage Pipe Testing Manual.* Canadian International Development Agency, Hull.

Peggs, I. D. (1987). Evaluating polyethylene geomembrane seams. Proceedings of the Geosynthetics of '87 Conference, New Orleans, pp. 505–18.

Peggs, I. D. & Charron, R. M. (1989). Microtome sections for examining polyethylene geosynthetics microstructures and carbon black dispersion. Proceedings of the Geosynthetics '89 Conference, San Diego, CA, pp. 421–32.

Peggs, I. D. & Little, D. (1985). The effectiveness of peel and shear tests in evaluating HDPE geomembrane seams. Proceedings of the Second Canadian Symposium on Geotextile and Geomembranes, Edmonton, Canada, pp. 141–6.

Peggs, I. D. & Rose, S. (1987). Practical aspects of polyethylene geomembranes seam welding. Geotechnical Fabrics Report, pp. 12–16.

Rollin, A. (1988). Evaluating field polyethylene geomembrane seams. Proceedings RILEM Meeting of TC-103 MHG Committee, Montréal.

Rollin, A., Vidovic, A., Denis, R. & Marcotte, M. (1989*a*). Evaluation of HDPE geomembrane field welding techniques; need to improve reliability of quality seams. Proceedings of the Geosynthetics '89 Conference, San Diego, CA, pp. 443–55.

Rollin, A., Vidovic, A., Denis, R. & Marcotte, M. (1989*b*). Microscopic evaluation of HDPE geomembrane field welding techniques. Geosynthetics: Microstructure and Performance, ed. I. D. Peggs. ASTM Special Technical Publication 1076, Orlando, FL, pp. 34–47.

4.5 Laboratory-Ageing of Geomembranes in Municipal Landfill Leachates

OLIVIER ARTIÈRES, FRANÇOIS GOUSSÉ & ERIC PRIGENT

CEMAGREF, Division Ouvrages Hydrauliques et Voirie Parc de Tourvoie, BP 121, F92185 Antony Cedex, France

INTRODUCTION

The time period required to obtain complete stabilization of landfilled waste is very long and may even involve several centuries (Stieff, 1989). During such a long period, chemical compounds in the leachate may react with the liner material, modifying its mechanical and hydraulic properties. The durability of synthetic geomembranes is considered to be one of the most critical points in ensuring long-term landfill performance. This chapter describes a long-term laboratory study undertaken to evaluate the possible modifications of the geomembrane characteristics during contact with leachates.

METHODOLOGY OF THE LABORATORY-AGEING

The aim of the study described below is to simulate the chemical compatibility of a selection of geomembranes in leachates. The laboratory-ageing must reproduce the chemical environment of a geomembrane placed at the bottom of the landfill, i.e. chemical, biological and mechanical stresses, temperature, flow, etc.

The general principle of the experiment is to immerse samples of various geomembranes in two different municipal landfill leachates and in distilled water at 20°C and at 50°C, and then to test these samples periodically to follow the variation of their characteristics.

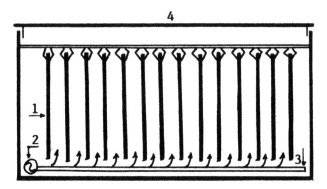

Figure 1. Diagram of the ageing tank. 1, Geomembrane sample; 2, immersed pump; 3, distribution pipe; 4, cover.

The size of the geomembrane samples is about 30 cm × 30 cm. The samples are submerged in polypropylene tanks (60 cm × 40 cm × 35 cm) filled with leachate or water to a depth of ~35 cm (Fig. 1). Each tank holds 17 samples of one type of geomembrane in one kind of fluid. There are approx. 21 litres of fluid per square metre of geomembrane area.

The local environment and stresses must be simulated as far as possible.

Flow and Dissolved Oxygen Rate

An immersed pump connected to a pipe distributes the fluid between the samples. This constant flow is necessary to renew the fluid at the sample surface avoiding mass transfer limitations of the chemical reactions. The flow is high enough to move the fluid along all the sample surfaces, but sufficiently low to avoid turbulence on the fluid surface, preventing its re-oxygenation. The anaerobic condition must be preserved to reproduce the real environment.

Temperature

A temperature of 50°C is the maximum value that may be reached in the leachate at the bottom of the landfill near the geomembrane. A value of 70°C has been reported elsewhere (Haxo & Haxo, 1988; Landreth, 1988).

However, 20°C can be considered as the average temperature at the bottom of a landfill when the degradation phase of the organic matter is almost finished or when the reaction core of the landfill is far from the geomembrane.

Knowledge of the ageing kinetics at these temperatures also allows the kinetics at intermediate or, if necessary, at higher temperatures to be extrapolated.

Light

The tank is protected from the light, thus avoiding the development of unsuitable bacteria or fungi, or acceleration of the ageing of the samples by photo-degradation.

Chemical Media

Leachates. It was decided that real leachate would be used instead of a synthetic leachate because of the wide range of compounds contained in a real leachate. To take the variability of leachates into account, the samples were immersed at room temperature (~20°C) in two leachates from different municipal landfills:

- the first leachate is sampled at the outlet of the leachate drain of a landfill with compacted domestic waste (Table 1);

TABLE 1. Physico-chemical Parameters of the Two Leachates from Municipal Landfills used in the Ageing Study (Prigent, 1990)

Chemical parameters in mg/litre	Leachate from compacted wastes	Leachate from crushed wastes
pH	6·93	7
Dissolved O_2	0·5	1·7
Redox potential (mV)	−70	−72
COD (mg O_2/litre)	12 700	1480
NH_4^+	1 200	66
PO_4^{3-}	0·8	1
Cl^-	14 960	328
Acetone	1·45	0·75
Butanone	0·053	0·025
Methylisobutylcetone	0·018	—
Hydrocarbon	<0·01	—

[a] Unless otherwise specified.

- the second leachate is sampled in the treatment pond of a landfill were the wastes are crushed, but not compacted (Table 1).

The concentration of dissolved oxygen in the two leachates is very low. The value is a little higher in the case of the crushed wastes because of the sampling in the treatment pond. But both conditions are anaerobic and it is of utmost importance that no air is introduced into the ageing tanks during the tests.

Reference medium. All the samples are also immersed at room temperature in distilled water, used as reference medium, in the same conditions as those immersed in the leachates.

The samples at 50°C are only immersed in the leachate from the compacted wastes landfill to simulate the extreme conditions.

All immersion media are usually renewed every 3 or 4 months. The physico-chemical properties (pH, conductivity, dissolved O_2) are checked every 2 weeks to follow their evolution.

Mechanical Stress

Chemical and mechanical stresses can combine to affect the polymer's service life. The combined stress on semicrystalline polymeric materials can cause stress cracks (Landreth, 1988). This phenomenon is now well known for polyethylene geomembranes and is one of the disadvantages of this material (Halse *et al.*, 1988; Peggs & Carlson, 1988). These failures often occur in the seam areas, due to residual stresses after seaming and also due to the overlapping geometry of the seams (Halse *et al.*, 1988). Cracks and strain due to residual stress increase the diffusion of chemicals in the polymer material leading to chain the breaks.

This acceleration of chemical damage due to synergy of stresses was also tested in this research programme. A device according to ISO 6252 'Determination of environmental stress cracking—Constant tensile force method' was therefore developed. Twelve test specimens of one liner material as described in Fig. 2 are loaded to a portion of their breaking stress by means of static loading. Then, four of them are immersed in crushed waste leachate, four in distilled water and four in air. The leachate and the samples are protected from light (Fig. 3).

The first tests were initially made on continuous samples and at

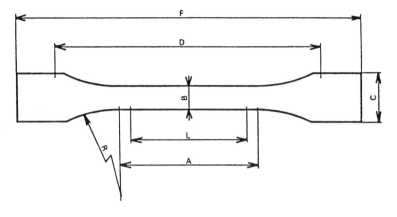

L	A	B	F	C	D	R
50 ± 0·5	60 ± 0·5	10 ± 0·5	150 min	20 ± 0·5	115 ± 3	60 min

Figure 2. Sample of ISO type for uniaxial tensile test. D = Initial distance between clamps; L = distance between the marks; A = constant width part. All the sizes in are in millimetres.

20°C. Further tests will involve seamed samples and higher temperatures.

SELECTED GEOMEMBRANES

One of the most frequently-used geomembranes is high-density polyethylene (HDPE). This polymer is commonly selected because of its chemical stability. But data about long-term ageing of other types of geomembranes in domestic waste landfill leachates are so few in the literature (Haxo & Nelson, 1984), that it was decided to test a representative range of the present international production of geomembranes.

Seven kinds of material constituting the geomembranes were selected (Table 2).

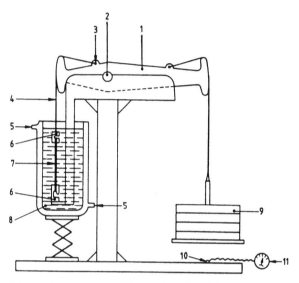

Figure 3. Schematic of stress cracking device after ISO 6252. 1, Beam; 2, pivot without friction or fulcrum; 3, anchor; 4, stainless steel cable; 5, thermoregulated fluid circulation; 6, clamps; 7, sample, 8, ageing fluid; 9, masses; 10, clock circuit breaker; 11, clock.

TABLE 2. Ageing Time and Media of the Selected Geomembranes (1 = since December 1989; 2 = stopped after 5 months; 3 = since December 1990; 4 = since November 1990)

Type of geo-membrane	Thickness (mm)	Room temperature (~20°C)			50°C
		Compacted waste leachate	Crushed waste leachate	Distilled water	Crushed waste leachate
Bituminous	3·9	(2)	(2)	(2)	
SBS/Bitumin	4	(1)	(1)	(1)	(4)
PVC/DOP	1·2	(1)	(1)	(1)	(4)
PVC/EVA	1·8				(3)
HDPE	2	(1)	(1)	(1)	(4)
co-PP	1·2	(3)	(3)	(3)	(3)
EPDM	1	(1)	(1)	(1)	(3)

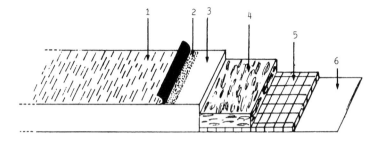

Figure 4. Structure of the bituminous geomembrane, 1, Non-stick film; 2, sanding; 3, asphalt impregnation; 4, bidim non-woven fabric; 5, glass fiber layer; 6, anti-perforation film.

Bituminous Geomembrane

The geomembrane is made up of a polyester non-woven geotextile and a glass voile arming the sheet, both impregnated with oxidized bitumen (Fig. 4). Bitumen is particularly sensitive to hydrocarbons, ether, xylene, benzene, etc. Polyester is sensitive to hydrolysis, especially over 50–70°C, but it has a good chemical stability to petrol and solvents (Rayne, 1990).

Modified Bituminous Geomembrane

The structure is the same as the previous one, but the impregnation is with bitumen modified with a copolymer of styrene–butadiene–styrene (SBS). SBS is a styrenic elastomer which gives some of its elastic properties to the mixture. SBS is sensitive to oxidation, hydrocarbons and solvents (Reyne, 1990).

Plasticized PVC Geomembranes

Polyvinylchloride (PVC) polymer has good stability to chemical compounds (oil, water, oxidizing agents) (Reyne, 1990). But it is a rigid material. A large proportion of plasticizers (between 20 and 50%) must be included to produce soft geomembranes. These plasticizers are very sensitive to oxidation (alcohols, hydrocarbons, etc.) because of their polarity and they migrate out of the material.

Other additives are then added to protect the plasticizers, i.e. anti-oxidizing and blocking agents (Reyne, 1990).

The type of plasticizers and additives is therefore very important for the long-term evolution of the geomembrane. Two PVC geomembranes were tested to evaluate the influence of the plasticizers:

- PVC plasticized with dioctylphtalate (DOP)
- PVC plasticized with a copolymer of ethylene and vinylacetate (EVA)

EVA, which belongs to the polyolefinic family (like PE or PP), is more resistant to oxidation than the phtalates.

Polyolefinic Geomembranes

Polyolefines come from the polymerization of ethylene and/or propylene. The resulting polymers have good resistance to chemical compounds because of the low chain ramifications and the crystalline structure, but they are sensitive to oxidation (especially UV) (Reyne, 1990). They contain very few additives ($<5\%$)—mainly anti-oxidizing agents.

Two kinds of polyolefinic geomembranes were tested:

- high-density polyethylene (HDPE)
- copolymer of ethylene and propylene (co-PP)

EPDM Geomembrane. EPDM is an elastomer terpolymer of ethylene–propylene–diene monomer. Like the polyolefines, EPDM is sensitive to oxidation, but also to aromatic solvents and chlorinated hydrocarbons. Some additives are used to protect it against oxidation.

TESTS

The selected tests intend to quantify and understand the ageing of the geomembranes over time. These tests may be mechanical, hydraulic or analytical.

Mechanical Tests

Uniaxial tensile test. The classic uniaxial tensile test is used to assess the mechanical values of the geomembranes, i.e. tensile strength and strain at break and/or yield points, secant tensile modulus at 10% strain. All these values are very common and easy to apply, even if the test does not describe in detail the real behaviour of the geomembrane once placed.

The sample size shown in Fig. 2 is of ISO type defined in the ISO/R527 standard. The thickness of the sample is that of the geomembrane.

The strain rate of the machine is 50%/min. Strain is measured with an optical extensometer following two lines previously drawn on the sample.

Biaxial tensile test. The principle of this test is to inflate the sample of geomembrane facing a circular opening by air pressure while clamps prevent it from shortening (Fig. 5). The air pressure is raised by steps of 10 kPa every 2 min and the geomembrane forms a spherical dome which grows in relation with the pressure up to bursting failure. The result of the test is a relationship between air pressure and increase of the dome.

The bursting test has the following advantages regarding the uniaxial tensile test:

- it generates a 2D tension which is very close to field conditions,
- it tests a larger area of material,

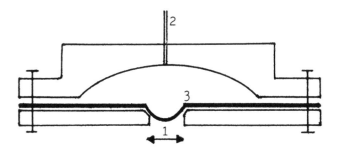

Figure 5. Bursting test apparatus. 1 = 10 cm; 2, air input; 3, geomembrane.

• it displays leaks in the geomembrane under strain (see below).

The two mechanical tests complement one another.

Hydraulic Test

The ageing process also has an affect on the impermeability of the geomembranes to leachates.

Mass transfer in geomembranes is due to the diffusion process. This mass transport is due to two driving forces: concentration and pressure gradients between the two sides of the sheet (Faure *et al.*, 1990). In landfill applications, the leachate level over the geomembrane is low (<1 m) and the pressure gradient is negligible. The diffusion due to concentration gradient is therefore the main cause of permeation of leachate through the geomembrane.

The sorption test is an easy method of quantifying diffusion by the concentration gradient. It may be performed with any kind of compound. Water is chosen as the medium to characterize the permeability of the aged geomembranes, because it is the main compound in the case of municipal landfill leachate.

The test involves monitoring with time the mass of absorbed water (Mt) in samples of geomembrane which are immersed in water until they reach mass stabilization (M_∞). A mathematical interpretation of diffusion process with the mass increase leads to the diffusivity (diffusion coefficient) in measuring the half-sorption time $t_{1/2}$ ($Mt_{1/2} = M_\infty/2$). This model is described in Faure *et al.* (1990). The relationship between diffusivity D (in m^2/s) and the half-sorption time $t_{1/2}$ (in s) is:

$$D = 0 \cdot 0492 \times T_g^2/t_{1/2}$$

where T_g is the thickness of the geomembrane in metres.

Analytical Test

All the above macroscopic tests are beneficial in assessing the general behaviour of the geomembrane after ageing, in the evaluation of the possible loss of characteristics and in comparing with the design values. However, these tests do not allow interpretation of the changes which happen at the molecular level.

An analytical test is useful to demonstrate the type of degradation and its extent. Many analytical tests are employed to characterize polymers (see, for instance, Van Langenhove, 1990; Verschoor *et al.*, 1990). Each of them describes one aspect of the polymer structure. But among these methods, the micro (Fourier transformation infrared) spectrophotometric technique applied to almost all the polymer materials tested, and gives the chemical and morphological changes in the polymer matrix and quantifies them. The basis of the method described in Jouan & Gardette (1987) involves sampling with a microtome small slices of geomembranes (with a thickness between 80 and 100 μm for PVC and HDPE; only 7 μm for EPDM which is very opaque because of its high black carbon content) normal to its surface and in analysing these films by IR light (Figs 6 and 7). The micro-(FTir) spectrophotometric measurement gives distribution profiles of the compounds which could appear during the ageing of the material in the thickness of the liner.

The opacity of the bituminous geomembranes is too high to allow micro-(FTir) spectrophotometric measurement. Photoacoustic

IR BEAM FOR micro-(F.T.i.r.) MEASUREMENT

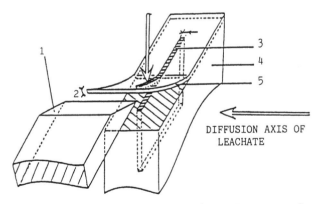

Figure 6. Cutting of a geomembrane slice to make an IR spectrum after CNEP report. 1, Microtome blade; 2, thickness of 80 or 7 μm; 3, geomembrane; 4, PE plates maintaining the sample; 5, slice cutting with the microtome blade.

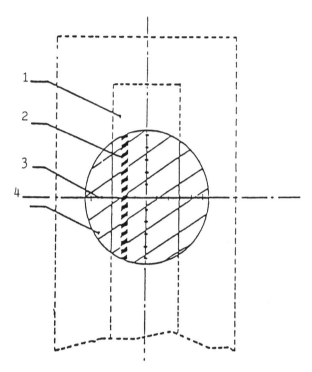

Figure 7. Micro-(FTir) spectrophotometric measurement on a microtomed slice after CNEP report. 1, Geomembrane; 2, measurement area (30 μm); 3, optical sighting mark; 4, IR beam.

spectroscopy is an alternative method which analyses surfaces of sheets with a high black carbon content.

These tests were performed at the Centre National d'Evaluation de Photoprotection (CNEP) in Aubières (France).

RESULTS

The following results describe the changes in geomembrane character-istics after ageing for about 16 months at 20°C and 3·5 months at 50°C.

Mechanical Changes

The uniaxial tensile tests showed no significant changes in the mechanical properties between unexposed and exposed geomembrane samples.

The bursting tests confirm this point. They are however a little more sensitive than the uniaxial tensile tests for the reasons explained on p. 402. The curves drawn in Fig. 8 indicate small variations between unexposed and aged samples for HDPE geomembranes. The latter are softer (lower modulus) and have greater pressure and elongation at break. For the bituminous/SBS geomembrane, the tendency

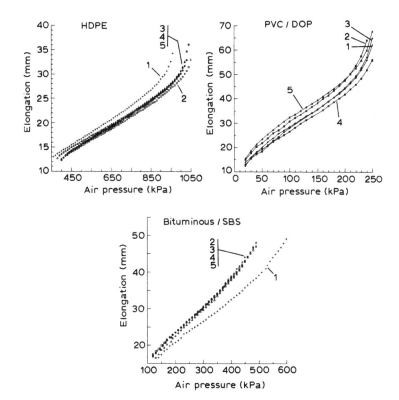

Figure 8. Bursting curves of some selected geomembranes. 1, Unexposed; 2, distilled water at 20°C; 3, crushed wastes leachate at 20°C; 4, compacted wastes leachate at 20°C; 5, compacted wastes leachate at 50°C.

is quite the reverse. The differences are very small for PVC and EPDM materials.

The bituminous geomembrane used was omitted from the initial liner selection, because the bursting test showed that this material was porous under little deformation (<6%) and low pressure (<200 kPa). It was assumed that such behaviour is inadequate for safe use in landfills.

For PVC/DOP and HDPE the uniaxial tests under constant stress showed neither failure nor differences between the three media after 3 months. Further tests must be conducted at 50°C to increase stress cracking, on samples with and without seams.

Finally, the most important result obtained from all these tests is that there is no noticeable difference between the exposed samples, i.e. there is no mechanical degradation due to leachate. These initial results agree with those presented in Haxo and Nelson (1984). In this study, low-density polyethylene, PVC, EPDM geomembranes were exposed to landfill leachate for 56 months at 10–20°C. There was only a slight change in the physical properties of the media.

Hydraulic Changes

Sorption tests were carried out on the HDPE and PVC/DOP samples after 16 months of ageing. Due to the long duration of the test (many months at 20°C), it is not possible to determine the diffusivity. But the comparison of the beginning of the curves for the PVC shows an acceleration of diffusion, which is higher for samples exposed to distilled water (Fig. 9). The differences are, however, very small.

Because of the very low diffusivity of HDPE, no changes were noticeable yet.

Changes at Molecular Level

These tests are more sensitive than the macroscopic tests.

The HDPE matrix shows no sign of oxidation, even in the superficial layer (0–24 μm), at 20°C or at 50°C. The HDPE geomembrane contains an anti-oxidizing agent containing an ester (maximum at 1736 cm^{-1}). This ester is hydrolysed producing OH groups in the hydroxyle zone (3100–3500 cm^{-1}).

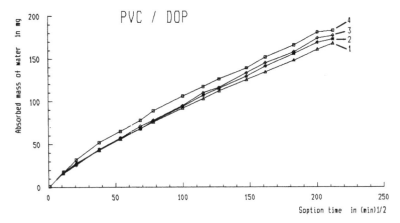

Figure 9. Absorption of water in PVC/DOP samples. 1 = Unexposed; 2 = compacted wastes leachate; 3 = crushed wastes leachate; 4 = distilled water.

The hydrolysis is very small at 20°C in distilled water, and still smaller with the compacted waste leachate at 20°C (<5%) (Fig. 10). It is located in the superficial zones (0–25 μm).

This reaction is higher with leachate at 50°C, where the ester is quite consumed in the first 30 μm but remains intact after 100 μm (Fig. 11). This last result conforms to the fact that hydrolysis increases with temperature.

Like HDPE, the PVC matrix is oxidized in any environment. On the other hand, the plasticizer (extremum at 1580 and 1600 cm^{-1} on Fig. 12) is hydrolyzed producing an acid of phtalic type and alcohols (range from 3100 to 3500 cm^{-1}) (Fig. 13). At 20°C, this reaction is greater in leachate than in distilled water, because hydrolysis is higher when the pH is other than 7, and faster at 50°C. In all cases, the loss of plasticizer is limited to the first 80 μm.

After 16 months exposure to leachate at 20°C, the EPDM matrix also shows no oxidation. On the other hand, there is a small oxidation in the bituminous/SBS samples which is faster at 50°C, due to the thermal evolution of SBS elastomer (Fig. 14). The water absorption in these two materials is high.

As a conclusion of these analytical tests, it was found that there is no significant oxidation of the geomembranes, whatever the environmental conditions. Hydrolysis of the additives occurs but is very limited in

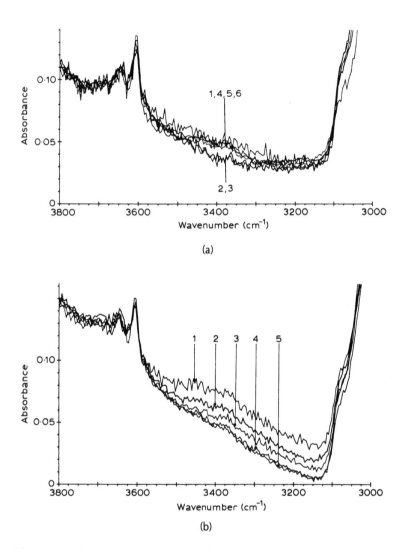

Figure 10. IR spectrum of the hydroxile zone for HDPE exposed during 16 months at 20°C: (a) compacted wastes leachate; (b) distilled water. $1 = 0/24 \, \mu m$; $2 = 24/48 \, \mu m$; $3 = 48/72 \, \mu m$; $4 = 72/96 \, \mu m$; $5 = 96/120 \, \mu m$; $6 = $ centre.

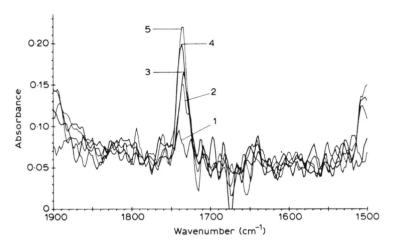

Figure 11. IR spectrum of the ester function of the anti-oxidizing agent for HDPE exposed during 3·5 months at 50°C in compacted wastes leachate. $1 = 0/28\,\mu$m; $2 = 28/56\,\mu$m; $3 = 56/84\,\mu$m; $4 = 84/119\,\mu$m; $5 = 119/185\,\mu$m.

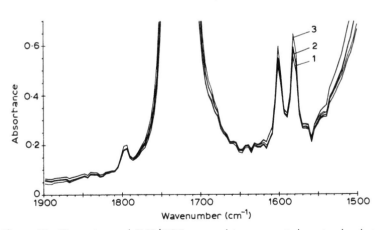

Figure 12. IR spectrum of PVC/DOP exposed to compacted wastes leachate during 16 months at 20°C. $1 = 0/24\,\mu$m; $2 = 24/48\,\mu$m; $3 = $ center.

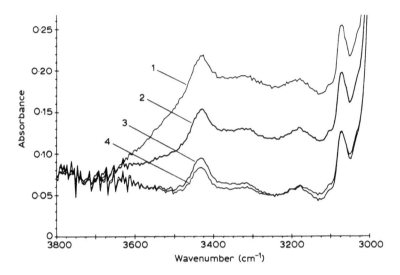

Figure 13. IR spectrum of hydroxile zone for PVC/DOP exposed to compacted wastes leachate during 16 months at 20°C. $1 = 0/24 \mu$m; $2 = 24/48 \mu$m; $3 = 48/72 \mu$m; $4 =$ centre.

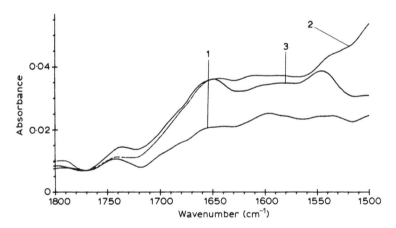

Figure 14. IR spectrum of the surface of bituminous/SBS samples. $1 = 16$ months exposure to distilled water at 20°C; $2 = 16$ months exposure to compacted wastes leachate at 20°C; $3 = 3 \cdot 5$ months exposure to compacted wastes leachate at 50°C.

quantity and thickness. All the facts observed are of minor importance.

CONCLUSIONS

With the increasing use of geomembranes in domestic waste landfills, the question of their chemical compatibility to leachate and their long-term durability becomes crucial.

Many experiments and tests have been undertaken to answer this question in the case of hazardous waste landfills (for instance the US EPA 9090 method). But the nature of domestic waste leachates, although they are of more complex composition, is certainly less aggressive towards geosynthetics than hazardous waste leachate.

The main selection criterion for a geomembrane must not be its chemical stability to leachate. Its long-term mechanical behavior, its suitability for laying and seaming, and its compatibility with the other elements of the impermeability system, are also of paramount importance and must be taken into account. It is therefore necessary to accord the chemical stability level of the materials sufficient but not excessive importance to ensure long durability, and to keep a sufficient material range to allow for choice.

The study described aims to provide some information on this point. The first results after an exposure of 16 months at room temperature and 3·5 months at 50°C show that all the tested geomembranes keep their initial characteristics. But this ageing time is too short to give any conclusion yet. The kinetics of ageing must be calculated over a longer period to assess the durability of the liner, when degradation, if it occurs, is greater.

It was shown that the ageing study is based on good definition and simulation of the environmental stresses and on judiciously selected tests. The macroscopic tests are used to compare the geomembrane characteristics to reference design criteria. The microscopic tests aim to describe the ageing processes to be interpreted and quantified, and to assess their durability.

REFERENCES

Faure, Y. H., Pierson, P., Artières, O. & Goussé, F. (1990). Tests of geomembranes water permeability. In *Proceedings of the 4th International*

Conference on Geotextiles, Geomembranes and Related Products, ed. G. den Hoedt. A. A. Balkema, Rotterdam, pp. 543–53.

Halse, Y. H., Koerner, R. M. & Lord, A. E. (1988). Laboratory evaluation of stress cracking in HDPE geomembrane seams. Proceedings of a Seminar on Durability and Ageing of Geosynthetics. GRI, Drexel University, USA.

Haxo, H. E. & Haxo, P. D. (1988). Environmental conditions encountered by geosynthetics in waste containment applications. In Proc. Seminar on Durability and Ageing of Geosynthetics, ed. K. M. Koerner. Geosynthetic Research Institute, Philadelphia, PA.

Haxo, H. E. & Nelson, N. A. (1984). Factors in the durability of polymeric membrane liners. In Proceedings of the International Conference on Geomembranes, Denver, ed. Industrial Fabric Association International. St. Paul, Minnesota, pp. 287–92.

Jouan, X. & Gardette, J. L. (1987). Development of micro(FTi.r.) spectrophotometric method for characterization of heterogeneities in polymer films. Polymer Communications, 28 (December), 329–31.

Landreth, R. E. (1988). Durability of geosynthetics in waste management facilities: needed research. Proceedings of a Seminar on Durability and Ageing of Geosynthetics. GRI, Drexel University, USA.

Peggs, I. D. & Carlson, D. S. (1988). Stress cracking of polyethylene geomembranes: field experiences. Proceedings of a Seminar on Durability and Ageing of Geosynthetics. GRI, Drexel University, USA.

Prigent, E. (1990). Contribution to the study of the chemical compatibility of geomembranes to landfill leachates. CEMAGREF. Mémoire de 3ème année de l'ENITRTS. Strasbourg 133 pp (in French).

Reyne, M. (1990). Plastics—application and transformations. Treaty of the New Technologies: Materials. Paris, 268 pp.

Stief, K. (1989). Technic of landfills in revolution. In Modern Technic of Landfills II. Report on Waste Management, Erich Schmidt Verlag, Bielefeld, pp. 7–31 (in German).

Van Langenhove, L. (1990). Conclusions of an extensive BRITE-research programme on ageing. In Proceedings of the 4th International Conference on Geotextiles, Geomembranes and Related Products, ed. G. den Hoedt. A. A. Balkema, Rotterdam, pp. 703–07.

Verschoor, K. L., White, D. F. & Thomas, R. W. (1990). An overview of practices used in the United States to determine the compatibility of geosynthetics with chemical wastes. In Proceedings of the 4th International Conference on Geotextiles, Geomembranes and Related Products, ed. G. den Hoedt. A. A. Balkema, Rotterdam, pp. 715–18.

4.6 Testing Program to Assure the Durability of Geomembranes

IAN D. PEGGS

*I-Corp International, 5920 North Ocean Boulevard,
Ocean Ridge, Florida 33435, USA*

INTRODUCTION

It is becoming standard practice to require independent construction quality assurance (CQA) activities during the installation of the geosynthetic components of landfill lining systems. Such activities require the preparation of comprehensive CQA plan and CQA specification documents to ensure that appropriate procedures and test protocols are followed, and that test results fall within acceptable ranges. Much has been written about the contents of these documents (Fluet, 1985; Giroud & Fluet, 1986) and no attempts will be made to expand on the nature of these documents. However, as practical experience is gained with the performance of geosynthetic systems, the types of testing required as part of the CQA program, and additional testing to support the CQA program, change. Until recently many of the tests required in CQA programs have simply ensured that the geomembrane is installed without apparent defects at the time of installation. Little thought has been given to those features of the liner that, although not a problem at the time of installation, could be used to help assure the durability of the geomembrane.

There are three basic components to a geomembrane CQA program (Giroud & Peggs, 1990):

- *Conformation verification*: verification that the geomembrane delivered to the site meets the specifications.

- *Integrity verification*: verification that the geomembrane is installed according to the design and installation specifications.
- *Survivability verification*: verification that the installed geomembrane has satisfactorily survived the rigors of installation and will perform as intended, i.e. it will be durable.

The following sections review the tests that have been found to be appropriate in the pre-construction and construction stages to assure the durability of the geomembrane.

The emphasis will be placed on high-density polyethylene (HDPE) geomembrane due to its predominant use in the USA, its developing use in Europe, and the fact that it is not the simple material it at first appears to be. The same philosophical approach can be used for other types of geomembrane, but the tests that relate to the stress cracking susceptibility of HDPE will not be necessary for most other materials.

PRE-CONSTRUCTION STAGE

Prior to constructing a landfill it is, or should be, necessary to show that the proposed geomembrane is capable of performing as intended and is capable of containing the specific landfill leachate over the required operating and post-closure periods. The geomembrane, therefore, requires both chemical compatibility with the leachate and mechanical durability.

Chemical Compatibility

The demonstration of leachate containment is typically achieved by performing a chemical compatibility test such as the US Environmental Protection Agency (EPA) Method 9090 'Compatibility Test for Wastes and Membrane Liners' (USEPA, 1986).

In EPA Method 9090 the geomembrane is exposed to the landfill leachate, or, in the case of a new landfill, a synthetic mixture that approximates the expected leachate, at 23 and 50°C for 120 days. Samples of geomembrane are removed from the leachate every 30 days and are measured to determine changes in the following properties:

- mass
- specific gravity

- hardness
- tensile strength
- tensile elongation
- modulus of elasticity
- tear resistance
- puncture resistance
- burst strength
- volatile content
- extractables content.

Note: Extra care should be taken when measuring elastic modulus since no polymers have linear elastic regions in the stress–strain curve. The slope of the steepest segment of the stress–strain curve should be measured.

The results are then analyzed by an expert computer program, FLEX (USEPA, 1990), to determine whether or not the geomembrane is considered to be compatible with the leachate.

These monitoring tests are relative to the long-term mechanical performance of the geomembrane and it is apparent to most engineers and regulators that the material is degrading if changes in these properties occur. However, there are a number of other appropriate materials properties that should be monitored to assess the compatibility of the geomembrane. Also omitted from the test program are tests that will assess the synergistic effects of stress and chemical environment. This synergism is particularly important for semi-crystalline HDPE because it is susceptible to environmental stress cracking. Because of potential extraction of plasticizers, PVC is also susceptible to embrittlement and stress cracking (not environmental stress cracking).

The USEPA does recommend that bent strip environmental stress cracking (ESCR) tests be performed on HDPE in the leachate, and that seamed samples be exposed to the leachate. It is essential that seam samples be exposed to the leachate and tested in peel and shear. However, the ESCR test will not provide useful information, as explained later.

All of the EPA Method 9090 monitoring tests reflect the bulk properties of the material and, except perhaps for break strength and elongation, are not responsive to surface changes. When exposed to the leachate the surface of the geomembrane will be the first to degrade, but this surface degradation may not be reflected in the bulk property changes since the surface layer is such a small percentage of the overall volume of the material. Many of the fracture processes that occur in materials are initiated at surfaces exposed to the environment; therefore, it is essential that, if material degradation is to be monitored, appropriate tests that evaluate surface changes be used.

The protocol for the EPA Method 9090 test requires that a large sample of geomembrane be exposed to the leachate and that, after exposure, specimens for the monitoring tests be die-stamped out of the sample. Surface degradation effects, at least on unsupported (unreinforced) geomembranes, may be better monitored if at least the tear specimens are pre-cut. The tear initiation characteristics of the ASTM D1004(C) specimens may then reflect surface degradation since the surface at which the tear commences would have been exposed to the leachate. As long as the tear specimens are removed from larger exposed samples the tear initiation will occur in unexposed bulk material, and the test results will not reflect the surface changes.

The inclusion of the environmental stress cracking test is a step in the right direction but this is not a severe test to assess stress cracking susceptibility since, with time, and at elevated temperatures, stress relaxation of the specimen will occur. The HDPE geomembranes presently manufactured rarely fail a bent strip stress cracking test. A more appropriate stress cracking test is the single edge notched, constant tensile load (CTL) test (Geosynthetic Research Institute, 1990; Peggs & Carlson, 1990) originally developed within the HDPE gas pipe industry. This test is described later.

When monitoring degradation, care should be taken not to discard an otherwise good geomembrane simply because it is seen to 'degrade'. It may well degrade, but to a level beyond which no further changes occur and at a level that still far exceeds the minimum level for that geomembrane to provide adequate containment.

Fingerprinting Techniques

It is recommended, when the EPA Method 9090 test is performed, that the geomembrane be fingerprinted by methods such as Differential Scanning Calorimetry (DSC), Thermal Gravimetric Analysis (TGA), and Infrared Spectrophotometry (IR) to define parameters and characteristics such as:

- DSC: melting temperature and range, crystallinity, oxidative induction time (OIT)
- TGA: composition
- IR: chemical structural groups

The geomembrane that is later delivered to the site for construction of the lining system will also be fingerprinted to ensure that it is the 'same' as the material that was tested and approved for construction. The difficulty, of course, is to define how, similar the fingerptints must be for the materials to be considered the 'same', and whether any differences indicate a gain or loss in appropriate properties. The experts that are capable of making such decisions are few in the core geosynthetics industry. These topics were discussed at a USEPA fingerprinting workshop in August 1991 (USEPA, 1991).

While the use of analytical methods for fingerprinting is still being debated, there is no question that the same analytical methods can be used to provide useful information on the degradation of geomembranes when exposed to leachates. The thermal analytical methods require smaller specimens, and, therefore, smaller volumes of leachates. Shorter exposure times are required to observe significant changes, and specimens can easily be cut and tested from surface layers, or from within the body of the geomembrane for more detailed and meaningful analyses of degradation processes.

For instance, it is practical to expose thin (microtomed) sections of geomembrane approximately 20 μm thick, or 50 μm on each side of a cube, to small volumes of leachate for 30 days, at 25 and 50°C and to effectively determine whether the geomembrane is compatible with the leachate. The analytical tests should be supported with density, tensile break strength and elongation, and tear initiation measurements on pre-cut specimens. In the tear test the tear strength should be recorded at the first peak on the curve representing the first surface tear in the specimen.

The above tests should be further supported with a CTL stress cracking test in the leachate.

Mechanical Durability (Stress Cracking Resistance)

Prior to the performance of a chemical compatibility test a predetermination of the one or two most mechanically durable HDPE geomembranes that will be subjected to the test should be made. This will ensure that not only is the geomembrane chemically durable in the leachate but also that the geomembrane is mechanically durable. It has been established by those working in the gas pipe industry, and a number of people working in geosynthetics, that HDPE pipe and geomembranes are susceptible to stress cracking: a brittle cracking phenomenon that can occur when the material is stressed at a constant load significantly lower than its yield stress. It can be envisioned as a monotonic fatigue process. A number of stress cracking failures in HDPE geomembranes have been observed (Peggs & Carlson, 1989) but all have occurred in uncovered liquid impoundment liners or landfill caps. There has been no evidence of stress cracking in landfill bottom liners.

All semicrystalline HDPEs are susceptible to stress cracking. Surveys (Peggs *et al.*, 1991) of the different geomembranes have shown that resistances to stress cracking can vary by factors up to 150. For maximum durability of a landfill liner, it is recommended that CTL stress cracking tests be performed on a number of geomembranes to determine those with maximum resistance to stress cracking. Based on this information a more quantitative estimate of the actual durability of the lining system may be generated using the following relationship (Hessel, 1990):

Actual geomembrane stress < Limit of tolerable stress

or:

$$(\sigma_{RW} f_{cm} \alpha \beta) < (K f_{CR\sigma} f_s)$$

where

σ_{RW} = tensile stress due to uniaxial relaxation in water,
f_{cm} = effect of waste expressed as a Chemical Relaxation factor,
α = excess tensile stress at different strain rates,
β = conversion factor to allow for field biaxial stress compared to laboratory uniaxial stress,
K = creep strength in water (determined from CTL test),
$f_{CR\sigma}$ = effect of waste expressed as a chemical Resistance Factor,
f_s = effect of seams expressed as a Long Term Welding Factor.

With a geomembrane resistant to stress cracking the lining system will be more tolerant of the seam notch geometries, excessive grinding during seaming, overheated seams, contraction stresses, folded wrinkles, and protruding stones that unavoidably occur, despite careful CQA, during installation.

The stress cracking test can be performed in a number of ways at different stages of the project. Initially a complete stress rupture curve should be generated in order to define the ductile–brittle transition time and stress for each selected geomembrane. This will allow the selection of the better material(s) to be subjected to the chemical compatibility test. When the final selection of a geomembrane has been made, conformance index (CTL) tests could be subsequently performed (instead of the bent strip environmental stress cracking tests) at a stress just below the ductile–brittle transition stress. The time to failure should be measured to ensure that it matches that given by the complete stress rupture curve. Unfortunately, with stress crack resistant resins, the time to failure may exceed several hundred hours. There are two possible ways, but only one practical way, to shorten the test time. The practical way is to perform the CTL test at a stress equal to approximately 40% of the room temperature yield stress for 50 to 100 h then to remove the specimen, prepare a thin microsection (Peggs *et al.*, 1991) and to compare the notch extension with that previously measured in the known reference geomembrane. This effectively compares the crack growth rates in the two materials.

The alternative method for accelerating stress cracking tests is to perform the test under a cyclic load rather than a constant load. Such procedures have been developed (Moet *et al.*, 1989) within the gas pipe industry but have not yet been adapted to geomembranes. With this technique, stress rupture curves will be generated within 2–3 days rather than over several weeks as presently required for the CTL tests on the better HDPE geomembrane resins.

It is essential to recognize that all HDPE geomembranes are not the same; some have far superior resistance to stress cracking than others. Resistance to stress cracking is the primary determinant of HDPE geomembrane durability.

Thermal Expansion

The Coefficient of Linear Thermal Expansion (CTE) should be measured over the full range of temperatures to be experienced by the geomembrane. These temperatures, under the sun, will be much higher than the ambient temperatures, and could reach more than

75°C. The CTE increases significantly with temperature and is generally higher (Giroud & Peggs, 1990) than the often quoted value of $1 \cdot 2 \times 10^{-4}/°C$.

The CTE is required in order to calculate the amount of slackness that needs to be built in to a liner during installation so that, at the lower temperatures (usually) of covering, there are no wrinkles in the liner that could be folded over during soil placement. If the liner is to remain uncovered for a long time the amount of slackness will need to be of an amount that the liner will not trampoline or bridge corners at the lowest temperatures of exposure prior to covering. Alternatively, the converse occasionally, but infrequently, occurs: the liner is laid at a lower temperature than that at which it is covered. To minimize the potential for large foldable wrinkles developing at the time of covering, the installation may need to be done in such a way as to limit the generation and distribution of wrinkles to small discrete areas. The development of wrinkles in a geomembrane is discussed by Giroud and Morel (1992).

Seaming

Pre-qualification seams should be made and tested by conventional peel and shear testing methods according to the CQA specifications. In addition, thin slice microsections, cut from across the seam, should be examined under a transmitted light microscope to detect the presence of residual stresses and crazes within, and adjacent to, the seam. Particular attention should be paid to the edge of the seam on the top of the bottom sheet and to the top (bevelled) edge of the top sheet in fillet extrusion seams. Crazes are, potentially, the precursors of stress cracks, and residual stress within the seam may compromise the durability of the seam and therefore the liner.

CONSTRUCTION STAGE

Conformance Testing

When the optimum geomembrane has been selected for mechanical and chemical durability and has been ordered for delivery to the landfill site it is necessary, according to most CQA plans, to review the manufacturers Quality Control (QC) documents for the specific rolls delivered to the site, and to perform a number of conformance tests (approximately every $10\,000\,m^2$) to ensure that the material meets project specifications. This is typically done by removing a full roll width sample from selected rolls delivered to the site and sending the sample to an independent testing laboratory. Unfortunately, if samples do not meet specifications it is time consuming for the owner, and expensive for the manufacturer, to replace the rejected material with acceptable material. It is more economical and more efficient to perform conformance tests before the material leaves the manufacturing plant. Barring a transportation accident, the material will then be known to be acceptable when it arrives on site and it can be immediately installed.

The functions performed by a CQA monitor in the plant should be as follows:

- Visually monitor production of rolls destined for the site.
- Monitor laboratory QC testing and sign the QC test reports to confirm that tests have been done according to appropriate standards and acceptable results have been obtained. If the monitor observes this procedure it will be unnecessary to perform some conformance tests.
- Remove samples and submit them to the independent laboratory for appropriate conformance tests.
- Monitor the loading of rolls for transport to the site. Rolls will only leave the plant when all QC and conformance test results have been received and approved.

The plant monitor must be familiar with laboratory testing procedures and the standards used.

A number of CQA plans and national standards require that the bent strip environmental stress cracking test and low temperature brittleness test be performed. At the present level of HDPE geomembrane technology it is not necessary to perform these tests since it is unlikely that any geomembrane will fail them. As previously discussed, the CTL test should be substituted for the bent strip ESCR test. If an HDPE geomembrane is to be left exposed to UV radiation for more than three months it must be confirmed that it contains between 2 and 3% of carbon black and, most importantly, that it is uniformly and finely dispersed. Therefore, the Carbon Black Dispersion should be assessed on microsections cut directly from the geomembrane. The geomembrane should not be heated, melted, and pressed to a thin section since this procedure may satisfactorily disperse agglomerates that were unacceptable in the geomembrane.

The tests that should be included in conformance testing programs are as follows:

- Density or specific gravity
- Melt Index
- Thickness
- Tensile strength/elongation at yield and break
- Tear initiation resistance
- Puncture resistance
- Carbon black content
- Carbon black dispersion
- CTL stress cracking resistance

CQA Plan and Specifications

During the installation of the geomembrane there are a number of items that should be added to the typical CQA plan, most of which are related to minimizing the potential for stress cracking in HDPE. Naturally, these items are not required for geomembranes fabricated from very low density polyethylene, PVC, polypropylene, and polypropylene/EPDM alloys that are not susceptible to stress cracking.

Slackness. The amount of slackness that is built into the geomembrane needs to be carefully monitored for the reasons described above. This is particularly important if any of the geotextile-encapsulated bentonite, or geomembrane/bentonite sandwich, composites are being used under geomembranes, since their long-term effectiveness depends on there being intimate contact between the underside of the geomembrane and the bentonite. Folds, wrinkles, and creases should not exist in the geomembranes so that the volume of hydrated bentonite under the liner is minimized if leaks in the liner occur where there is not intimate contact between the two components.

Grinding. Grinding that is required to prepare surfaces for extrusion seaming should be oriented normal to the direction of the seam, and grinding gouges, and any surface defects, should not exceed 10% of the thickness of the geomembrane in depth. The stress cracking resistance of the geomembrane area adjacent to the seam will, thereby, be minimally affected.

Seam repairs. All the stress cracking failures examined by the author have been initiated at overheated seams or (the majority) at seams that have been repaired by reseaming. It is, therefore, advised that repairing seams by extrusion reseaming not be permitted unless it can be shown that such procedures do not unduly reduce the stress cracking resistance of the geomembrane.

Peel testing. It is necessary, when double hot wedge seams are tested, that both tracks be tested in peel and that both tracks must meet the acceptance criteria. If the outer track (the one closer to the edge of the top sheet) is not tested it is necessary to take special precautions at intersecting seams to eliminate the potential for leakage along the edge of the bottom seam. If the double track seam has a free flap of material at the edge of the top sheet it is common practice to cut the flap back to the edge of the bonded outer track on the bottom seam to prevent leakage through the seam under the free flap. If the outer track is not tested in peel (even though the outer track may be placed right on the edge of the top sheet) the CQA staff cannot be assured that the outer track will not behave as a free flap. Therefore, at the point of seam intersection, the outer track and center annulus of the bottom seam should be cut back to the edge of the inner track. Such a procedure will be required at all patches placed on top of seams. An easier alternative would be to place the patch on the underside of the geomembrane and to extrude a fillet seam around the edge of the hole in the geomembrane. Such a patch can be placed if the hole in the geomembrane is a rectangle. A larger patch can be placed through the hole and under the geomembrane (minor axis of patch through major axis of hole) then turned through 90° to fill the hole in the geomembrane.

Vacuum box design. With the advent of textured geomembrane to increase friction angles it becomes extremely important to monitor vacuum box testing. It is more difficult to keep the surface of textured sheet clean, and there is a tendency for grinding debris to remain attached to the surface, thereby producing a seam with many small air voids on the interface. Such a seam will not produce a large soap bubble in the vacuum test but may produce a fine froth of very small bubbles. Depending on the pressure placed on the seam this froth may appear at different locations during different tests on the same area. It is therefore essential that sufficient time (15 s) be allowed for each vacuum test and that the viewing glass be clean enough that a fine froth can be defined.

Destructive Seam Testing The CQA Plan will probably require that destructive seam samples be removed from the installed geomembrane (approximately ever 150 m of seam) and sent to an independent testing laboratory for peel and shear testing. A good laboratory should be able to provide a printed report by telefax to the site shortly after noon on the day that samples are received. With Laboratory Information Management Systems such reports should include comparisons of the test results with specifications, and statements of acceptance or rejection of individual test specimens and the complete sample. Note that the CQA Specifications should provide acceptance criteria for the individual peel and shear specimens and for the sample as a whole, i.e. how many peel and/or shear specimens can fail before the complete seam sample is rejected.

Conventional seam specifications, usually based on National Sanitation Foundation (NSF) International Standard 54 'Flexible Membrane Liners' (NSF, 1991), require that seams meet minimum peel and shear strength criteria. Such criteria are adequate to demonstrate that the seam has satisfactory strength at the time of installation but indicate nothing about the durability of the seam. The same tests that are performed to generate strength data can also be used to generate data that assist in providing assurance that the seam also has resistance to stress cracking. Such procedures have been described elsewhere (Peggs, 1987; Giroud & Peggs, 1990) and include monitoring the ductility of the geomembrane adjacent to the seam in the shear test and the amount of separation during the peel test. The conventional criteria that, if the shear specimen meets a minimum strength value and fails in the geomembrane outside the edges of the seam it is acceptable, are not adequate. If the geomembrane fails in a brittle manner in the heat-affected zone of the seam, or at grinding gouges next to the seam, it will probably fail in the same areas during service, perhaps after only a few months. The same features that cause the low ductility break during seam testing may also induce stress cracking during service.

If, during a peel test, a specimen separates partially along the bonded interface before failing through the bottom geomembrane, but with adequate strength, it may conventionally be considered to be acceptable. However, if such a specimen is examined under a microscope it will be seen that the separated surfaces contain a large number of crazes. These crazes result in a large reduction in the stress cracking resistance of the geomembrane (Peggs & Charron, 1989; Peggs, 1990). Therefore, if, over time, a seam were to partially separate under the peeling action of an adjacent fold, or wrinkle, or due to subgrade settlement, a continued constant stress applied to the geomembrane may lead to stress cracking. Following this argument it is logical that no peel separation should be allowed. However, to achieve this in practice may require the seam to be overheated, thereby causing additional problems. As a compromise a 10% maximum peel separation is often specified, and, more often than not, causes seams to be fabricated with zero peel separation.

The peel separation can be specified as the area of separation expressed as a percentage of the originally bonded area (excluding the squeeze-out of hot wedge seams) or it can be expressed as the length of the most advanced front of separation as a percentage of the original width of the seam.

A tough, but achievable, set of seam specifications may be as follows:

- shear strength: >90% specified geomembrane yield stress
- shear elongation: >100% (smooth) > 50% (textured)
- failure in the geomembrane outside the seam
- peel strength: >75% specified geomembrane yield stress
- peel separation: <10%

For the complete sample of five peel and five shear specimens to be acceptable there should be no more than one peel and one sheer specimen failure.

There has always been a problem when testing PVC geomembrane seams in that time is required for the adhesive seams to cure, and that results cannot be obtained as quickly as for HDPE geomembranes. The PVC Geomembrane Institute in the US has shown that curing to within 80% of naturally cured values can be achieved by accelerating the curing process at 60°C for 16 h.

Nondestructive Seam Testing

Vacuum and air pressure testing. The most conventional methods of nondestructively testing geomembrane seams are air pressure testing for double track seams, and vacuum box testing for all seams. ASTM is developing standards for testing by these two methods. Typical procedures for these methods are:

- Air pressure:

    —pressurize to 160 to 200 kPa
    —leave for 5 min
    —pressure loss must not exceed 20 kPa

- Vacuum box:

    —evacuate to 35 kPa
    —no bubbles, or froth, within 15 s

When the air pressure test is finished the first clamp to be removed from the seam should be the one at the opposite end of the tested seam from the pressure gage. If air is not expelled from the seam annulus there is a blockage in the annulus and the seam must be retested.

The disadvantages of these techniques are that they are relatively time consuming, there is no hard copy of the test data, and they only detect complete penetrations through the liner.

Ultrasonics. Ultrasonic methods have been used, primarily in Europe, to detect defects, including those that are non-penetrating, in geomembrane seams, but while theoretically possible on ideal seams, the pulse-echo technique is difficult to apply on some field seams, and impossible to apply on others. The ultrasonic shadow technique in which a multi-frequency pulse is passed through the seam from one geomembrane panel to the other (Peggs *et al.,* 1985; Geosynthetic Research Institute, 1986) offers significant potential (even to assess the degree of bonding) but needs more development work. The author has experimented with infrared thermography but this method also requires further developmental work.

Electrical. There are two variants of the electrical method of leak location (Darilek *et al.,* 1989; Peggs, 1990) and ASTM is drafting a standard of practice to cover this technique. This technique has the advantage that it is capable of rapidly evaluating the complete surface of the liner, not only the seams, for leaks.

The method was developed to locate leaks in the geomembranes of uncovered liquid impoundments but has recently been extended to the detection of leaks underneath the soil covers on landfill liners. This has led to an interesting observation. When leaks are identified in a specific area under the soil cover the usual procedure is to remove the soil and to more accurately locate the leak with a vacuum box. It should be noted, however, that when a geomembrane has been placed correctly on a clay subgrade, and it has been forced into intimate contact with the clay by approximately 0·5 m of soil cover, a hole in the geomembrane may not be detected by a vacuum box. This confirms that the geomembrane has been placed correctly and is in intimate contact with the clay. However, the apparently conflicting observations between traditional (vacuum box) and new (electrical) methods can lead to some confusion. The conflict will not arise if the geomembrane has been placed on sand.

SUMMARY

A comprehensive testing program to assure optimum durability of the geomembrane should contain:

Design Stage (Prequalification Testing):

- Select the two best geomembranes based on stress cracking resistance of geomembrane and seams using the CTL technique.
- Select the better geomembrane of these two based on chemical compatibility testing.
- Chemical resistance evaluation must include thermal and chemical structural analytical techniques.

Pre-construction Stage (Conformance Testing):

- Monitor production and QC testing in the manufacturing plant.
- Perform conformance testing before geomembrane leaves the manufacturing plant.

Construction Stage:

- Perform destructive testing on seams, monitoring shear elongation and peel separation in addition to strength parameters.
- Perform nondestructive tests on seams using vacuum box, air pressure, ultrasonic, or electrical leak survey methods.
- Consider the feasibility of nondestructively testing the complete liner, prior to covering, by the electrical leak survey or infrared thermographic methods.

Post-construction Stage:

- Perform electrical leak survey to detect holes in geomembrane under the drainage/protective soil layer. This approach is not recommended as a CQA procedure but only as an investigative procedure as required. If the geomembrane is placed on a clay, subgrade vacuum testing may not confirm the presence of a hole in the geomembrane.

If such a testing program is performed, together with the implementation of a thorough CQA plan, to comprehensive CQA specifications, on a well-designed lining system, the most durable geomembrane liner will be installed.

It should be remembered that testing protocols for geomembranes are continuing to evolve. It should also be remembered that assuring durability of the geomembrane will be ineffective if proper waste handling and placement procedures are not instituted.

REFERENCES

Darilek, J. T., Laine, D. & Parra, J. O. (1989). The electrical leak location method for geomembrane liners: development and applications. In *Proc. of Geosynthetics '89*. pp. 454–62.

Fluet, J. E. (1985). Quality assurance of hazardous waste geosynthetic lining systems. *Proc. Second Canadian Symp. on Geotextiles and Geomembrane*. Edmonton, Alberta, pp. 147–51.

Giroud, J. P. & Fluet, J. E. (1986). Quality assurance of geosynthetic lining systems. *Geotextiles and Geomembranes*, **3**(4), 249–87.

Giroud, J. P. & Morel, N. (1992). Analysis of geomembrane wrinkles. *Geotextiles and Geomebranes*, **11**(3), 255–276.

Giroud, J. P. & Peggs, I. D. (1990). Geomembrane constructing quality assurance. In *Waste Containment Systems: Construction, Regulation and Performance*, Geotechnical Special Publication No. 26, ASCE, pp. 190–225.

Geosynthetic Research Institute (1986). GMI: seam evaluation by ultrasonic shadow method. *GRI Test Methods and Standards*, Philadelphia, PA, 25 pp.

Geosynthetic Research Institute (1990). GM5: ductile/brittle transition time for notched polyethylene specimens under constant stress. *GRI Test Methods and Standards*. Philadelphia, PA, 7 pp.

Hessel, J. (1990). Welding of PE-liners for waste disposal landfills. In *Advances in Joining Newer Structural Materials*. Pergamon Press, Toronto, Ontario, p. 187.

Moet, S. *et al.* (1989). Slow fatigue crack growth test for ranking polyethylene pipes. In *Proc. 11th Plastic Fuel Gas Pipe Conference*, American Gas Association *et al.* San Francisco, California, pp. 327–33.

NSF (1991). International Standard 54: Flexible membrane liners. National Sanitation Foundation, Ann Arbor, Michigan.

Peggs, I. D. (1987). Evaluating polyethylene geomembrane seams. In *Geosynthetics '87*. Industrial Fabrics Association International, St Paul, Minnesota, pp. 505–18.

Peggs, I. D. (1990). Detection and investigation of leaks in geomembrane liners. *Geosynthetics World*, 1(2), 7–14.

Peggs, L. D. & Carlson, D. S. (1989). Stress cracking of polyethylene geomembranes: field experience. In *Durability and Aging of Geosynthetics*. ed. R. M. Koerner. Elsevier Applied Science, London, pp. 195–211.

Peggs, I. D. & Carlson, D. S. (1990). The effects of seaming on the durability of adjacent polyethylene geomembrane. In *Geosynthetic Testing for Waste Containment*, ASTM Special Technical Publication 1081, Philadelphia, PA, 132–142.

Peggs, I. D. & Charron, R. M. (1989). Microtome sections for examining polyethylene geosynthetic microstructures and carbon black dispersion. In *Geosynthetics '89*. Industrial Fabrics Association, St Paul, Minnesota, pp. 421–32.

Peggs, I. D., Briggs, R. & Little, D. (1985). Developments in ultrasonics for geomembrane seam inspection. In *Proc. Second Canadian Symp. on Geotextiles and Geomembranes*. Edmonton, Alberta, pp. 153–6.

Peggs, I. D., Carlson, D. S. & Peggs, S. G. (1991). Understanding and preventing shattering failures of polyethylene geomembranes. In *Geotextiles, Geomembranes and Related Products*. Balkema, Rotterdam.

USEPA (1986). Method 9090: Compatibility test for wastes and membrane liners. US Environmental Protection Agency, Cincinnati, Ohio.

USEPA (1990). FLEX: Flexible Membrane Liner Advisory Expert System. US Environmental Protection Agency, Cincinnati, Ohio.

USEPA (1991). Workshop on chemical fingerprinting of geosynthetics. US Environmental Protection Agency; Cincinnati, Ohio.

4.7 Performance Changes in Aged In-Situ HDPE Geomembrane

ANDRÉ L. ROLLIN,[a] JACEK MLYNAREK,[b] JEAN LAFLEUR[b]
& ALEXANDRE ZANESCU

[a] *Department of Chemical Engineering, Ecole Polytechnique de Montréal, CP 6079, Succursale A, Montréal, Canada H3C 3A7*
[b] *Department of Civil Engineering, Ecole Polytechnique de Montréal, CP 6079, Succursale A, Montréal, Canada H3C 3A7*

INTRODUCTION

The performance of in-situ synthetic lining materials is generally monitored on the basis of the quantity of the leachate collected above the liner or of the environmental effects of leachate escaping from the lining system (Haxo & Haxo, 1989).

Some efforts have been made in placing specimens in landfill sumps and measuring property changes in specimens withdrawn at different times. Nevertheless the operational conditions of the in-situ material and the specimen are not the same with particular regard to mechanical stress (Matrecon Inc., 1988).

The most interesting source of information is the investigation of material collected at dismantled landfills.

This chapter reports on the results of an investigation carried out on an aged HDPE geomembrane used at a landfill site for contaminated solids. Samples of geomembranes were collected from the bottom, slopes and cover of cells that were emptied. These geomembranes were in contact with leachates and contaminated soils for seven years. A testing programme was performed to evaluate and compare the characteristics of the collected samples to the original characteristics of the sheets to assess the ageing process.

LANDFILL HISTORY

In 1983, 60 000 m^3 of contaminated soils were recuperated and stored in three retaining cells located in Ville LaSalle, Montréal, Canada. These soils had been contaminated in 1979 by an industrial leak of approximately 13 000 000 litres of effluent containing waste water, oils and pyrolitic tars from a coke production plant. Soil analysis indicated that a layer as deep as 1·8 m contained cyanides, phenolic compounds as well as heavy hydrocarbon fractions. Maximum concentrations for some of the contaminants found during the investigation programme are presented in Table 1 (Zanescu & Demers, 1990, unpublished).

Before filling the cells, the contaminated soils were overturned to accelerate the oxidation and drying process while the cells were built. As shown schematically in Fig. 1, each of the cells was constructed using three impermeable layers: (a) 300-mm-thick compacted clay; (b) 2·0-mm-thick high-density polyethylene geomembrane; and (c) 600-mm compacted silty clay. Typical dimensions of the cells were: 64 × 81 m at the bottom; 80 × 97 m at the top; and a depth of 2·7 m. Liquid pumping as well as venting of gas facilities were also installed. Each cell was covered using a 1·6-mm HDPE geomembrane.

The geomembranes sheets were welded using a fillet extrusion technique and the clay layers were compacted. The characteristics of the clays and their compaction level as the Proctor standard are presented in Fig. 1.

During the 1984–1990 period, the leachate levels, the phenolic content and the cyanide content in the leachate retained in each cell were monitored as shown in Figs 2–4. Since initially these cells were

TABLE 1. Characteristics of the Contaminated Soil at the Villa LaSalle Landfill, Montreal

pH	>1·21
Sulphate (mg/kg)	<328 000
Copper (mg/kg)	<295
Zinc (mg/kg)	<534
Arsenic (mg/kg)	<385
Benzo-a-pyrene (mg/kg)	<1 300
Oil (mg/kg)	<9 000

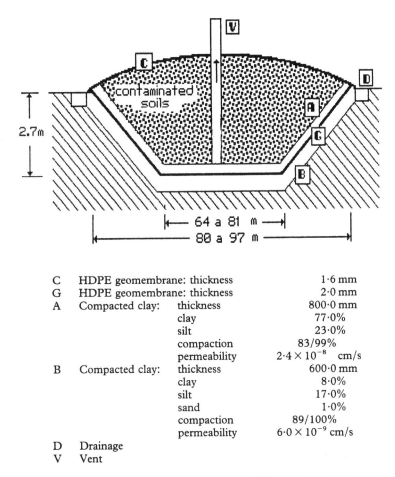

C	HDPE geomembrane:	thickness	1·6 mm
G	HDPE geomembrane:	thickness	2·0 mm
A	Compacted clay:	thickness	800·0 mm
		clay	77·0%
		silt	23·0%
		compaction	83/99%
		permeability	$2·4 \times 10^{-8}$ cm/s
B	Compacted clay:	thickness	600·0 mm
		clay	8·0%
		silt	17·0%
		sand	1·0%
		compaction	89/100%
		permeability	$6·0 \times 10^{-9}$ cm/s
D	Drainage		
V	Vent		

Figure 1. Scheme of a cell at the Villa LaSalle Landfill site and characteristics of the lining system.

guaranteed for a period of 10 years, it was decided in 1990 to empty the cells and to transfer the contaminated soils into a double lined cell. This offered a unique opportunity to evaluate visually the liners' condition and to collect and analyse the geomembranes after seven years of service. Special care was given to collect good and potential faulty seams as well as geomembrane sheets that were under strain and in contact with leachates.

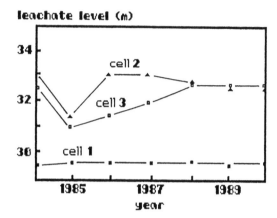

Figure 2. Leachate level in each cell.

In the following paragraphs, the measured characteristics of the collected geomembranes are compared to their initial characteristics. Basically, values of the tensile resistance of the membranes, the tensile and peel resistances of the seams, the brittleness of the sheets and seams, and the determination of micro-cracks by microanalysis are presented to evaluate the geomembranes behaviour.

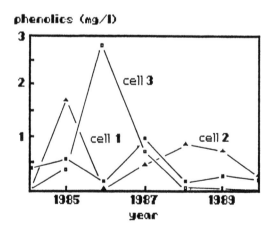

Figure 3. Concentration of phenolics in leachate.

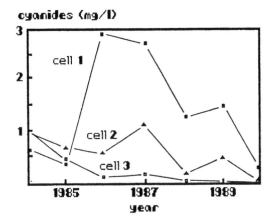

Figure 4. Concentration of cyanides in leachate.

GEOMEMBRANE SPECIMENS

Quality Control in 1983

HDPE 2·0-mm-thick geomembranes were used at the bottom and on the slopes of each cell and HDPE 1·6-mm-thick geomembranes were used as cell covers. In 1983, a quality control check was performed on sheets shipped to the site. A total of 34 samples of 2·0-mm-thick membranes were tested using ASTM D-1593 standard for thickness and ASTM D-638, type IV, 50 mm/min for tensile strength. Similarly 11 samples were tested for the 1·6-mm-thick membranes.

The tensile results were gathered from Gundle technical reports and the following correlations were established as a function of the sheet thickness (*T*):

- the tensile strength of the membranes:

—strength at yield (N/m) = (500 (N/m) × *T* (mm)) − 12 (N/m)
—strength at break (N/m) = (676 (N/m) × *T* (mm)) + 238 (N/m)

- the average elongation at break:

—for the 1·6-mm sheet: elongation = 846%
—for the 2·0-mm sheet: elongation = 836%

TABLE 2. Description of samples

Sample no.	Cell no.	Location	Remarks	Thickness (mm)
1	1	Cover	Normal	1·47
2	1	Cover	Folded sheet	
3	1	Slope	Cracked sheet	
4	1	Slope	Normal	1·88
5	1	Bottom	Folded sheet	
6	1	Bottom	Normal	1·92
7	2	Cover	Folded sheet	
8	2	Cover	Cracks at seams, open seams	1·22
9	2	Cover	Repaired sheet	1·52
10	2	Cover	Folded sheet, repaired sheet	1·62

Quality Control in 1990

In 1990, samples of geomembranes were collected after the cells were emptied. A total of 6 samples were collected from the bottom, slopes and cover of cell no. 1, and 4 samples were collected from the cover of cell no. 2. Descriptions of the collected samples are given in Table 2.

SHEET PROPERTIES

In a first phase of the programme, sheets collected in 1990 were tested and the results compared to the original sheets analysed in 1983. The data presented in Table 3 are averaged results of tests performed on five specimens.

The comparison of the 1990 and the 1983 sheet tensile strengths indicates that the yield strength of the 7-year-old membranes is slightly higher than the yield strength of the original membranes.

On the other hand, both the tensile strength at rupture and the elongation at rupture decreased greatly with average decreases of 16% for samples collected on slopes, 25% for samples collected from covers and 60% for samples collected from the base of the cells. These data are an indication of the ageing of the geomembranes: increased yield

TABLE 3. Tensile Strength of Sheets

Sample no.	Strength at yield (N/m)			Strength at break (N/m)			Elongation at break (%)		
	1983	*1990*	*%*	*1983*	*1990*	*%*	*1983*	*1990*	*%*
1	724	754	4	1232	847	31	846	661	22
8	599	648	8	1063	770	28	846	655	23
9	749	748	0	1265	509	16	846	454	46
10	799	815	2	1333	1 078	19	846	756	11
4	929	986	6	1509	1 263	16	836	703	16
6	939	963	3	1522	532	65	836	346	59

strength, decreased strength at rupture and lower elongation at rupture.

The tensile strength curves for five sheet specimens from sample no.9 are presented in Fig. 5.

SEAM PROPERTIES

The seams of collected geomembranes were tested to measure their tensile and peel strengths using the ASTM D-368 and the ASTM D-413 as modified by NSF-54. The obtained results are presented in Tables 4 and 5.

As shown in Table 4, the average tensile strength of seams are acceptable as per the strength ratio used as acceptance criteria and defined as the ratio of the yield strength of the seam to the yield strength of the sheet. It is recommended that the ratio be greater than 80% (Rollin *et al.*, 1991*a*).

On the other hand, the seams' elongation at break are much less than the sheet itself with values lower by 79% to 96%. These results indicate that ageing has occurred along the edges of the seams.

The peel yield strength ratio calculated from data collected during the testing programme is shown in Table 5. It is recommended as an acceptance criterion, that the ratio of the yield strength of the seam to the sheet yield strength should be greater than 70% (Rollin *et al.*, 1991*a*). In this study all values measured are lower than 60%, which indicates that the adhesion between the sheets does not meet today's standards.

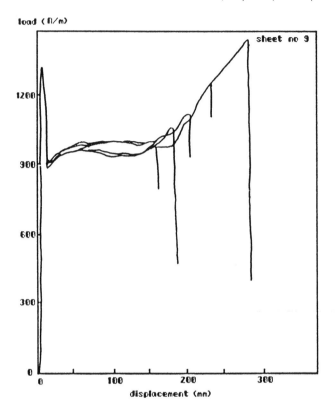

Figure 5. Tensile strength curves of geomembrane sample no.9.

TABLE 4. Tensile Strength of Seams

Sample	Thickness (mm)	Yield strength		Elongation			Strength ratio %
		Seam (N/m)	Sheet (N/m)	Yield %	Break %	Sheet %	
1	1·59	672	754	11	27	661	89
8	1·10	508	648	9	102	655	78
9	1·46	629	748	14	134	454	84
10	1·56	765	815	11	159	756	94
4	1·98	885	986	12	55	703	90
6	2·01	783	963	11	66	346	91

TABLE 5. Peel Strength of Seams

Sample	Thickness (mm)	Yield strength		Strength ratio (%)
		Seam (N/m)	Sheet (N/m)	
1	1·62	279	754	37
8	1·15	327	648	51
9	1·52	217	748	29
10	1·61	162	815	20
4	1·76	308	986	31
6	2·04	565	963	59

The tensile strength and peel strength curves for specimens of sample no.9 are presented in Figs 6 and 7.

MICROANALYSIS OF GEOMEMBRANES AND SEAMS

Seam cross sections were analyzed using the microanalysis technique as developed by Rollin and co-workers (Rollin *et al.*, 1989, 1991*b*; Rollin & Peggs, 1991) to support the tensile and peel results. Care was taken to analyse specimens having properties that had been altered during the service life of the geomembranes.

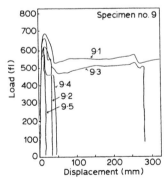

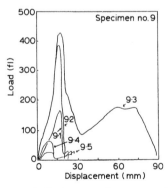

Figure 6. Tensile strength curves for seams.

Figure 7. Peel strength curves for seams.

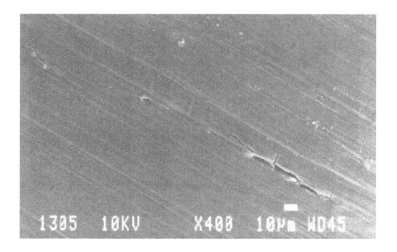

Figure 8. Cross section of a seam: unbonded area.

Unbonded Areas

As presented in Fig. 8, a cross section of a fillet extrusion weld from sample no.9 can be observed at a magnification of 400×. A horizontal interface can be visualized, indicating an unbonded area between the upper sheet and the lower sheet. The adhesion of this seam (see curve 9·5 on Fig. 7) was performed and the peel strength was measured at 47 N/m as compared to 575 N/m to meet the acceptance criteria. Evidence presented in this photograph suggests very poor adhesion during the peel test.

Stress Cracks

Residual stress cracks can be identified easily on microscopic photographs as shown in Fig. 9. This micrograph with a magnification of 400× shows the presence of two transversal microcracks in the upper sheet as well at the interface of both sheets. Their presence can be related to ageing of the geomembrane as it was noticed for specimen no 9·5 (see Fig. 6) with an elongation at rupture of 341% for the sheet and of 21% for the seam. The elongation of the specimen has decreased drastically during the seven years of service since the elongation at rupture of the original sheet was measured to be 846%.

Figure 9. Cross section of a seam: stress cracks.

Seam Examination

The cross section of the fillet extrusion weld observed in Fig. 10 (specimen no. 6) shows that the lower sheet was altered in thickness during the welding process. The thinning of the sheet explains the low yield strength (450 N/m) and the inacceptable elongation at rupture (12%) of the seam as compared to the sheet properties in 1983 (yield strength of 940 N/m and an elongation of 836%).

CONCLUSIONS

Analysis of the geomembranes collected in cells containing contaminated soils for 7 years indicated that ageing had occurred. The results indicated that the ageing is more severe for the samples recuperated from the bottom of the cells compared to the samples recuperated from the slopes and covers. The ageing was detected from an increase in the yield strength, a decrease in the tensile resistance at rupture and a lowering of the elongation at break.

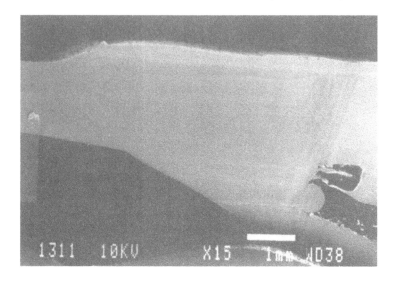

Figure 10. Cross section of a seam: thinning of the sheet.

It is important to note that the yield strength of all the seams tested was acceptable with a strength ratio ranging from 80% to 95%. This can be associated with the visual observation of the emptied cells where no seam was found unbonded and only one or two cracks were observed in the sheets. The cells were found to be in very good condition.

Tensile and peel tests were performed during this programme to obtain results that could be compared to the original characteristics of the sheets, and microscopic analysis was used to gather useful information which helped to assess the sheets and seams quality and to identify faulty seams.

REFERENCES

Haxo, H. E. & Haxo, P. D. (1989). Synthetic lining systems for land waste disposal facilities. Proceedings Sardinia '89, Second International Landfill Symposium, Alghero (Italy), October 9–13 1989.
Matrecon, Inc. (1988). Lining of waste containment and other impoundment

facilities. SW-870, 2nd revised edition; US Environmental Protection Agency, EPA/600/12-88/052, Cincinnati, OH.

Rollin, A. L. & Peggs, I. D. (1991). Microanalysis of polyethylene geomembrane seams. In *Geomembranes: Identification and Performance Testing*, ed. A. L. Rollin & J. M. Rigo. Chapman & Hall, London, Chapter 12.

Rollin, A. L., Vidovic, A. & Ciubotariu, V. (1989). Assessment of HDPE geomembrane seams. *Proceedings of Sardinia 89, Second International Landfill Symposium*. Sardinia, pp. XIX-1-11.

Rollin, A. L., Fayoux, D. & Benneton, J. P. (1991a). Non-destructive and destructive seam testing. In *Geomembranes: Identification and Performance Testing*, ed. A. L. Rollin & J. M. Rigo. Chapman & Hall, London, Chapter 4.

Rollin, A. L., Lefebvre, M., Lafleur, J. & Marcotte, M. (1991b). Evaluation of field seams quality by the impact test procedure. In *Proceedings Geosynthetics '91*. IFAI, Atlanta, pp. 223-39.

5. GEOTEXTILES AND GEOGRIDS

5.1 Efficiency of Geotextiles and Geocomposites in Landfill Drainage Systems

DANIELE CAZZUFFI,[a] RAFFAELLO COSSU[b] & M. CRISTINA LAVAGNOLO[c]

[a] ENEL–CRIS, Via Ornato 90, 20162 Milan, Italy
[b] Department of Land Engineering, University of Cagliari, Piazza D'Armi, 09123 Cagliari, Italy
[c] CISA, Environmental Sanitary Engineering Centre, Via Marengo 34, 09123 Cagliari, Italy

INTRODUCTION

Geotextiles and geocomposites are among the most widely used groups of geosynthetics and they perform many important functions in various fields of civil engineering.

Geotextiles consist of synthetic fibers arranged in different ways. They can be manufactured using a normal weaving loom (woven geotextiles) or processes which give rise to a random structure (non-woven geotextiles). The non-woven geotextiles can be made from different types of polymeric fibers, the main types being continuous (mono or multi) filament (CF) and staple fibers (SF). For CF geotextiles cohesion between fibers is normally obtained by thermal (melt or thermo-bonded, TB), mechanical (needle-punched, NP) or chemical (resin-bonded, RB) treatment. The most widely used polymers for manufacturing geotextile fibers are polypropylene (PP) and polyester (PES), although polyethylene (PE) and polyamide (PA) are sometimes used.

Geocomposites consist of a combination of different geosynthetic materials (geotextile + geonet, geotextile + geomembrane, geonet + geomembrane, geotextile + clayey soil, etc.).

In sanitary landfills geotextiles and geocomposites can perform several functions. The main functions are:

- filtration and drainage
- separation of soil with different grain size
- mechanical protection of synthetic liners
- reinforcement
- lining

The application of course requires specific properties of the individual product. In this chapter geotextiles and geocomposites for filtration (cross-plane flow) and drainage (in-plane flow) function are considered.

A drainage system represents the key factor in landfill design, management and environmental impact. The function of an efficient drainage system is, in fact, to allow a quick flow and collection of leachate, avoiding formation of waterhead above the landfill bottom and of perched water tables. When these phenomena occur the following risks arise:

- leachate diffusion along the slopes or through lining failures,
- difficulty in biogas abstraction,
- stability problems, particularly when landfilling is operated with hill method or on a slope.

These remarks imply that materials for drainage systems should be carefully considered and studied. The utilization of geotextiles and geocomposites in landfill drainage systems has very often been handled without proper consideration of the performance of these materials in landfill environments. Geotextiles were seen to clog, sooner or later, due to both mechanical and biological phenomena (Cazzuffi & Cancelli 1987; Cancelli *et al.*, 1988; Koerner & Koerner, 1988). Therefore, specific design criteria have been proposed (Rollin & Denis, 1987). The study described here was carried out in order to investigate the hydraulic behavior to leachate of several geotextiles and filtering geocomposites, which are representative of the most widely used types of synthetic material proposed for landfill drainage systems.

METHODOLOGY

The research was carried out by performing permittivity tests on different kinds of geotextiles and geocomposites commercially available in Italy. Three experimental runs were planned according to the program presented in Table 1.

Testing Apparatus

All tests were carried out by means of a permeameter which had been specifically designed for geotextiles and built at ENEL-CRIS laboratories in Milan (Francia, 1983). Most parts are made of plexiglass: in this way direct observation of all phenomena occurring during the permeability tests is possible. The hydraulic scheme of the apparatus is shown in Fig. 1; the basic elements being the permeameter, the feeding group, and the flow measuring group.

The permeameter is formed of three coaxial cylinders with an

TABLE 1. Testing Programme

Run	Tested geotextiles	Test methodology
I	Non-woven geotextiles	Short-term permittivity test
II	Woven geotextiles and geocomposites	Long-term permittivity test
III	Non-woven and woven geotextiles; geocomposites	Short-term permittivity test after long-term leachate exposition: a = 2 months b = 3 months c = 5 months d = 8 months

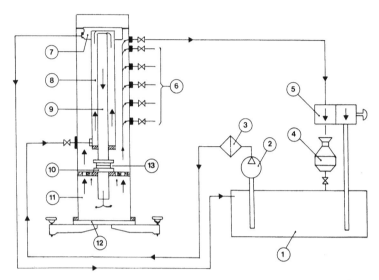

Figure 1. Hydraulic scheme of the closed circuit, constant head permeameter for permittivity tests: (1) backwater tank, (2) feed pump, (3) filter, (4) calibrated vessel, (5) two-ways valve, (6) outlets, (7) overflow system, (8) constant level cylinder, (9) feed cylinder, (10) specimen bearing, (11) external cylinder, (12) base, (13) geotextile specimen.

internal diameter of 284, 124 and 51 mm respectively. The internal cylinder holds at its base the geotextile specimen to be tested, and simultaneously acts as feeding cylinder for the permeability test. The intermediate cylinder acts as a constant level cylinder; the water exceeding the spillway is directed, by means of an overflow pipe, to the feeding tank. The external cylinder is the support of the entire apparatus; six different values of the hydrostatic load, h, can be applied (h is the difference between the spillway level and the selected outlet and is maintained constant during a single test). Each opening has a cross section of 100×12 mm^2.

The feeding apparatus consists of a pump which sucks water form a feeding tank; this tank also represents the backwater tank at the end of the test circuit. The flow measuring group is essentially formed by a calibrated reservoir that can be fed or excluded by means of a two-way valve.

This constant head permeameter has a closed hydraulic circuit so that the same permeability test can be performed by using the same small quantity of liquid (demineralized de-aerated water or leachate).

In performing normal permeability tests on geotextiles, a particular problem is caused by the validity limits of Darcy-Ritter's law. In view of this, an upper limit value was computed for the apparent water velocity: $v_{max} = 0.031$ m/day (Gourc *et al.*, 1982; Francia, 1983). This means that a minimum value of the specimen thickness, H, related to the hydrostatic load, h, should be maintained. For the purpose of this research study, when testing materials with water, the minimum value of H was obtained by superimposing different layers of the same geotextile; the number of layers ranged from 3 to 6, depending on the thickness Tg of the geotextile or of the geocomposite to be tested. When using leachate, laminar flux conditions were observed and only one layer of material was used. The coefficient of normal permeability K_n and the permittivity Ψ of the geotextile are computed by the formulae:

$$K_n = \frac{V \cdot H}{A \cdot h \cdot t} = \frac{V \cdot x \cdot Tg}{A \cdot h \cdot t} \tag{1}$$

and

$$\Psi = \frac{V \cdot x}{A \cdot h \cdot t}$$

where

V = calibrated volume = 1450×10^{-6} m^3,
H = specimen thickness (m),
x = number of geotextile layers forming a specimen,
h = hydrostatic load (m),
t = time required to fill calibrated volume V during a single test (s).
Tg = geotextile or geocomposite thickness,
A = specimen surface = 2043×10^{-6} m^2.

The testing apparatus, with simple modifications, also allows testing under normal compressive stress. However, in this study only tests in the absence of normal stress have been performed in order to investigate the less critical conditions.

TABLE 2. General Characteristics of Materials Tested in Different Runs (See Table 1) (μ = area mass, Tg = thickness, K_w = permeability to water)[a]

Run	Sample reference	Name	Type	Material	Geo-textile process	μ (g/m²)	Tg (mm)	K_w (m/s)
I	A	Bidim U34	non-woven	PES	NP, CF	290	2·9	$4·1 \times 10^{-3}$
I	B	Drefon S45	non-woven	PES	NP, SF	200	2·2	$4·5 \times 10^{-3}$
I	C	Tecnofelt FAG	non-woven	PES	NP, SF	300	3·4	$5·9 \times 10^{-3}$
I	D	Geodren PE/S	non-woven	PES	NP, SF	300	2·8	$1·2 \times 10^{-3}$
I	E	Polyfelt TS 750	non-woven	PP	NP, CF	370	3·0	$2·8 \times 10^{-3}$
I	F	Drefon SIA 200	non-woven	PP	NP, SF	200	2·9	$5·0 \times 10^{-3}$
I, III	G	Terram 1000	non-woven	PP–PE	TB, CF	140	0·8	$1·7 \times 10^{-3}$
I	H	Typar 3807	non-woven	PP	TB, CF	280	0·7	$1·7 \times 10^{-4}$
I	I	Drefon SIA 400	non-woven	PP	NP, SF	400	5·2	$5·0 \times 10^{-3}$
I	L	Stratum	non-woven	PP	NP, SF	450	5·4	$1·2 \times 10^{-3}$
II	M	Terram W/3-3	woven	PP	WN	206	0·8	3×10^{-2}
II	N	Terram W/7-7	woven	PP	WN	485	1·4	$2·5 \times 10^{-3}$
II, III	O	Terram W/12-12	woven	PP	WN	794	2·1	$1·1 \times 10^{-3}$
II	P	Flotex	woven with open structure	PP	WN	277	0·8	$1·3 \times 10^{-1}$
II, III	Q	Superdrain 600	non woven with open structure	PP	NP, SF	600	8·9	$1·2 \times 10^{-1}$
II	R	Tecnodrain	geocomposite (non-woven + geonet)	PP–PVC	TB, CF	863	6·4	5×10^{-3}
II	S	Tigerdrain	geocomposite (non-woven + geonet)	PP–PE	NP, SF	1 177	17·5	$7·3 \times 10^{-3}$
II	T	Filtram 1B1	geocomposite (non-woven + geonet + non-woven)	PP–PE	TB, CF	1 256	6·1	$5·4 \times 10^{-4}$
II	U	Filtram 1BZ	geocomposite (non-woven + geonet + non-woven)	PP–PE	TB, CF	1 321	12·5	$3·6 \times 10^{-4}$
II	V	Terram 7M7/700	geocomposite (non-woven + woven)	PP–PE	TB, CF	739	2·2	4×10^{-4}
III	Z	Terram W/20-4	woven	PP	WN	594	1·8	$2·2 \times 10^{-4}$
III	W	Polyfelt TS 700	non-woven	PP	NP, CF	306	2·4	$7·1 \times 10^{-3}$
III	Y	Terbond A-300	non-woven	PES	NP, SF	419	3·8	$3·6 \times 10^{-2}$

[a] PES = Polyester; PP = polypropylene; PE = polyethylene; PVC = polyvinylchloride; NP = needle-punched; TB = thermo-bonded; CF = continuous filament; SF = staple-fibre; WN = woven.

Materials

The materials used in the experiments include geotextiles, geocomposites, leachate and demineralized de-aerated water.

Geotextiles and geocomposites. The complete list of geotextiles and geocomposites tested is reported in Table 2, together with the most significant physical properties. Mass per unit area (μ) and thickness (Tg) were measured and did not differ significantly from the values reported by the manufacturer. Permeability to water was also measured and values were often lower than those reported by the manufacturer.

The list includes non-woven and woven geotextiles. The non-woven ones were mostly monofilament, both needled and thermobonded and some staple-type Italian products. The mass per unit area ranged from 140 to $600\,\text{g/m}^2$ and the nominal thickness ranged from 0·7 to 8·9 mm for non-composite geotextiles. The list also includes composites made by joining woven and non-woven geotextiles or geotextiles and geonets. The mass per unit area and the nominal thickness of these materials are reported in Table 2.

Leachates. The leachate used for the tests was collected at the sanitary landfill of Mariano Comense, Como, Italy. The tests were carried out in different periods and the leachate composition reflects typical variations for this wastewater. In Table 3 the range of the

TABLE 3. Analytical Composition of the Different Leachates used in Different Runs of the Experimental Tests (I–IIId; See Table 1)

Parameter	L1 (I)	L2 (II)	L3 (IIIa)	L4 (IIIb)	L5 (IIIc)	L6 (IIId)	L7 (imm)
pH	6–6·3	7·2–7·3	7·1–7·4	8·1–8·9	7·7–8·4	8·1–8·3	7·1–8·1
COD (g/liter)	28·8–38·5	13·6–35·4	31·9–34·4	7·3–11·7	19·1–20·3	5·4–6·0	8·8–32·0
BOD (g/liter)	3·0–10·4	5·0–11·0	13·0–14·5	2·0–3·7	6·4–7·8	0·7–1·0	3·4–13·0
TS (g/liter)[a]	12·4–20·5	11·6–21·3	20·2–22·7	12·1–12·4	16·2–16·5	10·0–10·3	9·1–22·7
TVS (g/liter)[b]	5·0–9·7	4·6–10·6	10·2–11·7	4·9–5·4	7·6–7·9	3·4–3·7	3·8–11·7

[a] TS = Total solids.
[b] TVS = Total volatile solids.

analytical composition of leachates used in the different experimental runs is reported.

Leachate L1 is a typical acidic phase leachate, while the others are typical of an unstable and stable methanogenic phase. Leachate L7 was used for the long-term exposition to leachate of geotextiles before conducting the permittivity test, in research phase III.

Experimental Procedure

The material for the test was prepared by cutting a geotextile foil in order to obtain circular specimens with a diameter of 68 mm. The thickness of the specimen was measured under $\sigma = 2$ kPa. The values in Table 2 are the averages of 7–10 measurements. The sample was weighed after drying at 80°C, with precision of ±0·1 mg.

Test cycles I and II. The samples of geotextiles were tested on the instrument previously described. Leachate was circulated through the permeameter until a relevant and stable permeability decrease was reached. In the cycle I, when non-woven geotextiles were studied, the test-duration was in the range 50–200 min. Permeability measurements during the test were carried out periodically according to the clogging dynamic of the individual geotextile. The higher measurement frequency was adopted for less permeable geotextiles. After each test, the leachate was changed and a fresh one was used.

Test cycle III. In the third test cycle, specimens of geotextile and geocomposite were immersed in leachate, under anaerobic conditions, in order to observe the effects on clogging of the biological growth.

The specimens were immersed in a stainless steel vessel with a volume of 50 liters, and hung up to the vessel top. The vessel was thermo-insulated at 30°C. Specimens of the same geotextile or geocomposite had the same weight. Leachate in the vessel was changed periodically. Four immersion periods were adopted: (a) 2 months, (b) 3 months, (c) 5 months, and (d) 8 months.

After immersion the specimen was tested on the permeater previously described. The same test was carried out on a virgin specimen of the same characteristics (material, weight and thickness). Contrary to the I and II test cycles, the same duration (600 min) was established for all tests. Following leachate circulation the specimens were analyzed in order to measure the trapped amount of total and volatile solids.

In the leachate immersion vessel rectangular specimens (300×170 mm in size) of the different materials were also exposed. These specimens were used for mechanical testing. Mechanical testing for tensile strength and strain at failure were performed using two pieces of equipment, operating at two different maximum load: JJ Instruments, London, model M5K (maximum load, 5 kN) and Galdabini model PMA/20 (maximum load, 200 kN).

RESULTS AND DISCUSSION

The results of the permeability tests on non-woven geotextiles are summarized in the graph K_n–time (Fig. 2). All tested materials show a marked decrease in the coefficient of permeability K_n with time elapsed from start of the test.

The ratio between initial and final permeability range between 100 and 1000 times. By comparing the values of permeability to leachate at the end of the test with permeability to water of the different geotextiles, it can be seen that the effect of leachate is to reduce normal permeability to less than 1/100 000 of the original value. The thermo-bounded geotextiles (G and H; see Table 2) proved to clog more than needle-punched geotextiles. Staple-fiber, needle-punched geotextiles generally showed better performances, particularly the thicker ones (I and L).

The results obtained with woven geotextiles and with geocomposites are reported graphically in Fig. 3 (a) and (b). Generally, woven geotextiles showed a good resistence to clogging—higher than that of non-woven geotextiles. Only one geotextile (M) with a close weave showed a very rapid and intense clogging.

The composite geotextiles, formed by a sandwich of two thermo-bonded staple-fiber non-woven geotextiles and geonet, showed low K_n values with the same behavior as geotextiles alone (Fig. 2). The geocomposite made by joining non-woven needle-punched geotextile and geonet showed good hydraulic behavior (S and R) and showed a clogging effect later than the other tested geotextiles. The best hydraulic behavior was displayed in cycle II by a special non-woven

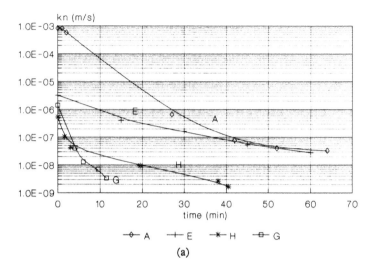

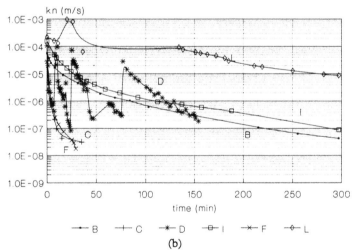

Figure 2. Permeability to leachate versus time observed in the permittivity tests on non-woven geotextiles (experimental cycle I): (a) continuous filament, (b) staple fibre. For the meaning of the geotextile references (A–L), see Table 2 (modified from Cancelli *et al.*, 1988).

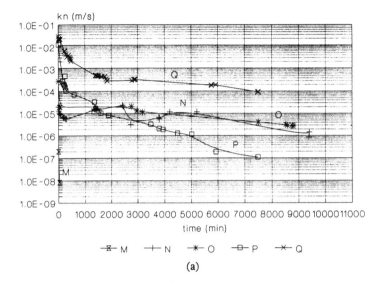

(a)

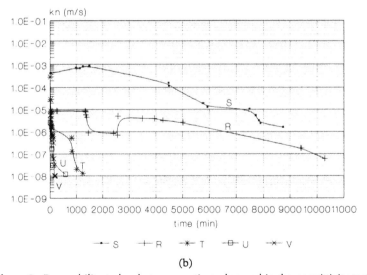

(b)

Figure 3. Permeability to leachate versus time observed in the permittivity tests on (a) woven geotextiles and mats and (b) geocomposites, in experimental cycle II. For the meaning of the geotextile references (M–V) see Table 2.

geotextile (Q) with a very open structure and relevant thickness (8·9 mm). This open structure is supported by a kind of geonet.

The results of cycle III are reported in Fig. 4. All the tests were performed for a duration of 600 min, as previously mentioned. Only for the geotextile Q (non-woven with an open structure) was the test duration of 1200 min, because it took longer to reach stabilization.

The graphs in Fig. 4 refer to different immersion times in the leachate (T2, T3, T5 and T8) and to the corresponding non-immersed geotextile specimen (T02, T03, T05 and T08). From these results it is possible to observe a permeability decrease of all geotextiles when compared with the permeability to water (Fig. 5). Here again, as in the cycle II, the woven geotextiles (O and Z) and the special non-woven geotextile with an open structure (Q) showed the best hydraulic behavior. The permeability decrease is influenced both by the content of solids in the tested leachate and by the immersion time in the leachate (biological clogging). Generally the permeability to leachate of the immersed samples was lower than the permeability of the non-immersed sample, with some exception regarding mainly the woven geotextiles (O and Z) (Fig. 6). The greatest clogging effect is observed, once again, for non-woven needle punched geotextile (W and Y).

A linear relationship between permeability decrease and immersion time in leachate cannot be clearly observed as the content of solids in the tested leachate varied in the different runs and was particularly high for the run with a two-month immersion time (Table 3). This is a general problem when performing this long-term test as leachate changes dramatically in quality if stored at either low temperature or during production in the same landfill.

The graphs in Fig. 7 clearly show how the specific permeability (ratio of permeability to leachate and permeability to water) depends on the content of solids in leachate and how this effect is emphasized by biological growth which occurs during immersion in leachate.

Figure 8 reports the accumulation of solids which occurred in the different samples after different immersion periods. The accumulation of solids proved to be highest in the non-woven geotextile with an open structure (Q) which is also the material with greater thickness.

The clogging phenomena appears to involve both surface and volume of the geotextile. The geotextile with higher specific volume generally have a higher solid accumulation capacity and consequently higher resistance to clogging phenomena. From another point of view

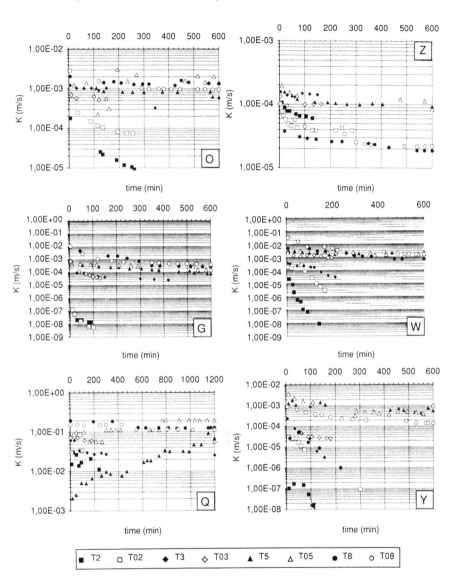

Figure 4. Permeability to leachate versus time observed in the permittivity tests on different kind of geotextiles, after 2, 3, 5 and 8 months of immersion in leachate (T2, T3, T5 and T8) and on corresponding samples of the same materials without immersion (T02, T03, T05 and T08) (experimental cycle III). For the meaning of the geotextile references (G–Y), see Table 2.

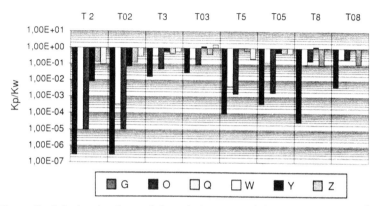

Figure 5. Calculated values of the relative permeability (or permittivity) for the different geotextiles used in experimental cycle III. For the meaning of the geotextile references, see Table 2: k_p = permeability to leachate; k_w = permeability to water of a virgin sample; $T0_i$ = non-immersed sample tested after i months; T_i = immersed sample tested after i months.

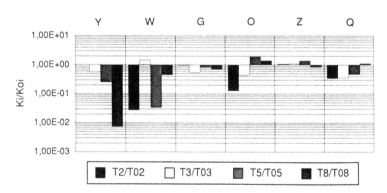

Figure 6. Ratio between permeability to leachate of immersed (k_i) and non-immersed samples (k_{oi}) after i months for the different geotextiles tested in experimental cycle III. For the meaning of the geotextile references, see Table 2.

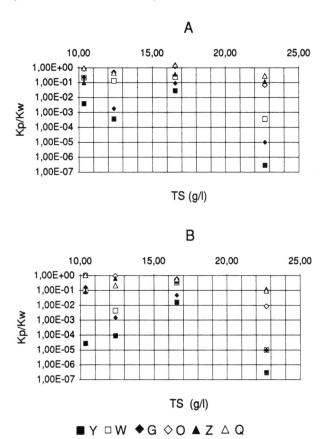

Figure 7. Relative permeability (k_p/k_w) of (A) non-immersed and (B) immersed samples versus total solids content in the testing leachates used in the different runs of experimental cycle III. For the meaning of the references, see Table 2 and Fig. 4.

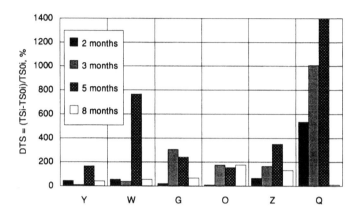

Figure 8. Relative accumulation of Total Solids (DTS) in the geotextiles specimen, after different immersion times. For the meaning of the geotextiles references, see Table 1. $TS0i$ = total solids trapped in the non-immersed samples after the permittivity tests with leachate; TSi = total solids trapped in the immersed samples after the permittivity tests with leachate.

the open structure of some geotextile with low specific volume (as with some woven material) show lower accumulation but also a lower filtering effect so that their clogging resistance is higher too. Generally however it is possible to observe a relationship between the relative

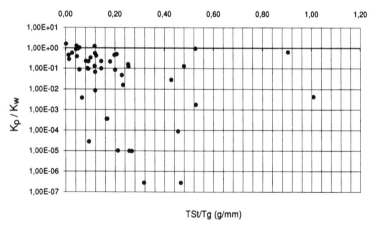

Figure 9. Relative permability (k_p/k_w) versus specific solid accumulation (TSt/Tg). k_p and k_w are defined in Figure 5. TSt = total solids trapped in the specimen; Tg = geotextile thickness (mm).

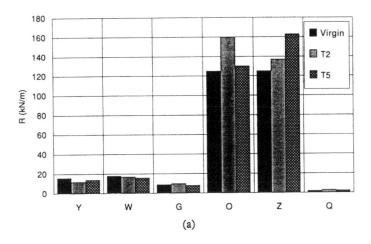

(a)

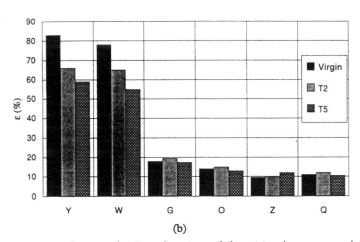

(b)

Figure 10. Tensile strength (*R*) and strain at failure (*ε*) values measured in mechanical tests on different geotextiles after 2 and 5 months (T2 and T5) of leachate immersion, and on virgin samples of the same materials. For the meaning of the geotextile references, see Table 2.

permeability and the specific solids accumulation (ratio of total trapped solids/geotextile thickness) as shown in Fig. 9.

Figure 10 shows the results of the mechanical tests on the immersed and non-immersed geotextiles after permittivity tests. There is an evident effect of immersion on tensile strength parameters, and particularly the non-woven needle-punched geotextiles show a marked decrease of resistence (R) and strain at failure (ε). On the contrary, woven geotextiles show an increase in resistance.

CONCLUSIONS

The results of three cycles of tests on permeability to leachate of different kinds of geotextiles lead to the following conclusions:

- Sanitary landfill leachate causes a marked decrease in permeability in every kind of geotextile, under the combined effects of surface deposition and volumetric accumulation of solids, both those contained in the leachate itself and those formed by biological growth.
- The best hydraulic behaviour is generally shown by woven geotextiles, and non-woven geotextiles with a very open structure.
- Among the non-woven geotextiles, the thermo-bonded showed permittivity values lower than the needle-punched ones.
- The performance of geocomposites varies according to the above consideration on the individual materials.

ACKNOWLEDGMENTS

The authors gratefully acknowledge the precious cooperation in the experimental study of Ludovica Ferruti, Dario Pessina and Pierpaola Trapani, and of the Waste Management Company which provided the leachates used in the tests. Furthermore the helpful assistance of all the technicians of ENEL-CRIS Special Materials Laboratory is gratefully acknowledged.

This study has been supported by the Hatian Ministry for Universities and Research (MUR).

REFERENCES

Cancelli, A., Cossu, R., Malpei, F. & Pessina, D. (1988). Permeability of different materials to landfill leachate. In *Proceedings ISWA 88 Congress.* Copenhagen, Academic Press, London, pp. 115–22.

Cazzuffi, D. & Cancelli, A. (1987). Permittivity of geotextiles in presence of water and pollutant fluids. Proceedings Geosynthetic '87 Conference, New Orleans, pp. 471–81.

Cazzuffi, D., Cossu, R., Ferruti, L. & Lavagnolo, C. (1991). Efficiency of geotextiles and geocomposites in landfill drainage systems. In *Proc. Sardinia '91: 3rd Inter. Symposium.*, CISA, Cagliari, 759–80.

Francia, L. (1983). Studi sperimentali sui geotessili per applicazioni in campo geotecnico ed idraulico (Experimental studies on geotextiles in geotechnical and hydraulic applications). Masters Thesis, Faculty of Engineering, University of Bologna.

Gourc, J. P., Faure, Y., Rollin, A. & La Fleur, J. (1982). Standard test of permittivity and application of Darcy's formula. Proceedings 2nd International Conference on Geotextiles, Las Vegas, pp. 149–54.

Koerner, G. R. & Koerner, R. M. (1988). Biological clogging in leachate collection system. Proceedings 2nd G.R.I. Seminar 'Durability and aging of geosynthetics', Philadelphia, Drexel University, 1–18 December 1988.

Rollin, A. & Denis, R. (1987). Geosynthetic filtration in landfill design. Proceedings Geosynthetic '87 Conference, New Orleans, pp. 456–70.

Trapani, P. (1992). Efficiency of Landfill Drainage Systems. Ph.D. Thesis, Technical University of Milan.

5.2 Permeability Models for Non-woven Geotextiles

JACEK MLYNAREK,[a] JEAN LAFLEUR,[a] ANDRÉ L. ROLLIN[b]
& J. BOGUMIL LEWANDOWSKI[c]

[a] Department of Civil Engineering, Ecole Polytechnique de Montréal,
CP6079, Succursale A, Montréal, Canada H3C 3A7
[b] Department of Chemical Engineering, Ecole Polytechnique de
Montréal, CP6079, Succursale A, Montréal, Canada H3C 3A7
[c] Department of Land Reclamation, Agricultural University of Poznan,
ul. Wojska Polskiego 73A, 60–625 Poznan, Poland

INTRODUCTION

An important field of geotextile research is the determination of the coefficient of permeability. The movement of water through granular porous media is a basic part of hydrodynamics. The geotextiles indeed, are very special porous media. Their thickness is usually very small $(0.2 < T_g < 10\,\text{mm})$ while their porosity (n) could be very high—up to 95%.

The purpose of this chapter is to present the modification of drag models to determine the permeability of geotextiles used in leachate drainage systems in landfills.

MODELS OF FLOW THROUGH GEOTEXTILES

The principal models of fluid flow through any porous media (Scheidegger, 1974) are:

- analytical
- capillary tube
- hydraulic radius

- drag
- statistical

The first four have been used to investigate the water flow through geotextiles.

Analytical Model

The flow through any porous medium can be described by the Navier–Stokes equations, with the following assumptions:

- the porous medium is homogeneous and isotropic on a macroscopic scale,
- the filtering fluid is incompressible, homogeneous and isotropic, and
- the electric, thermodynamic and chemical effects are insignificantly small.

With the above assumptions, the Forchheimer equation (1) can be obtained from the Navier-Stokes equations (Ahmed & Sunada, 1969)

$$i = aV + bV^2 \tag{1}$$

where i = hydraulic gradient, V = bulk velocity, and a, b = constants.

Mlynarek (1987) examined the possibility of applying the Forchheimer equation to describe the water flow through geotextiles. In the case of laminar flow, the nonlinear term of eqn (2) is neglected due to the too low velocities, and it can be changed to

$$V = \frac{1}{a} \cdot i$$

which corresponds to the empirical Darcy's formula

$$V = k \cdot i \tag{2}$$

Darcy's equation allows the experimental determination of the permeability of geotextiles (k) through measurements of the bulk velocity (V) and the hydraulic gradient (i). This equation is commonly used by geotextiles investigators.

Capillary Tube and Hydraulic Radius Models

Researchers have begun to set up other models of a geotextile structure to study the correlation between the permeability and other parameters such as porosity, fibre diameter d_f, specific area, tortuosity, etc. The result is that the obtained correlations are valid for some specific cases but can hardly be applied to all geotextiles.

Pivovar *et al.* (1980), Mlynarek (1983) and Gourc (1983) have used the same mixed model of the hydraulic radius and the capillary tube to describe the flow through geotextile structures. They assumed that a geotextile is represented by a tube model with the equivalent diameter equal to the hydraulic radius. The formula for the permeability (k) obtained from their considerations is

$$k = \frac{g}{v} \cdot S \cdot \frac{n^3}{(1-n)^2} \cdot d_f^2 \tag{3}$$

where g = gravity, v = viscosity, n = porosity, d_f = fibre diameter, S = constant.

The main difference between these considerations is related to the geotextile structure constant S. According to Pivovar *et al.* (1980), this constant depends on the porosity and the fibre diameter and it could take the following value:

$$S^P = \frac{1}{32} \text{ to } \frac{1}{320}$$

For geotextiles with a porosity between 0·81 and 0·90, Mlynarek (1983) recommended the following experimental constant:

$$S^M = \frac{1}{80} \text{ to } \frac{1}{160}$$

while the Gourc's (1983) constant S^G, which depends upon the tortuosity and the capillary tube cross-section, is equal to

$$S^G = \frac{1}{175}$$

These results can be compared with the Carman–Kozeny solution for a uniform granular porous media with a porosity equal to 0·3 and the experimental constant would then be equal to

$$S = \frac{1}{80}$$

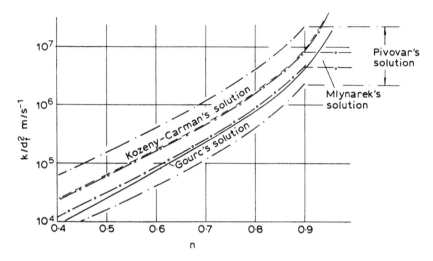

Figure 1. The different solutions of the hydraulic radius model.

A comparison of the above solutions is presented in Fig. 1 in terms of porosity versus the k/d_f^2 ratio.

Drag Model

The drag model is presented schematically in Fig. 2. Rollin *et al.* (1982) have suggested using this model to obtain a satisfactory understanding of the water flow through thick geotextiles. From the

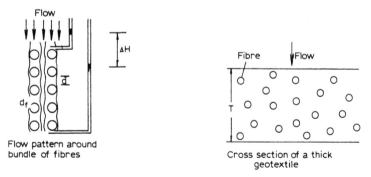

Figure 2. Schematic cross-section of a thick geotextile—the drag model.

analysis of the internal structure of the geotextiles, they concluded that the model of flow would corresponds to the flow around a bundle of cylinders. They obtained the following equation

$$k = \frac{g}{v} \cdot \frac{2}{A} \cdot \frac{\pi}{4} \cdot \frac{\sqrt{3}}{(1-n)} \cdot d_f^2 \tag{4}$$

to calculate the geotextile permeability, assuming that the flow is laminar. According to experimental data, the experimental constant A is between 65 and 76 for geotextiles, having a porosity ranging between 0·85 and 0·95.

MODIFIED DRAG MODEL

The drag model is based on the assumption the flow through thick geotextiles corresponds to the flow around a single bundle of many fibres in one direction.

The modification is based on the development of the friction term which is a function of the fibres' layout in the three dimensions. Consider a thick geotextile sample with a porosity between 0·80 and 0·95, area A_i, and thickness T_g. The sample consists of many fibres with diameter d_f and length L_f. The characteristics volumes (ω) are:

total volume $\quad \omega_T = A_i \cdot T_g$
volume of solids $\quad \omega_S = (1-n) \cdot A_i \cdot T_g$
volume of voids $\quad \omega_V = n \cdot A_i \cdot T_g$

The volume of the fibres is:

$$\omega_f = \left(\sum L_f \right) \cdot \frac{\pi d_f^2}{4}$$

and it is equal to the volume of solid. From the above equations, the sum of the length of the fibres can be calculated as follows:

$$\sum L_f = 4(1-n) \cdot A_i \cdot \frac{T_g}{\pi \cdot d_f^2}$$

and the area of fibres inside a given geotextile sample can be expressed as

$$a = \frac{4(1-n) \cdot A_i \cdot T_g}{d_f}$$

This so-called interfacial area a characterizes the resistance of the geotextile structure and the influence of the fibres on each other.

By determining the drag force F_D on one fibre:

$$F_D = C_D \cdot a \cdot \tfrac{1}{2} \cdot \rho \cdot V_f^2 \tag{5}$$

where C_D is the drag coefficient, ρ is the density and the total number of rows of fibres,

$$N = \frac{T_g}{\bar{d} + d_f} \tag{6}$$

where $\bar{d}$ is the average distance between fibres, the following equation can be derived:

$$F_D = \frac{T_g}{\bar{d} + d_f} \cdot \frac{4(1-n)A_i \cdot T_g}{d_f} \cdot C_D \cdot \frac{1}{2} \cdot \rho \cdot V_f^2 \tag{7}$$

The resistance force of a geotextile sample is expressed as follows:

$$F_R = \rho \cdot g \cdot i \cdot A_i \tag{8}$$

Substituting eqn (8) into eqn (7) leads to the following expression for head loss:

$$i = \frac{4T_g^2(1-n)C_D \cdot V_f^2}{2d_f(\bar{d} + d_f) \cdot g} \tag{9}$$

The Reynolds number is defined as:

$$R_e = \frac{V_f \cdot d_f}{v}$$

and the resistance coefficient as:

$$C_D = \frac{C}{R_e}$$

Combining these definitions with Darcy's law eqn (2) and eqn (9), the coefficient of permeability can be obtained:

$$k = \frac{g}{v} \cdot \frac{2}{C} \cdot \frac{\bar{d} + d_f}{4(1-n)T_g^2} \cdot d_f^2 \tag{10}$$

The average distance between the fibres can be expressed as follows

$$\bar{d} = d_f\left(\frac{1}{2}\sqrt{\frac{\pi\sqrt{3}}{1-n}} - 1\right)$$

and eqn (10) can be transformed into the final expression for the permeability:

$$k = \frac{g}{v} \cdot \frac{1}{C} \cdot \frac{d_f\sqrt{\pi\sqrt{3}}}{4(1-n)^{3/2} \cdot T_g^2} \cdot d_f^2 \tag{11}$$

From experimental data obtained on non-woven geotextiles, the material constant C was found to range between $0\cdot13$ and $0\cdot52$ with an average value of $0\cdot274$ for $0\cdot8 < n < 0\cdot94$.

The variation of the k/d_f^2 ratio as a function of the porosity, the geotextile's thickness, the fibre diameter and the material constant is presented in Fig. 3.

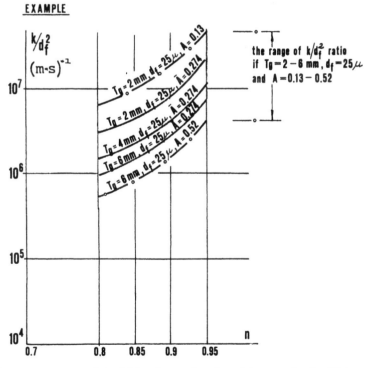

Figure 3. Variation of the k/d_f^2 ratio as a function of the porosity, the thickness and the structural constant calculated using the modificated drag model formula; the fibre diameter $d_f = 25 \ \mu$m.

CONCLUSIONS

The analysis performed show the following:

* The permeability of geotextiles can be determined experimentally by using the simplest form of analytical model—the Darcy equation for the laminar flow of water through geotextiles.
* The analysis of the different empirical formulae shows that the results obtained with these formulae are valid for some specific cases but cannot be applied to all types of geotextiles.
* The modified drag formulae are suggested to estimate the values of permeability of geotextiles with $0.85 < n < 0.95$.

Finally, the estimated values of geotextile permeability have to be treated as approximations, in the same manner as the soil permeability determined by empirical correlations.

REFERENCES

Ahmed, N. & Sunada, D. K. (1969). Nonlinear flow in porous media. *J. Hydraul. Div. ASCE*, **95** (HY6), 1847–57.

Gourc, J. P. (1983). Quelques aspects du comportement des géotextiles en mécanique des sols. PhD thesis, University of Grenoble, France.

Mlynarek, J. (1983). Geotextile as protective filter of an earth dam drainage. PhD thesis, Technical University of Gdansk, Poland.

Mlynarek, J. (1987). Determination of permeability of geosynthetics. In *Proc. of Geotechnical Institute of Technical University of Wroclaw*, (No. 51) Conference No. 23. Technical University of Wroclaw, Poland pp. 139–45.

Pivovar, N. G., Bugaj, N. G. & Ryczkow, W. A. (1980). *Drainage with Geotextile Protective Layer*. Izdatielstvo Naukowa Dumka, Ukrainian Academy of Sciences, Kiev, 216 pp. (in Russian).

Rollin, A. L., Masounave, J. & Lafleur, J. (1982). Pressure drop through non-woven geotextiles: a new analytical model. In *Proc. of the 2nd Int. Conf. on Geotextiles*, vol 1. IFAI, St Paul, USA, pp. 161–6.

Scheidegger, A. E. (1974). *The Physics of Flow through Porous Media*. University of Toronto Press, Toronto, 353 pp.

5.3 Use of Geotextiles for the Protection of Geomembranes

REINHARD KIRSCHNER[a] & ROBERT WITTE[b]

[a] Huesker Synthetic GmbH & Co., Fabrikstrasse 13–15, Postfach 1262, D-4423, Gescher, Germany
[b] Amtliche Materialprüfanstalt (AMPA), Appelstrasse 11A, D-3000, Hannover 1, Germany

INTRODUCTION

Increased environmental awareness and problems of waste disposal, such as preventing groundwater contamination from leachate, have led in recent years to new technical developments in German landfill construction aimed at reducing interference to the environment. Amongst others, the basic idea of the multibarrier concept was developed, requiring several impermeable barriers to make the landfill compatible with the environment (LAGA-pamphlet M 3, 1989).

One of the major elements of this concept is the so-called combined sealing system, consisting of a mineral sealing layer of defined character and minimum thickness and a geomembrane. A coarsely-graded drainage gravel (often with gradation 16/32) is applied above the geomembrane to draw off the collecting leachate. To prevent damage of the geomembrane by the gravel, a protective layer must be placed on the geomembrane.

FUNCTION OF THE PROTECTIVE LAYER

The short-term dynamic loads affecting the geomembrane in the construction phase, through installation of drainage gravel or from site

475

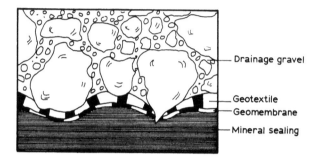

Figure 1. Point loading of a geomembrane and possible impacts (Saathoff, 1989).

traffic, can lead to damage of the unprotected lining membrane. Thus, point loading of the geomembrane occurs which is caused by the aggregates of the drainage layer.

There is a long-term risk of damage from single gravel grains under the static waste load. Potential damage to the geomembrane is thereby dependent on the amount of the surcharge, its long-term effect on the lining membrane, on the size and shape of the gravel aggregates and on the supportive action of the mineral sealing layer. As a result of poor protective layers being used between geomembrane and drainage gravel, base sealing systems placed above the geomembrane using graded 16/32 gravel can often lead to damage of the geomembrane itself (Fig. 1).

TYPE AND FORM OF PROTECTIVE LAYERS

The following design alternatives can be adopted:

- a layer of cohesive soils ($k \leq 10^{-7}$ m/s), free of sharp-edged aggregates and aggregates larger than 20 mm, installed to a thickness of at least 30 cm;
- a geotextile/aggregate combination consisting of a non-woven minimum 1200 g/m² geotextile and a minimum 15 cm layer of low grade calcium carbonate sand or chippings 0/8 mm;
- a geotextile and/or geocomposite.

Commercial geotextile protective layers can also be used:

- mechanically-bonded spun fibres or non-woven filament;
- composites of non-woven geotextile with a fabric inlay, and a weight per unit area up to $3000 \, \text{g/m}^2$;
- non-woven filament with a rigid, three-dimensional geonet placed above geotextile;
- composites with a bentonite filling (bentonite mat).

OPERATING MECHANISMS OF PROTECTIVE LAYERS

The load-distributing effect of conventional aggregate protective layers has already been described. Thus, in view of the relatively high thickness and the finely-graded structure, there is no need to demonstrate the mechanical protective effect in this case. On the other hand, experiments are required to prove the equivalent protective effect of the comparatively-thin geotextile protective layers.

According to Sehrbrock (1989), the effect of the geotextile protective layer is based on two mechanisms: firstly, the single grains of the drainage layer are surrounded by individual fibres of non-woven materials; consequently point loading which occurs due to the geometry of the single grain, is absorbed (Fig. 2(a)). Secondly, when two adjacent single grains sink into the geotextile, the intermediate spaces elongate thus redistributing load due to the tensile strength of single non-woven fibres (Fig. 2(b)).

In the case of bentonite mats, the increased protective effect in comparison to a simple non-woven material, is based purely on the improved absorbing action.

A marked rise in load distribution can be achieved if the non-woven fibre is combined with a fabric inlay. This is due to the higher secant

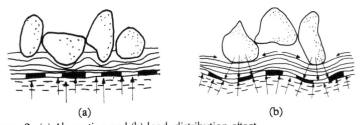

(a)   (b)

Figure 2. (a) Absorption and (b) load–distribution effect.

TABLE 1. Devices for Testing Protective Efficiency

Testing Institute	Test methods	Experimental set-up	Explanatory notes
Amtl. Material-prüfanstalt AMPA, TU Hannover	Applying load with pressure stamp and pressure plate on simulated gravel drainage layer. Base: hard foam		Design of sealing layer and load direction, true to existing system.
Inst. für Grundbau und Bodenmechanik, TU Braunschweig	Applying load with pressure stamp and pressure plate on simulated gravel drainage layer. Base: earth/clay		Design of sealing layer and load direction, true to existing system.
Franzius-Institut, TU Hannover	Simulation of point loading with stamp loading and pressure points. Base: hard and soft materials		Laboratory testing methods, not directly comparable with sealing design (basic research).

Dr.-Ing. Steffen Ingenieurges. mbH, Essen	Applying load by pressing the geomembrane into the drainage gravel. Base: none	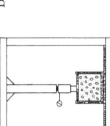	Type of load contrary to existing sealing design.
Grundbauinstitut der LGA Bayern, Nürnberg	Applying load with pressure stamp and pressure plate on simulated gravel drainage layer. Base: earth/clay	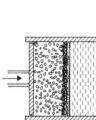	Sealing layer design and load direction, true to existing system.
Naue-Fasertechnik GmbH/Dr.-Ing. Knipschild, Lübbecke	Applying load with pressure stamp and pressure plate on simulated gravel drainage layer. Base: moulded fibre elastomer, etc.		Design of sealing layer and load direction, true to existing system.

modulus of fabric compared to non-woven materials whereby relatively high tensile forces can be mobilised even at low strain. In order to obtain maximum strain in the fabric, this should be placed as far away as possible from the geomembrane, in the top-third of the non-woven layer.

A better protective effect can be obtained using relatively thick, rigid geonets above a non-woven fabric. Apart from the thickness and rigidity of the geonet, the width of its opening (which should be smaller than the smallest single grain of drainage gravel) seems to be important. Thus, in addition to the load–distributing effect, a blocking effect is also achieved.

METHODS OF TESTING PROTECTIVE EFFICIENCY

Object-related experiments to calculate the protective efficiency particularly of geotextile protective layers have been sporadically carried out for some years on test fields and in test laboratories. However, to date, an objective comparison of individual results has not proven possible as testing methods and evaluation criteria differ, at times considerably. A simplified outline of the laboratory test methods applied to date is shown in Table 1. In most cases, this is a modified plate-bearing test to show the protective efficiency of a geotextile under long-term, static surcharge.

Status of development

A workshop in 1990 entitled 'Quo vadis protective layers', with representatives from specialist institutes, consultants, manufacturers and government agencies, pointed out the discrepancies in testing methods and evaluation criteria. Based on the workshop's studies, the Amtliche Materialprüfanstalt of the TU Hannover therefore developed a provisional test concept to estimate the protective effect.

In this concept the following questions should be addressed in evaluating protective suitability:

- Does the geomembrane show signs of buckling, depressions and/or nicks after the plate bearing test?

- Can perforations of the geomembrane be seen?
- How large was local deformation of the geomembrane after the experiment had expired?
- Is there an increase in local deformation and/or overall deformation depending on the stressing time?
- Does the number of local geomembrane deformations increase with progressive stressing-time?
- With a time-dependent increase of local and/or overall deformation, can asymptotic behaviour of the deformation with increased stressing time be established?
- To which ultimate value does time-dependent deformation aspire?
- Can characteristic damage be identified in the protective systems applied which could lead to an incalculable drop in protective efficiency with a longer stressing time?

To address these questions an experiment was set up as in Fig. 3 with a cylindrical vessel of 550 mm diameter. A high-density 2·5 mm-thick polyethylene lining membrane, as used in landfills, was installed as a geomembrane, or alternatively, the geomembrane used in the particular project under construction. As a base, to simulate the mineral sealing layer, a 20-mm-thick elastomer slab with a Shore-A hardness of 50 units was used, having the advantage of reproducibility, or alternatively the mineral sealing layer used in the particular project under construction. The research showed that no measurable differences were recorded in geomembrane deformation by using an

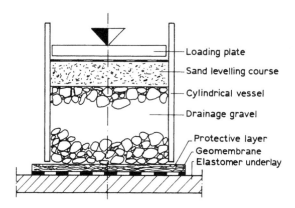

Figure 3. Test facilities of AMPA Hannover.

elastomer or mineral base. For the record a tin foil is laid on the geomembrane.

In order to estimate the temporary progression of eventual deformations or damage of the geomembrane and the protective layer both efficiently and at a low cost, loads were applied at time-staggered intervals of 10, 100 and 1000 h.

Besides a visual and thus to a certain extent subjective assessment of the geomembrane's appearance after dismantling from the cylindrical vessel, protrusions were measured using an automatic coordinate measuring instrument.

The deformation test data were plotted over the stressing time. Damage to the protective geotextile can be estimated by comparative tensile tests based on the DIN 53 857 standard on a loaded and unloaded geotextile.

EXPERIMENTAL RESULTS OF PROTECTIVE EFFICIENCY TESTS

The plate bearing tests described here with an experiment set up as in Fig. 3 were carried out at staggered intervals (in 10-, 100- and 1000-hourly stages). The mineral sealing layer under the geomembrane in this case was itself simulated by a hard foam base irreversibly deforming under loadings. After the end of each stressing time and removal from the mould, the surface of the hard foam base was first scanned in two diagonals of maximum deformation with the coordinate measuring instrument, and a photographic record was kept of the components of the experimental set-up for a lasting comparison.

Example A

Figure 4 shows a diagram of the scanner curves from a loading program with the following parameters:

Protective system 1:   PP spun non-woven geotextile, nominal
                        weight per unit area 1400 g/m², with an
                        incorporated fabric inlay (225 g/m²) and

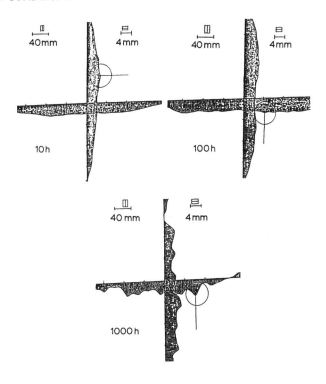

Figure 4. Deformation profiles of hard foam base.

	a gravel-faced overlay of an extruded flexible geonet
Surcharge:	$1300 \, kN/m^2$
Drainage gravel:	round, graded $16/32 \, mm$

In a visual inspection of the stressed geomembrane the increase of deformation with longer stressing time can be observed immediately. However, a quantitative and objective assessment is not possible on the basis of this finding alone. If one considers the scanner curves, the initial visual impression is clearly confirmed. While the course of the surface lines after 10 and 100 h loading does not present deformations worth mentioning, after 1000 h severe deformations were observed, corresponding to the visual inspection.

As a follow-on evaluation, the maximum local deformation (the item

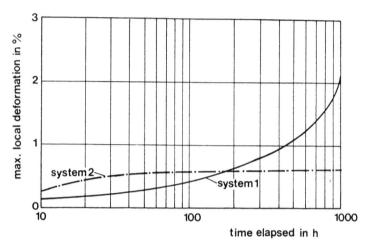

Figure 5. Deformation–time curves.

encircled in Figure 4) was derived from the individual scan diagrams. The course of maximum local deformation is plotted in Fig. 5. The continued course of time-dependent deformation can be estimated from this deformation–time diagram. In terms of reliable extrapolation of long-term protective behaviour, an asymptotic approximation of the deformation curve to an ultimate value is desirable.

The example shown evidentiates that asymptotic approximation of local deformation after 1000 h has not occurred. In order to estimate the long-term protective behaviour in this case, further tests over a period of time would be necessary.

Example B

Figure 5 shows the deformation–time curve of the following loading programme:

Protective system 2:   PP spun non-woven geotextile, nominal weight per unit area 1200 g/m^2 with gravel-face overlay of an extruded rigid geonet

Surcharge: 800 kN/m^2

Drainage gravel: round, graded $16/32 \text{ mm}$

Despite an initial higher deformation value, the increase per time interval compared with Example A is markedly less and clearly aspires in asymptotic terms to an ultimate value. The relatively rigid geonet obviously leads to good load distribution in the protective layer.

NEW DEVELOPMENTS

New developments can be observed both in the materials and testing direction. Official standards of mineral or mineral/geotextile protective layers have been stipulated and all future material developments will have to be measured against these ('equivalence'). There are marked advantages with regard to semi-finished, geotextile composites with defined qualities, which cause fewer problems particularly for installation in sloping areas. However, the question of long-term deformation of the geomembrane due to the combined action of stress from individual grains and possible subsoil settlement has still to be resolved.

With regard to testing, there are still several unresolved questions, the major ones being the simulation of a largely consolidated mineral base, the simulation of service life (time acceleration by increased surcharge, raised test temperature, etc.), and criteria and methods for the evaluation of test results.

The development of new methods to show protective efficiency should be directed at:

- defining protective categories as a function of waste surcharge and drainage material;
- the development and application of time acceleration methods in testing techniques;
- design of test underlays from synthetic semi-finished products with defined characteristics which better simulate the deformation behaviour of mineral layers;
- objective measuring of deformations occurring in the geomembrane, e.g. with coordinate measuring instruments as opposed to a subjective, visual assessment.

REFERENCES

LAGA-Pamphlet M 3 (1989). The orderly depositing of waste. Draft of 25 November 1989 (in German).

Saathoff, F. (1989). Geomembranes with and without protective non-woven geotextiles under pointloading. In *Proc. 5th Special Conference on The Safe Landfill.* Süddeutsches Kunststoff-Zentrum, Würzburg, pp. 122–35 (in German).

Sehrbrock, U. (1989). Protective effect of geotextiles in landfills. Journal No. **30,** Institute for Foundations Engineering and Soils Mechanics, Braunschweig (in German).

Schicketanz, R. & Witte, R. (1990). Proof of protective effect as part of the qualifying test for the use of protective layers in landfill sealing systems. Seminar on Advances in Landfill Techniques 1990. ed. K.-P. Fehlau & K. Stief. Haus der Technik e.V., Essen (in German).

5.4 Geogrids: Properties and Landfill Applications

VICKY E. CHOUERY-CURTIS & STEPHEN T. BUTCHKO

Tensar Environmental System Inc., 1210 Citizens Parkway, Morrow, Georgia 30260, USA

INTRODUCTION

The construction of landfills over areas having the potential to exhibit localized subsidence and/or differential settlement can generate excessive deflection on the lining systems. This can result in potential creep failure or rupture of the geomembrane, geonets and other components of the lining system.

Critical situations can be encountered in the following site conditions:

- karst terrain
- abandoned mines
- compressible foundation
- fault zones
- areas susceptible to mass movement or seismic impact zones
- weak and unstable soils that do not have the ability to support foundation loadings as a result of expansion, shrinkage and differential settlement (expansive and/or sensitive clays, organic soils, loess, etc.)
- waste material deposits

Optimizing the scarcely-available space in landfills can be easily accomplished by vertically expanding an existing waste deposit, and laterally expanding along existing side slopes.

Construction of a new landfill over existing refuse can lead to potential localized and differential settlement problems. In the outlined situations it could prove necessary to support and limit the deflection of the lining system in order to assure a long-term integrity of the materials, keeping within accetable limits the maximum allowable strain and the resultant tensile stresses. The solution could be a structural lining system stabilized by using a high-strength and low-deformation geogrid. These geosynthetics have been developed to provide tensile reinforcement and their use is steadily increasing in landfill construction.

COMPONENTS OF A STRUCTURAL LINING SYSTEM

The various components of a typical structural lining system are identified in Fig. 1 in the case of a vertical expansion of an existing waste containment facility.

A structural lining system can consist (from the base upwards) of:

- compacted subbase layer
- structural geogrid layer
- secondary containment zone (geomembrane)
- leak detention zone (geonet)
- primary containment zone (geomembrane)
- leachate collection zone
- protective soil cover layer

As a geomembrane is primarily designed for containment and a geonet for drainage, these materials generally show an inability to sustain long-term tensile loads.

In the following paragraphs the tensile properties of geomembranes and geogrids are discussed.

Geomembranes

When a depression forms below the lining system, the membrane will elongate as it attempts to bridge the depression. However, the stress–strain response of a geomembrane under biaxial loading conditions is different from the response under a uniaxial loading condition. Figure 2 illustrates typical stress–strain responses of various geomembranes.

Based on their polymeric characteristics, geomembranes cannot

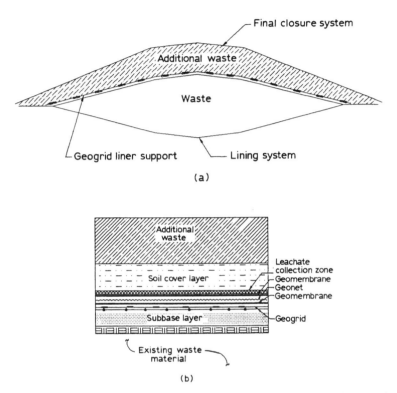

Figure 1. (a) Vertical expansion of an existing waste deposit and (b) typical lining system cross-section.

carry high tensile stresses under sustained loading conditions. Induced stresses in geomembranes result from several phenomena, such as non-uniformity and discontinuity in the liner thickness, differential settlement of the underlying materials, thermal expansion–contraction, and installation-induced damage. As a result of induced and concentrated stresses in the geomembrane, the average biaxial elongation at yield (or at failure) is significantly reduced and is in the range of (10–13%) (Giroud, 1984). This phenomena is important because the elongation at yield governs the ability of the geomembrane to withstand field-induced stresses and strains. As stated by Giroud (1984), the maximum allowable elongation that should be considered for final design should be equal to the elongation at yield divided by a

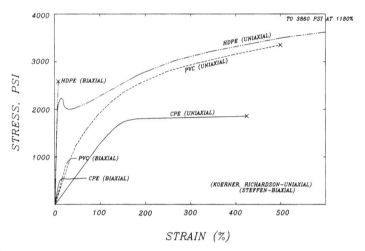

Figure 2. Typical stress–strain performances of various geomembranes (after Landreth *et al.* 1988).

factor of safety. Several countries have adopted factors of safety of 1·5 and up to 2·0.

Other primary factors that can further reduce the value of the uniaxial elongation at yield are long-term sustained loading conditions (i.e. creep behavior) and the effect of biaxial field conditions.

Koerner *et al.* (1990) studied the effect of a three-dimensional, axi-symmetric tension test on the stress–strain behavior of various geomembranes. The test results show a marked contrast to the usually-performed one-dimensional tension tests and a substantial difference in the behavioral trends. Figure 3, adapted from Koerner *et al.* (1990), illustrates typical stress–strain responses of a series of tests performed with various types of HDPE geomembranes. It should be noted that this particular testing program did not account for creep behavior. A thorough knowledge of the mechanical behavior of geomembranes is necessary for generating adequate final designs and, in general, geomembranes should not be used as tensile elements.

Geogrids

Geogrids were first introduced into North America in the early 1980s. According to ASTM definition, a geogrid is any planar structure formed by a regular network of tensile elements with apertures of

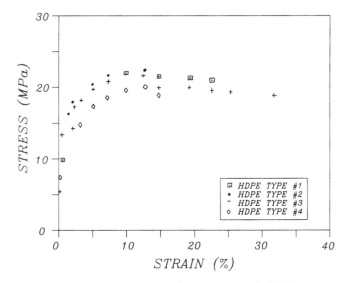

Figure 3. Responses of various HDPE (after Koerner *et al.,* 1990).

sufficient size to allow interlocking with the surrounding soil, rock, earth or any geotechnical material to perform the functions of reinforcement. Geogrids are produced in biaxial and uniaxial load configurations.

Uniaxial geogrids are manufactured from high-density polyethylene (HDPE) resins. Even within the families of HDPE resins, there are many differences among their products when used in critical load bearing applications. There are several resin characteristics that will determine processing, end-use properties, and performance of a finished product. These resin parameters are: density, melt index, molecular weight, molecular weight distribution, degree of orientation, and chain branching or entanglement.

Tensile properties and creep strain are significantly affected by the type and composition of the various resins. Testing of various HDPE resins with similar melt indices and densities indicated variations of up to 100% greater creep strain at designated loads after only 1000 h of creep testing. This is crucially important when critical structures are designed assuming a particular creep strain. The molecular weight and its distribution, as well as the number of chain branches control the creep characteristics. As recognized by polymer chemists, creep testing over 10 000 h is the only reliable way to assure that long-term performance is acceptable. It has been shown that short-term tensile tests do not predict long-term rupture performance. Figure 4 illustrates typical stress–strain responses of structural geogrids.

It is important for the geogrid supporting a landfill lining system to develop its working strength at strain levels that are compatible with the maximum allowable strains of the various lining system components. As a structural geogrid there are two primary design properties that need accurate definition. These properties are: *long-term stress–strain behavior* and *junction efficiency*. The key word here is long-term. Geosynthetics are polymer-based materials which means that they are viscoelastic, i.e. their stress–strain characteristics are time and temperature sensitive.

This structural classification means that the long-term physical properties of geogrids must be accurately defined before the geogrid can be specified for final design. The long-term stress–strain relationship should account for time or creep, temperature, installation damage, biological durability, chemical durability, and structural connections. Table 1 outlines typical chemical compatibility test results performed, per EPA 9090 test method, on an HDPE geogrid immersed in what was classified as the most severe leachate environment encountered in municipal solid waste facilities. Test methods and a standard of practice to define the long-term load-carrying capacity of a geosynthetic have been established by the Geosynthetic Research Institute (GRI) at Drexel University.

The GRI Standard of Practice GG4a 'Determination of the Long-Term Design Strength of Stiff Geogrids' addresses all the above-noted variables which should be accounted for in determining long-term load carrying capacity. Furthermore, it references specific test methods or provides procedures for definition of these variables. Minimum creep testing of 10 000 hours is required, per GRI:GG3a 'Tension Creep Testing of Stiff Geogrids'. The 10 000-h creep testing should be performed through the junctions as opposed to on the ribs only. This data may be extrapolated one order of magnitude, per polymer engineering rule of thumb (reference ASCE manual 63), to a design life of approximately 10 years. Creep performance for longer design lives may be defined with accepted polymer extrapolation techniques such as time–temperature superposition for up to two orders of magnitude shift from actual test data, thus leading to a design life in excess of 100 years (Bonaparte, 1987; Wrigley, 1988).

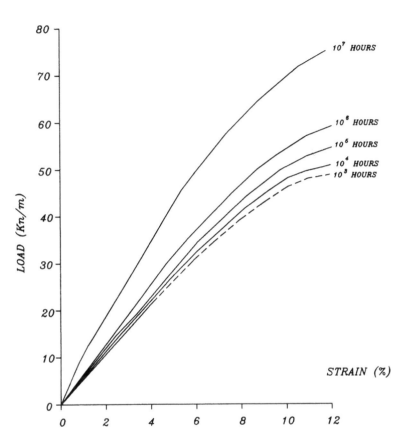

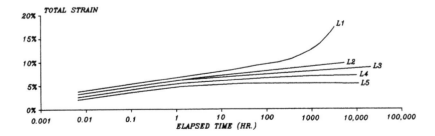

Figure 4. Typical structural geogrid response: (a) isochronous load–strain curve; (b) creep strain versus time.

TABLE 1. Summary of Typical Laboratory Test Data Performed on Structural HDPE Geogrid using EPA Test Method 9090

	Immersion temperature	Units	Initial value	30-Day Value	30-Day % Change	120-Day Value	120-Day % Change Initial	120-Day % Change Previous
Mechanical property								
Wide width tensile	23	kg	2 468	2 579	4·50	2 440	−1·13	−5·65
	50		2 468	2 455	−0·53	2 677	8·47	4·86
Rib strength	23	kg	303·7	323·5	6·52	316·8	4·31	0·73
	50		303·7	318·8	4·97	322·3	6·12	2·06
Node strength	23	kg	268·6	293·5	9·27	288·3	7·33	−5·29
	50		268·6	283·5	5·55	284·5	5·92	−6·04
Physical property								
Dimensions	23				0·00		0·00	
(Average both dir)	50				0·00		0·00	
Thickness	23				0·00		0·00	
	50				0·00		0·00	
Mass	23				0·17		−0·12	
	50				−0·44		−0·13	
Specific gravity	23	kg/liter	0·965 8	0·971 9	0·63	0·972 7	0·71	
	50		0·965 8	0·973 1	0·76	0·968 9	0·32	
Density	23	kg/liter	0·963 4	0·969 5	0·63	0·971 5	0·84	
	50		0·963 4	0·970 7	0·76	0·966 5	0·32	
Volatiles	23	kg/liter	0·32	0·23		0·74		
	50		0·32	0·20		0·23		

Figure 5 (a) and (b) illustrates the potential of creep failures after only 1000 h of creep testing of an HDPE geomembrane and a high-strength woven geotextile, respectively. The maximum applied constant load was approximately only 40% of the ultimate tensile load for all the geosynthetics considered. In contrast, creep testing of an HDPE structural geogrid (see Fig. 5(c)) shows an adequate response even after 10 000 h of testing. Notice the total strain variation between the structural geogrid and the other geosynthetic.

The junction efficiency of structural geogrids ensures adequate load transfer from the ribs through the junction to the adjacent ribs. It is crucial to ensure full tensile strength transfer from the ribs through the junctions to the adjacent ribs. Failure to do so will require reduction of the allowable long-term tensile strength as specified in GRI-GG4a.

Connections in geosynthetics are normally the weak link in the chain. Uniformity of the geogrid structure is essential to prevent stress concentration and provide continuity of the geogrid's tensile strength through the connections utilized at roll ends. Typically, a mechanical joint (referred to as Botkin connection) is normally used to connect structural geogrids at roll ends. Failure to ensure adequate connections will require further reduction of the allowable tensile strength used for the final design.

APPLICATIONS

Innovative use of structural geogrids provides a means of establishing a new structurally-secure landfill constructed atop an existing facility. Several applications will necessitate the use of structural lining systems, including those described below.

Vertical Expansion

A vertical expansion, also referred to as piggybacking, is the construction of a new facility on top of a closed landfill. Differential settlement associated with the decomposition of the buried waste can potentially damage the new lining system. However, the use of a structural geogrid secures the new liner system and provides extended service operation at a fraction of the projected expense of a new facility or

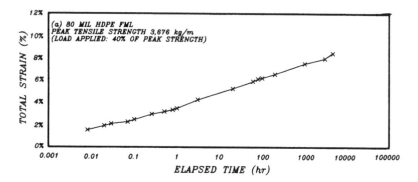

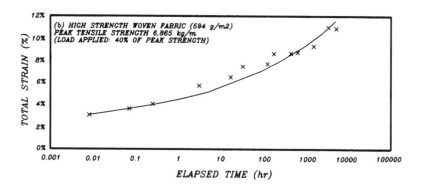

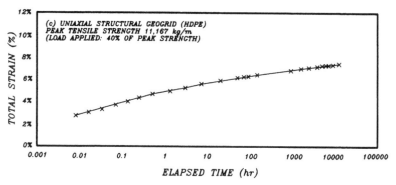

Figure 5. Typical creep data of (a) HDPE geomembrane, (b) high-strength woven fabric, and (c) HDPE structural geogrid.

Figure 6. Installation of structural geogrids on existing waste facilities.

such stop-gap measures as interstate exporting and hauling. Figure 6 illustrates the installation of a structural geogrid placed underneath the new lining system of a vertical expansion of an existing landfill.

Vertical expansions are rapidly becoming a conventional practice in landfill design. Proper design of a structural lining system will ensure that both existing and new waste are adequately separated and that both facilities act independently. Vertically-expanding landfills typically offer significant economic and permitting advantages in lieu of construction on a new site. Building a vertical expansion will eliminate the complicated siting process because it utilizes a previously permitted site. Therefore the public is less likely to object.

Lateral Expansion

Lateral expansion is the construction of a new cell alongside an existing facility. Standard construction procedures require side slopes of perimeter berms to be built at conventional angles (3H:1V or 4H:1V). Between adjacent cells (i.e. between the outer side slopes of two adjacent perimeter berms) there is generally an unused, although permitted area of land. This lost space is referred to as 'air space'. The use of this air space for storage will increase the life of the facility and maximize the effectiveness of the original permit.

Ash Storage

The operation of waste-to-energy plants requires ensuring adequate storage of the resultant ash. Cells for storing ash have been constructed within existing permitted municipal solid waste facilities through a structurally-supported lining system. New ash storage facilities have been constructed directly atop and within the existing landfill by incorporating a structural lining system to ensure long-term stability of the containment system.

Cover Systems

Due to the inherent instability of the waste, it is preferable to incorporate a structural geogrid in the design of a lining system for a final cover. The use of the structural geogrid beneath a cover system has several benefits: (a) long-term integrity of the system; (b) minimization of ponding of water thereby precluding water infiltration into the closed cell; and (c) a structural foundation for a potential vertical expansion. The last benefit is particularly important to the long-term sucess of the integrated waste management program. All facilities that are required to close may consider the use of a structurally-secured cover system. This will allow reuse of these facilities through vertical expansion in the future.

DESIGN OF A STRUCTURAL LINING SYSTEM

Structural geogrids have been used to support lining systems constructed over areas having the potential to exhibit differential settlements and/or localized subsidence. When a depression forms below a layer of geogrid supporting a landfill lining system, the tensile element deflects into the depression. This deflection has two effects: the materials (soil and/or waste) overlying the geogrid bend, which generates soil arching; and the structural geogrid strains which mobilizes a portion of the geogrid's tensile strength.

The most commonly-used design methodology was developed by combining the arching theory (for the fill material) with the tensioned membrane theory (for the structural element). This method has been progressively developed by Giroud (1981, 1982), Bonaparte and Berg (1987), and Giroud *et al.* (1988). Due to soil arching, part of the applied load is laterally transferred to the sides of the depression. As a result the vertical stress over the void is smaller than the applied overburden pressure (Giroud, 1988). Based on the tensioned membrane theory, the geogrid is assumed to behave as a structural element able to transmit tensile stresses but not shear stresses. Figure 7 illustrates the soil arching effect and the redistribution of the vertical stresses over the depressions.

Another important design criteria is the size of the depression. A circular depression with a certain radius *r* is the most common assumption used for the design of a structural support. Subsidence of the underlying subgrade is assumed to be caused by localized collapse. Conditions causing subsidence include collapse of Karstic subgrade, collapse of waste, such as barrels or household appliances. The 'rusted

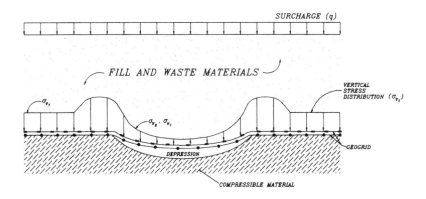

Figure 7. Effect of soil arching on vertical stress distribution.

refrigerator' theory, which assumes a circular depression with a 2 m diameter, is commonly used to analyze lining and cover systems placed over municipal solid waste landfill.

CONCLUSION

Expansions of existing landfills can be accomplished with the use of high-density polyethylene (HDPE) structural geogrids. The geogrid can provide the required support that will ensure the long-term integrity and prevent creep failure and/or rupture of any components of a lining and/or cover system.

REFERENCES

ASCE Structural plastics selection manual. Manuals and Reports on Engineering Practice No. 66.

ASCE Structural plastics design manual. Manuals and Reports on Engineering Practice No. 63.

Bonaparte, R. & Berg, R. R. (1987). The use of polymer geosynthetics to support roadways over sinkhole prone areas. Proceedings, Second Multidisciplinary Conference on Sinkholes and the Environmental Impacts of Karsts, Orlando, FL, February 1987, pp. 437–45.

Bonaparte, R. & Berg, R. R. (1987). Long-term allowable tension for geosynthetic reinforcement. Proceedings, Geosynthetics '87, vol. 1, New Orleans, LA, February 1987, pp. 181–92.

Geosynthetic Research Institute GRI Test Methods GG3a: Tension creep testing of stiff geogrids.

Geosynthetic Research Institute (1990). GRI Test Method GG4a: Determination of the long-term design strength of stiff geogrids (Revised). Drexel University, Philadelphia, PA.

Giroud, J. P. (1984). Analysis of stresses and elongation in geomembranes. Proceedings of International Conference on Geomembranes, vol. II, Denver, CO., pp. 481–4.

Giroud, J. P., Bonaparte, R., Beech, J. F. & Gross, B. A. (1988). Load-carrying capacity of a soil layer supported by a geosynthetic. *Proceedings of Kyushu International Symposium on Theory and Practice of Earth Reinforcement*, Japan, October 1988.

Koerner, R., 17, Koerner, G. R. and Hwu, B. L. (1990). Three dimensional, axysynmetric geomembrane tension test. ASTM STP on Geosynthetic Testing for Waste Containment Applications.

Landreth, R., Hokanson, S. A., Daniel, D., Richardson, G. N. & Koerner, R.

M. (1988). Seminars—Requirements for Hazardous Waste Landfill Design, Construction and Closure. US Environmental Protection Agency.

Tensar Inc. (1987). High density polyethylene resin characteristics which affect performance of products bearing high sustained loads in soil reinforcement applications. Tensar Corporation, Morrow, Georgia, October 1987, Tensar Technical Note PT13.0.

Tensar Inc. (1989). Design of TENSAR geogrid reinforcement to support landfill lining and cover systems. Tensar Corporation, Morrow, Georgia, Tensar Technical Note: WM3, May 1989.

Wrigley, N. E. (1988). The Durability and aging of geogrids. Proceedings, Durability and Aging of Geosynthetics, Geosynthetic Research Institute, Drexel University, Philadelphia, PA, December 1988.

5.5   Performance Tests for Geosynthetics

YVES HENRI FAURE, JEAN PIERRE GOURC & PATRICK PIERSON

IRIGM, University Joseph Fourier, BP 53X, Grenoble Cedex, France

INTRODUCTION

Geosynthetic lining systems can involve different products with different functions (Chapter 1). This chapter presents performance tests developed to determine the efficiency of some of these products, based on the following parameters:

- water permeability and diffusion coefficients of the geomembrane;
- friction coefficients of the soil/geocomposite interface or at the interface of two different geocomposites;
- transmissivity coefficients (drainage capacity) of geocomposites.

GEOMEMBRANE PERMEABILITY TEST

Multipurpose testing equipment has been developed at IRIGM (University Josep Fourier, Grenoble). This equipment is used to take permeation measurements (diffusion of a fluid through a membrane under an energy gradient) under a pressure and concentration gradient. This type of test is not suitable for determining the diffusion coefficient, as in the case of Fick's law (Park, 1986), but only permeability coefficients. The test is therefore accompanied by an absorption test. Permeability and diffusion coefficients characterise a permeant/geomembrane pair. The results presented here concern the water/LDPE and water/PVC pairs under hydraulic pressure gradient.

Measurement of the Water Permeability of a Geomembrane

This test involves measuring the flow rate of water through the geomembrane under the action of a hydraulic pressure gradient. The measured permeability coefficient K is defined by Darcy's equation (Faure *et al.*, 1990):

$$\Delta V/\Delta t = K \cdot \Delta h/T_g$$

where ΔV is the volume of water flow during time Δt, under a hydraulic pressure difference Δh through a membrane of thickness T_g. In the case of geomembranes, K is a function of the upstream and downstream pressures.

The IRIGM equipment can measure the K coefficients of different membranes (Faure *et al.*, 1989), and allows for efficient monitoring of temperature and lateral leakage of permeant. For this purpose, the equipment was placed in a chamber thermostatically controlled to $0 \cdot 1°C$. A peripheral chamber (Fig. 1) collects any permeant that leaks laterally so that it can be measured. The geometry of the apparatus prevents deterioration of the membrane that might result from excessive tightening and enables precise control of upstream and downstream pressures and trapped air. The equipment can be used to measure K for any geomembrane subjected to an upstream/downstream pressure difference of 0 to 800 kPa, at a temperature of 20 to 45°C.

Figure 2 gives the results of the permeation test of water through LDPE at 25°C. Later tests showed that the correct value of the permeability coefficient is close to the upstream value: $K_{upstream} = 6 \cdot 7 \times 10^{-15}$ m/s. The difference with $K_{downstream}$ ($= 11 \cdot 5 \times 10^{-15}$ m/s) is due to the fact that the regime was not yet perfectly stationary: the stationary state is achieved when the upstream flow rate is equal to the downstream flow rate; the membrane is then in equilibrium with the mechanical and physico-chemical conditions to which it is subjected. Tests on PVC conducted at 23°C showed that it was necessary to wait several months to obtain the steady state needed to estimate K. Figure 3 gives the measurements obtained for PVC

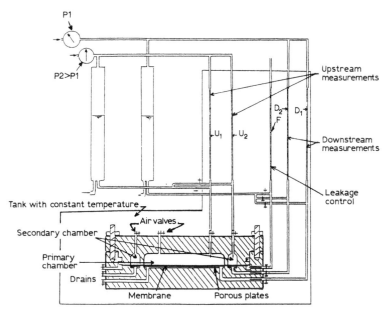

Figure 1. New apparatus used at IRIGM to determine geomembrane permeability.

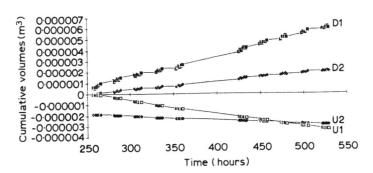

Figure 2. Results of the permeation test of water through LDPE at 25°C with 100 kPa as pressure difference; D1: $K = 1 \cdot 1 \times 10^{-14}$ m/s; D2: $K = 1 \cdot 1 \times 10^{-14}$ m/s; U2: $K = 5 \cdot 8 \times 10^{-15}$ m/s; U1: $K = 6 \cdot 2 \times 10^{-15}$ m/s.

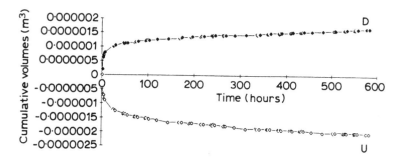

Figure 3. Results of the permeation test of water through PVC at 23°C with 100 kPa as pressure difference; D (downstream rooms with 12·5 cm³ of air): $K = 8·8 \times 10^{-16}$ m/s; U (upstream rooms with 5·5 cm³ of air): $K = 7·4 \times 10^{-16}$ m/s.

with a pressure difference of 100 kPa after a first test lasting 990 h in the same conditions: $K = 8·8 \times 10^{-16}$ m/s averaged over the two downstream surfaces and $K = 7·4 \times 10^{-16}$ m/s over the two upstream surfaces. The difference between these two K values here is due to an air flow through the membrane (to be subtracted from the water flow). The start of the curves shows the end of the mechanical deformation of the geomembrane. Last, it should be mentioned that with a pressure difference of 250 kPa, $K = 5 \times 10^{-16}$ m/s downstream and $K = 8 \times 10^{-16}$ m/s upstream (the leakage upstream explaining the high value of K).

Absorption Test

The absorption test is performed essentially to determine the diffusion coefficient and the maximum absorption rate of the permeant/geomembrane pair. Fick's second law is integrated to express the permeant concentration in the membrane as a function of time. It has been shown (Park, 1986; Faure *et al.*, 1989) that at the start of the process this relationship is linear with respect to $\sqrt{t}$ and can be simplified as follows:

$$D = (M(t)/\sqrt{t})^2 . \Pi . T_g^2/(16 . M(\infty)^2)$$

where M is the volume of water contained in the sample, t is the immersion time and T_g is the thickness of the membrane. An example

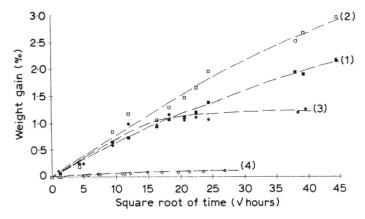

Figure 4. Results of the absorption test with water at 25°C: (1) PVC1; (2) PVC2; (3) bitumen; (4) LDPE.

of the values obtained on samples of PVC, bitumen and LDPE at 25°C is presented in Fig. 4, giving the coefficients: $D_{\text{LDPE}} = 3 \cdot 5 \times 10^{-13}$ m/s, $D_{\text{bitumen}} = 2 \cdot 4 \times 10^{-12}$ m/s, $D_{\text{PVC1}} = 9 \times 10^{-14}$ m/s and $D_{\text{PVC2}} = 2 \cdot 2 \times 10^{-13}$ m/s. It should be noted that the absorption tests take at least one month for LDPE, two months for bitumen and close to six months for PVC (the coefficients given for PVC are thus approximate, but of an order of magnitude confirmed by the current tests conducted with a strict operating method and with stable, controlled immersion conditions). Certain results are fairly irregular: this is due to temperature control problems and other problems originating in the sample (bitumen) (Fig. 4).

FRICTION TEST ON GEOMEMBRANES

Waste disposal sites must be capable of storing a maximum volume of waste and for this reason designers are considering increasingly steeper slopes for excavating and for covering the site. In such conditions—and this has been confirmed by incidents at actual sites—it is of the utmost importance to study the sliding resistance of the soil/geosynthetic composite (geotextile, geogrid and geomembrane), all types of interface being possible by combining the above components in pairs.

At IRIGM, a special apparatus (Gballou, 1991) based on the principle of the shear boxes used in soil mechanics has been designed to measure interface friction. This apparatus has particularly interesting experimental features:

- the device can be used for tests on interfaces covering a large area ($S = 0.40$ m $\times$ 0.25 m);
- the normal compression load N is distributed by means of a jacket membrane filled with pressurised fluid ($\sigma_N = N/S$);
- the geosynthetics can be placed in different positions.

A few applications are presented as examples in the following paragraphs:

Soil–Geomembrane Friction Measurement

The test presented concerns friction between sand and a bituminous geomembrane, this type of geomembrane being used frequently as a protective cover against rain water at waste disposal sites.

The experimental apparatus shown in (Fig. 5) was designed to show

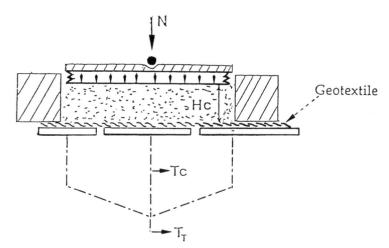

Figure 5. Friction unit with apparatus for measuring central frictional stress.

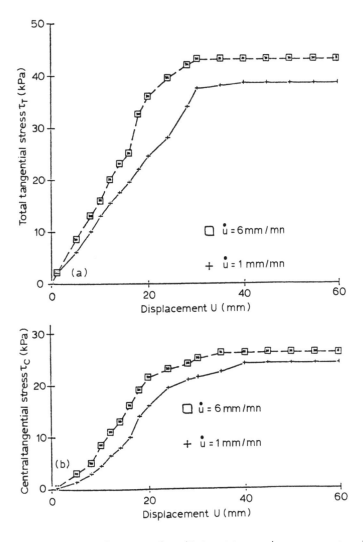

Figure 6. Sand–geomembrane interface friction: (a) central measurement and (b) total measurement. Influence of shear speed $\dot{u}$.

that edge effects can have a considerable influence on results, when grains get stuck between the edge of the unit and the geomembrane or when the deformation conditions are changed near the edges. The central tangential $T_c = \tau_c S_c$ on the central part of the interface ($S_c = S/2$) and the total tangential $T_T = \tau_T . S$ are measured separately. Comparison of the average tangential stresses τ_c and τ_T shows that the usual total measurement greatly overestimates the interface friction: $\tau_T \gg \tau_c$ (Fig. 6).

Figure 6 also shows the considerable influence of the tangential shear rate $\dot{u}$, due to the viscosity of bitumen at a temperature of $\theta = 21°C$. If $\bar{\phi}_g$ is the average interface angle of friction obtained for four normal stresses σ_N with $\tan \phi_g = \tau_{max}/\sigma_N$

$\dot{u} = 6$ mm/min $\bar{\phi}_{gT} = 43°$ $\bar{\phi}_{gc} = 26°$

$\dot{u} = 1$ mm/min $\bar{\phi}_{gT} = 29°$ $\bar{\phi}_{gc} = 19°$

Geonet–geotextile or geomembrane friction measurement

The experimental apparatus is shown in Fig. 7. The geogrid is placed on a metal plate and fixed from above. The geosynthetic material,

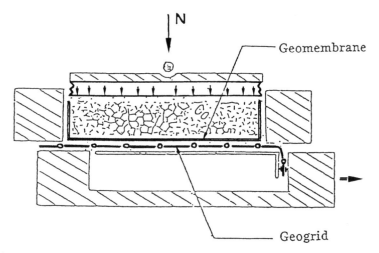

Figure 7. Influence of sheer speed on the sand–geomembrane interface friction.

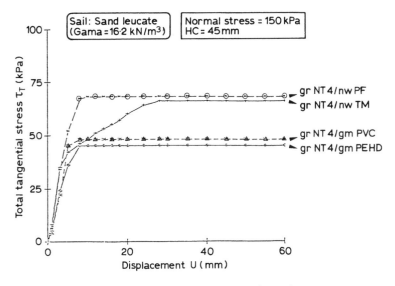

Figure 8. Geonet–non-woven geotextile or geomembrane friction.

placed in the upper part, is wrapped around a volume of sand in order to create a flexible contact between the geogrid and the geomembrane or geotextile, and to enable the geosynthetic to penetrate the geogrid mesh under the effect of compression. This constitutes a more realistic simulation than the standard test involving placing the geotextile or geomembrane bonded to a metal plate.

The test was carried out on a polyethylene geogrid (gr NT4) with a diamond-shaped mesh (10×10 mm^2) interfaced with two non-woven geotextiles (one needle-punched (nw PF) and one heat-bonded (nw TM)) and two geomembranes (gm) (one PVC and one PEHD) (Fig. 8).

The influence of the surface state is demonstrated (there is greater friction with the non-woven geotextiles than with the geomembranes), as well as the flexibility under bending (the PVC geomembrane is more flexible than the PEHD geomembrane).

nw PF: $\phi_g = 23°$ nw TM: $\phi_g = 23°$

gm PVC: $\phi_g = 20°$ gm PEHD: $\phi_g = 17°$

The values are relatively low for geogrid–geomembrane friction.

DRAINAGE CAPACITY TEST ON GEOCOMPOSITES

In order to determine the drainage capacity (transmissivity) of geocomposites, a special permeameter was designed (Fig. 9) to test large size samples: 250×550 mm (Suryolelono, 1991).

The geosynthetic test piece is placed inside a flexible latex membrane $0·5$ mm thick. This membrane separates the geocomposite from a pressurised fluid which exerts a uniform compressive stress on the two sides of the test piece of up to 400 kPa. This system results in deformation of the filter, which penetrates the voids of the draining part, the core, in the same way as when *in situ* under pressure from the soil. When the geocomposite core is a rigid structure with surface relief (grid, grooved or waffle-shaped sheet films) the filter is stretched and yielded between the rough parts of the core. These deformations reduce the voids between the solid parts of the core and decrease the drainage capacity of the composite.

Figure 10 shows the influence of the tensile modulus of the filter on the drainage capacity of the composite: gr NT4 net + geotextile, as a function of compressive stress. The drainage capacity of the net alone, compressed by rigid interleaved plates between the latex membrane and the grid, shows a distinct reduction in drainage capacity due to the filter.

On another more rigid gr NT5 net, a net heat-bonded geotextile was heat-bonded to the grid. Figure 11 shows the best behaviour of this gr NT5 net and also the best behaviour of the geotextile: the adherence of the geotextile to the nets filaments reduces puncturing and deformation of the geotextile.

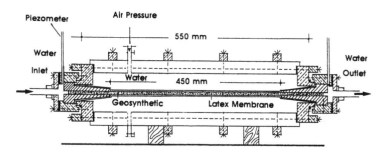

Figure 9. Transmissivity meter diagram.

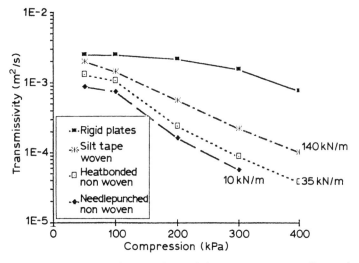

Figure 10. Influence of filter tensile modulus: non-woven needle-punched filter, 10 kN/m; non-woven heat-bonded filter, 35 kN/m; woven strips, 140 kN/m (average).

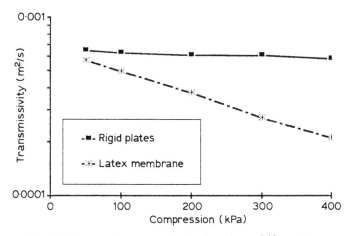

Figure 11. Gr NT5 net with a geotextile fixed on the grid filaments.

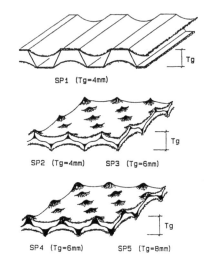

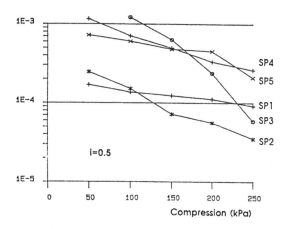

Figure 12. Influence of the core structure shape.

Other structures with surface relief were also tested (Fig. 12):

- a ribbed structure: SP1
- structures with sharp points: SP2 and SP3
- structures with rounded points: SP4 and SP5

Despite the fact that it is not very thick, the ribbed structure was more effective under strong compression than structures SP2 and SP3, which performed very poorly. The rounded points (SP4 and SP5) puncture the geotextile less than the sharp points and are therefore preferable.

The drainage capacity of the geocomposites under pressure thus depends not only on the shape and size of the empty space in the core, but also and above all on the mechanical behaviour of the filter stretched on its support. This also has a long-term effect on the behaviour (Faure, 1989).

It is thus important to test the geocomposites with apparatus that will clearly reveal the behaviour of the core-filter (or geomembrane) system, in order to determine the most suitable draining structure.

REFERENCES

Faure, Y. H. & Suryolelono, B. K. (1989). Comportement hydraulique des drains verticaux préfabriqués. Conference No. 10, Troisièmes Entretiens Jacques Cartier, ENTPE, Lyon.

Faure, Y. H., Pierson, P. & Delalance, A. (1989). Study of a water tightness test for geomembranes. Sardinia '89, Second International Landfill Symposium, Vol 2. CIPA. S.R.L., Milano. pp. B.III.1–11.

Faure, Y. H., Pierson, P., Artieres, O. & Gousse, F. (1990). Tests of geomembranes water permeability. 4th International Conference on Geotextiles, Geomembranes and Related Products. The Hague 1990, 28 May–1 June, vol. 2, pp. 543–8.

Gballou, Z. J. (1991). Renforcement des sols—Optimisation de l'essai d'interaction sol-geosynthétique. Thesis, IRIGM, Université de Grenoble 1, France.

Park, G. S. (1986). Transport principles—solution, diffusion and permeation in polymer membranes. In *Synthetic Membranes: Science, Engineering and Application,* ed. P. M. Bangay *et al.* Reidel Publishing Company, pp. 57–107.

Suryolelono, B. K. (1991). Etude du comportement drainant des géosynthétiques. Thesis, IRIGM, Université de Grenoble 1, France.

6. LEACHATE DRAINAGE AND COLLECTION

6.1 Design Criteria for Leachate Drainage and Collection Systems

PETER LECHNER

Department of Waste Management, Vienna University of Engineering and Technology, Karlsplatz 13, A 1040 Vienna, Austria

INTRODUCTION

A study of publications on the subject of landfilling in recent years reveals conspicuous shortcomings in the area of leachate drainage systems. This is all the more surprising in view of the fact that damage to drainage and leachate collection systems is one of the reasons for the contamination of groundwater by landfills.

In the early days of landfill engineering, leachate collection was managed by means of a drainage system meeting the requirements of agricultural drainage; in other words, a main and lateral drain system set in a gravel bed. However, the disadvantages of this technique very soon became apparent. Blockages occurred as a result of the ingression of fine materials and the formation of ocherous deposits. It was impossible to remedy these conditions due to the lack of accessibility caused by small pipe diameters and junctions.

An attempt was subsequently made to transfer the experience gained in wastewater engineering to the construction of leachate collection systems. A larger pipe diameter was now used, and every junction was provided with a shaft, often with a diameter of only 0·8 m. However, on completion of the landfill the shafts were no longer vertical, or had in many instances been destroyed. Even if such shafts were still in order today, they could not be accessed due to their small diameter and the landfill gas which they contain. It is thus impossible to inspect and clean a large part of the drainage system. If the leachate collection system does not function properly, there is a build-up at the base of the landfill, and hence an increase in the hydraulic gradient. Depending on the hydraulic gradient. k_f, leachate begins to percolate through the mineral liner, drastically reducing the protective function of this bottom liner.

The possibility that has become available in recent years of inspecting piping by means of video systems has revealed the following types of damage (Bothmann, 1989):

- deformation due to the imposed load and high temperatures (synthetic pipes);
- arching of the base of the pipes and in the case of plastic tunnel pipes;
- longitudinal and radial cracks (earthenware pipes), including distortion;
- fracturing, breakage and displacement of the pipes;
- silting-up with fine material;
- growth of bacterial sludge;
- recrystallization.

The main causes of damage are as follows (Weitzel, 1985; Bothmann, 1989):

- incorrect assumptions for static computations;
- construction deficiencies, such as local settlement, absence of or incorrect pipe bedding, inadequate pipe covering, unsuitable pipe materials, unsuitable gravel types;
- operating faults, such as driving vehicles over pipes with too little cover, displacement of pipes during landfilling activities, etc.

The renovation of such damaged pipe systems is generally extremely difficult, as the necessary conditions are not usually met:

- the pipes must be accessible from outside the landfill area,
- the lengths of piping must be suitably short,
- the piping must run in a straight line,
- there must be no junctions or inaccessible intermediate shafts,
- the pipe diameter must be sufficiently large.

The logical conclusion that can be drawn from this is that pipe systems must not only ensure the collection and rapid removal of leachate, but that it must also be possible to inspect, maintain and repair them.

FREE LEACHATE FLOW

The necessity for free leachate flow—the leachate must be able to exit from a landfill following a natural gradient—is now accepted (Lechner & Pawlick, 1988; Koerner & Koerner, 1989), and is already anchored in technical specifications. If this requirement is met, it is logically no longer permissible to fill a pit completely. As an alternative, the Austrian recommendations provide for partial filling with sufficient space for a leachate storage basin.

In the case of free leachate flow, the hydraulic gradient results only from the controlled flooding of the drainage system according to the hydraulic requirements for the runoff of the leachate. The hydraulic gradient will not generally exceed a value of 1·5. The value of $k_{i=1·5}$ relevant for the actual transmission through the base liner is thus much lower than $k_{i=30}$, which is the value usually used for the determination of the coefficient of permeability in the laboratory. At small hydraulic gradients i, mineral materials of low permeability $(k_{i=30} < 10^{-9}\,\mathrm{m/s})$ exhibit an exponential relationship between filter speed v and i. In other words, their resistance to the transmission of leachate is virtually infinitely large at a low hydraulic gradient (see Fig. 1). This exponential relationship is explained by the fact that the

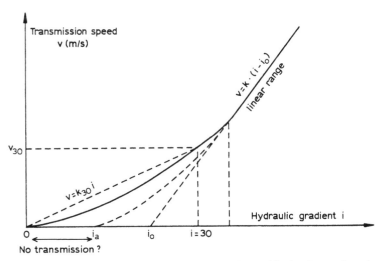

Figure 1. Correlation between transmission speed v and hydraulic gradient i.

adsorption water only contributes to the flow at an increasing hydraulic gradient. Only then does the cross-sectional area of flow—and with it permeability—increase. As long as this is not the case, a mineral barrier liner of low permeability can therefore be described as 'technically impermeable' with respect to flow.

If a pit is completely filled, in other words, if there is no free leachate flow, there is a build-up of leachate as soon as the pumping system fails, even if the leachate collection system is optimally constructed. In an extreme case, at the relevant depth of filling or height of build-up, all the free water in the mineral base liner begins to permeate the liner. Permeation is now governed by the relationship $v = k(i - i_0)$ (see Fig. 1). In other words, there is laminar flow through the liner. The mineral barrier liner is thus no longer 'technically impermeable'. This is the reason for the decisive importance attached to free leachate flow.

LEACHATE DRAINAGE MEASURES

Leachate drainage measures are designed to conduct the free water in a landfill area quickly to a leachate collection system. However, this requirement would appear to be inconsistent with the mode of operation of the reactor landfill.

In order to accelerate anaerobic decomposition in a reactor landfill, waste should have a suitably high water content, and should be highly compacted. Although this results in the desired high gas production, on the other hand, the high volume weight leads to severe compression of the lower layers of the landfill. High 'virtual water tables' can develop as a result. Such high water tables inside the landfill have been reported from many landfills. Stegmann (1990) gives the following theoretical explanation for this phenomenon.

The formation of gas causes the pores to become filled with gas, as a result of which the water is partially displaced. In view of the fact that the movement of landfill gas and leachate is in opposite directions, there is a further reduction in the permeability of the lower region of the landfill. This causes the formation of a local build-up of gas and leachate at relatively high pressure.

The problem now is to institute effective measures to drain off this leachate. Timely removal of the landfill gas by itself does not appear to offer a solution to this problem, as shown by investigations at Vienna Technical University's trial landfill (Lechner & Riehl-Herwirsch, 1993). According to Stegmann (1990), the provision of vertical wells to enable the leachate to be pumped off is not a solution either, because such wells apparently affect only a small area due to the influence of the gas.

One possible solution might be the arrangement of inclined, plane drainage layers. These would have to be connected with the base drainage by means of vertical drainage slits. However, this would mean joining gas and leachate systems. The problem that may possibly result from this—high gas pressures in the area of the base drainage—must be taken into consideration when designing a leachate collection system.

The dewatering of the base of the landfill should take the form of a wide-area drainage. This is one of the latest techniques in landfilling. McEnroe (Chapter 6.2) developed a formula for estimating the transmission of leachate through a mineral barrier liner with a drainage layer lying above it. This analysis assumes that the drainage layer is fully drained, and that the barrier layer is saturated, but that the initial saturated depth is much smaller than the barrier thickness. In this case the leakage volume depends on:

- the inflow volume
- the barrier slope
- the distance of the leachate collection pipes (drainpipes)
- the porosity of the drainage layer
- the hydraulic conductivity of the drainage layer
- the hydraulic conductivity of the barrier layer.

There are technical restrictions with regard to a reduction in the permeability of a mineral barrier liner and to large increase in slope. Furthermore, it is also impractical to install pipes at small intervals, so that the most practical solution would seem to be the use of a coarse filter gravel with a high porosity.

On the basis of calculations. Ramke (1989) also comes to the conclusion that if the permeability of the filter gravel is too low, e.g. only 10^{-4} m/s, the hydraulic gradient will depend primarily on the gradient at the base of the landfill and the distance between pipes. Only if the filter gravel has a high permeability of $k = 10^{-2}$ m/s will the effect of the gradient and the distance between the pipes be of secondary importance. As mentioned above, the hydraulic gradient should not exceed a value of $i = 1 \cdot 5$.

In practice, therefore, it is necessary to use filter gravel material with the highest possible permeability and porosity. It is a proven fact that the use of a fine-graded material and is not to be recommended for reasons of filter stability, and because of the increased propensity for incrustation (see Chapter 6.2). There is no danger of ingression by fine materials from the lowest layer of waste due to the slow seepage speeds encountered in municipal waste. Even in instances where fine-grained, uncompacted special waste is to be disposed of, the provision of an additional layer of well-graded material would appear to be necessary only in exceptional cases.

The most important requirements in designing a base drainage layer are therefore as follows:

- A filter gravel of round grain 16/32 mm, with sufficient grain strength, resistant to weathering and with a low carbon content should be used.

  If coarser grains are used, there are complications in driving vehicles over the drainage layer. If the grain strength is insufficient, the grains may break and cause damage to the synthetic liner. A heavy geotextile can be used to provide additional protection for the synthetic liner.

  CO_2 from biochemical processes releases calcium with the formation of hydrogen carbonate. A fall-off in partial CO_2-pressure then leads to calcium crystallization (precipitation) in the area of the drainage pipes.

  The installation of a base drainage system should be carried out with the aid of a backloader. If the gravel layer is graded using a bulldozer, there is a danger of folds forming in the synthetic liner (Weitzel, 1985).

- The drainage layer thickness should be a minimum of 50 cm. If the gravel layer has an insufficient thickness, there are problems with driving over it, and the base liner may be damaged by the compacting vehicles. There is also a risk that too thin a drainage layer may reduce the functional life of the drainage system.

- The gradient should be a minimum of 2%. A lower gradient than this cannot compensate sufficiently for subsequent settlement, and there may be a local build-up of leachate. A sufficient gradient in the base drainage system coupled with the material properties listed above should ensure adequate drainage at the base of the landfill even if the collection pipes should become blocked.

LEACHATE COLLECTION

There do not as yet exist set standards or recognized methods for the structural calculation of leachate collection pipes and shafts. For this reason, the possible interactions between the inhomogeneous wastes and the construction components are often incorrectly estimated or not even accounted for. Moreover, the leachate collection system must periodically be verified and maintained. Thus, the system must be designed fulfilling its primary function of landfill drainage while being accessible over the long term. In the past, this aspect was often neglected.

Leachate Collection Pipes

Suitable pipes are made of either HDPE (high-density polyethylene) or rigid earthenware. Thin-walled pipes with insufficient profile stiffness are not appropriate. Since drainpipes must remain functional over a long period of time, the appropriate wear factors must be accounted for. These include factors for temperature, surrounding environment and time. The long-term modulus of elasticity must also be determined independently of the above mentioned factors (Hoch, 1989).

The conditions around the pipes, i.e. the type and material of the pipe covering and pipe bed also significantly influence the load-bearing capabilities of the pipe, as have been shown in the investigations of Hoch (1989) and Reuter and Prein (1989).

A state of the art application is illustrated in Fig. 2. in this case, the leachate collection pipe (drainpipe) is laid in a rounded pipe bed in order to ensure an as even as possible pressure distribution in the area surrounding the pipe. The covering, i.e. the fill surrounding and covering the pipe, consists of filter gravel. Even though the calculated contact angle between the pipe and its bed is 90°, the uncertainties in the calculations mean that an actual angle of 120° should be used.

Still being examined is the question of whether one should for

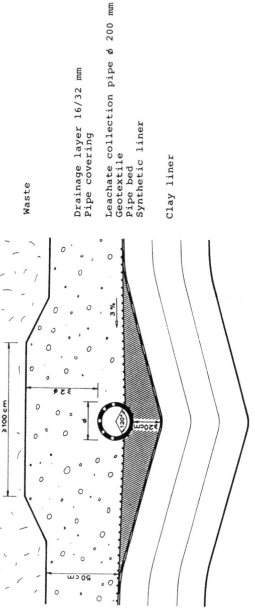

Waste

Drainage layer 16/32 mm
Pipe covering

Leachate collection pipe ø 200 mm
Geotextile
Pipe bed
Synthetic liner

Clay liner

Figure 2. Leachate collection pipe—example of an application.

structural reasons use sand (having for instance a grain size of 0/4 mm) as a pipe bed, or whether one should utilize materials high in clay content. The now-suspected softening of these latter materials through exposure to leachate could not be confirmed in the studies of Reuter and Prein (1989). Their investigations showed that materials high in clay content, compressed as they would be under a landfill site, did not reach a fluid state. Instead, they display a deformation modulus which can be determined through compression tests. Clearly, the use of materials high in clay content as pipe bedding should not be discounted entirely.

It is important to note that the material selection of the drain pipe must be made independently of the available gravel grain size. The reverse must also be true. Large grain sizes require either pipes with high pressure capabilities or ones made of a malleable material. Since gravel with a grain size of 16/32 mm is widely used, the application of earthenware pipes should be subject to thorough investigation.

While the depth of the covering over the top of the pipe is only of minimal importance in the case of HDPE pipes, it is of some significance in the case of earthenware pipes. In recent years, so-called deformation mats have been layed over earthenware pipes to accommodate their rigid deformation characteristics. The effectiveness of these mats in disposal sites deeper than 10 m is now being questioned.

Leachate drainage pipes must be regularly verified and maintained. Furthermore, it must be possible to clean and remedy them. The selection of pipe diameter, length and layout must take this into account.

Shafts and Tunnels

Shafts and tunnels are used to collect leachate and to allow for periodic maintenance of the drainage system. Shafts are particularly exposed to physical and chemical stresses when they are built into the landfill body itself. There are two problems with such positioning. Firstly, the necessary values for structural calculations are unknown for a landfill of different waste materials, and secondly, the shafts and tunnels should be safely accessible over a long time period. At present, it is the task of the structural engineer to identify possible load conditions. For this reason it is useful to locate shafts at the edges or tunnels even below the landfill (see Fig. 3).

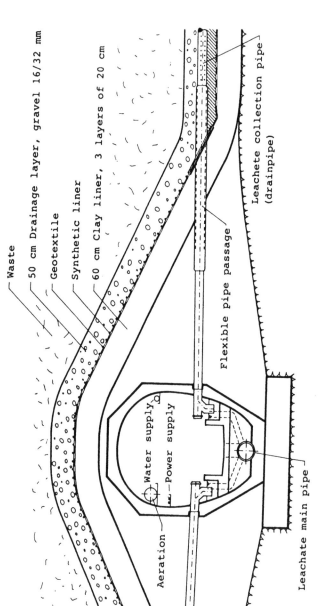

Figure 3. Tunnel for leachate collection below a landfill.

Waste

50 cm Drainage layer, gravel 16/32 mm

Geotextile

Synthetic liner

60 cm Clay liner, 3 layers of 20 cm

Leachete collection pipe (drainpipe)

Flexible pipe passage

Water supply

Power supply

Aeration

Leachate main pipe

It is difficult or impossible to quantify even in general terms values for the soil mechanics of garbage, especially over the long term. A covering consisting of material with known mechanical properties such as gravel evens out the loading and facilitates calculation. HDPE shafts must be able to shift in the horizontal—something which must be taken into account in choosing the type and thickness of the surrounding covering. Special attention should be paid to areas of particular weakness, such as where leachate collection pipes pass through the shaft wall.

Shafts must always rest on an appropriately dimensioned foundation below the landfill barrier layer. The design and construction of the sealed connection between the periphery of the shaft and the barrier layer requires particular care.

The most important criteria for the calculation and layout of the leachate collection systems are:

- The structural calculation of individual system components must be made with knowledge of the possible load conditions. If required, appropriate laboratory tests should be carried out.
- Leachate collection systems must be designed to allow for both verification and repair. This means that the pipe diameter must be no smaller than 200 mm, and that the pipe itself should be laid straight and in such a manner that its entire length can be cleaned.
- The leachate collection pipes should be laid in a suitable bedding with an appropriate covering material.
- Shafts for maintenance and for leachate collection should be located at the pipe ends. Shafts within the landfill body should be avoided if at all possible. Should shafts within the landfill body be necessary, the trouble-free removal of gasses and supply of fresh air must be ensured. The shaft diameter must be chosen to allow for periodic verification and maintenance of the collection system. This requires shafts with a minimum diameter of 2·5 m.

REFERENCES

Bothmann, P. (1989). Demands for the repair of leachate collection systems in landfills—overview on repair systems. In *Abfallwirtschaft in Forschung und*

Praxis, Vol. 30; ed. S. Fehlau. Erich Schmidt Verlag, Berlin, 49–64 (in German).

Hoch, A. (1989). Structural analysis of shafts and leachate collection pipes. In *Publications of the Institute of Foundations of the Landesgewerbeanstalt Bayern,* Vol. 54; ed. E. Gartung. LGA, Nürnberg, 225–57 (in German).

Koerner, G. & Koerner, R. (1989). Biological clogging in leachate collection systems. In *Proc. of 2nd Inter. Landfill Symposium.* Cagliari, B.XI-1–B.XI-18.

Lechner, P. & Pawlick, R. (1988). Directions for Landfilling Household Waste, ed. Ministry of Environment, Youth and Family/Ministry of Agriculture and Forestry. Vienna, 1988 (in German).

Lechner, P., Riehl-Herwirsch, G., 1993. The Test Landfill 'Breitenen', Institute of Geology and Institute of Water Quality and Waste Management, Interim Report. TU Vienna (in preparation), (in German).

Ramke, H. G. (1989). Leachate collection system. In *Sanitary Landfilling—Process, Technology and Environmental Impact,* ed. T. H. Christensen, R., Cossu, & R. Stegmann. Academic Press, London, pp. 343–64.

Reuter, E. & Prein, T. H. (1989). Investigations into the stability of drainage pipes in landfills. In *Publications of the Institute of Foundations of the Landesgewerbeanstalt Bayern,* Vol. 54; ed. E. Gartung. LGA, Nürnberg (in German).

Stegmann, R. (1985). Landfill gas extraction. In *Sanitary Landfilling—Process. Technology and Environmental Impact,* ed. T. H. Christensen, R. Cossu, & R. Stegmann. Academic Press, London, pp. 167–74.

Weitzel, H. (1985). Contribution to the water balance and stability of landfills. In *Müll and Abfall,* 17(4), 126–8 (in German).

6.2 Hydraulics of Leachate Collection and Cover Drainage

BRUCE M. McENROE

School of Engineering, Department of Civil Engineering, University of Kansas, Lawrence, 66045 Kansas, USA

INTRODUCTION

In designing a drainage system for a bottom liner or final cover of a landfill, one must consider two measures of hydraulic performance: the maximum saturated depth over the barrier, and the amount of leakage through the barrier. The hydraulic behavior of a drainage system depends upon the amount and temporal pattern of percolation into the drainage layer. Where rainfall is abundant, it is reasonable to assume a continuous steady inflow, at least as a first approximation. In arid locations, periods of significant inflow are likely to be infrequent and relatively brief, so the system's response to an isolated instantaneous inflow is more relevant.

This chapter presents two tools for preliminary evaluation of drainage-system designs. The first is a set of equations that define the saturated-depth profile for steady drainage over an impervious barrier. The second is a simple algebraic formula for estimating the leakage resulting from a single instantaneous inflow to a drainage layer over a compacted-clay barrier.

SATURATED-DEPTH PROFILE FOR STEADY INFLOW

This section presents an analytical solution for steady drainage over an impervious sloping barrier. Figure 1 defines some of the relevant

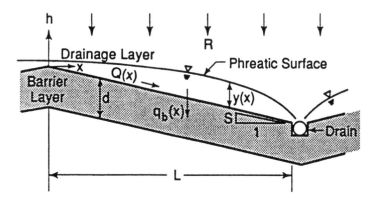

Figure 1. Definition sketch for landfill drainage system.

variables. The upstream boundary may be either a drainage divide, as shown, or an impervious sidewall.

Analytical Solution

Based on the Dupuit assumption for unconfined flow, the differential equation governing steady drainage on a sloping barrier is

$$K_d y\left(\frac{dy}{dx} - S\right) + Rx = 0 \tag{1}$$

where x is the horizontal distance from the upstream boundary, y is the saturated depth over the barrier, S is the dimensionless slope of the barrier, K_d is the hydraulic conductivity of the drainage layer, and R is the vertical inflow rate. Equation (1) is solved for a specified saturated depth, Y, at the downstream boundary.

This problem and its solution are expressed more concisely in dimensionless form. The following dimensionless variables are defined: $x_\star = x/L$, $y_\star = y/L$, $Y_\star = Y/L$, and $R_\star = R/K_d$. The dimensionless form of eqn (1) becomes separable through a change of variables. Defining $u_\star$ as $y_\star/x_\star$, substituting $u_\star x_\star$ for $y_\star$, and then separating variables leads to the equivalent form

$$-\frac{dx_\star}{x_\star} = \frac{u_\star\, du_\star}{R_\star - Su_\star + u_\star^2} \tag{2}$$

Equation (2) can be integrated directly. The constant of integration is determined by the condition $u_\star = Y_\star$ at $x_\star = 1$. The result depends upon the relative magnitudes of $R_\star$ and S. For the case $R_\star < S^2/4$, the result is

$$x_\star = \left[\frac{R_\star - SY_\star + Y_\star^2}{R_\star - Su_\star + u_\star^2}\right]^{1/2}\left[\frac{(2Y_\star - S - A)(2u_\star - S + A)}{(2Y_\star - S + A)(2u_\star - S - A)}\right]^{S/2A} \tag{3}$$

where $A = (S^2 - 4R_\star)^{1/2}$. For the special case $R_\star = S^2/4$, the result is

$$x_\star = \frac{S - 2Y_\star}{S - 2u_\star}\exp\left[\frac{2S(Y_\star - u_\star)}{(S - 2Y_\star)(S - 2u_\star)}\right] \tag{4}$$

For the case $R_\star > S^2/4$, the result is

$$x_\star = \left[\frac{R_\star - SY_\star + Y_\star^2}{R_\star - Su_\star + u_\star^2}\right]^{1/2}$$

$$\times \exp\left\{-\frac{S}{2}\left[\tan^{-1}\left(\frac{2u_\star - S}{B}\right) - \tan^{-1}\left(\frac{2Y_\star - S}{B}\right)\right]\right\} \tag{5}$$

where $B = (4R_\star - S^2)^{1/2}$. The solutions for the cases $R_\star = S^2/4$ and $R_\star > S^2/4$ were first presented by Demetracopoulos and Korfiatis (1984). The solution for the more important case, $R_\star < S^2/4$, is new.

Practical Applications

Saturated-depth profile. One obtains the coordinates $(x_\star, y_\star)$ of a single point on the saturated-depth profile by first assuming a value of $u_\star$ and then using eqn (3), (4) or (5) to determine the corresponding value of $x_\star$. The dimensionless saturated depth, $y_\star$, at this dimensionless location is the product of $u_\star$ and $x_\star$. By repeating this procedure for a number of $u_\star$ values over the appropriate range, one can define the entire profile. Several different types of profiles are possible, depending upon the relative magnitudes of S, $R_\star$ and $y_\star$.

If the drain system is working properly, the water level in the drain trench will be below the top of the barrier layer, and will have no effect on the saturated-depth profile over the barrier. This is termed the

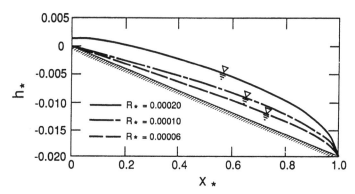

Figure 2. Dimensionless saturated-depth profiles for a barrier slope of 0·02.

free-drainage condition. Under this condition, the hydraulic gradient at the brink of the drain is approximately -1, and consequently $Y_\star$ is approximately equal to $R_\star$. If the drain system is undersized or partially obstructed, $Y_\star$ may exceed this value.

With free drainage at the downstream boundary, two types of upstream boundary conditions are possible. Where $R_\star$ is less than or equal to $S^2/4$, the saturated depth is zero and the hydraulic gradient is nonzero at the upstream boundary. Where $R_\star$ is greater than $S^2/4$, the saturated depth is nonzero and the hydraulic gradient is zero at the upstream boundary. In both cases, no flow occurs across this boundary. Figure 2 shows example saturated-depth profiles for $R_\star$ values less than, equal to, and greater than $S^2/4$ under the free-drainage condition (i.e. $Y_\star = R_\star$). The point of maximum saturated depth moves upstream as the ratio of the inflow rate to the hydraulic conductivity of the drainage layer increases.

Maximum saturated depth. The maximum saturated depth along the profile, y_{max}, is of particular interest. Regulatory agencies commonly require that this maximum saturated depth not exceed a certain value. For example, US Environmental Protection Agency regulations for hazardous-waste landfills limit the leachate depth over the bottom liner to 30 cm. Figure 3 shows the relationship between the dimensionless maximum saturated depth, $y_{max\star}(=y_{max}/L)$, and the dimensionless inflow rate, $R_\star$, for barrier slopes of 1, 2, 5 and 10%, assuming free drainage at the downstream boundary. This graph can

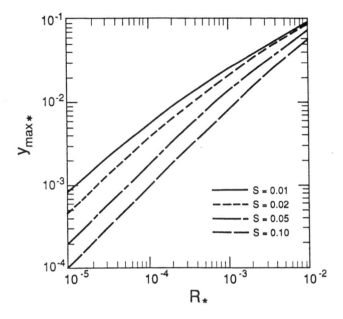

Figure 3. Dimensionless maximum saturated depth.

be used to estimate the maximum saturated depth under a steady inflow for most drainage-layer designs.

Example

Problem. A proposed drainage system has a barrier slope of 2% and a maximum drainage distance of 30 m. The drainage-layer material is a fine sand with a hydraulic conductivity of 10^{-3} cm/s. Regulations limit the maximum saturated depth over the barrier to 30 cm. Find the steady inflow rate that will produce this saturated depth.

Solution. For this design, $y_{max\star}$ equals 0·01. Figure 3 shows that the corresponding maximum value of $R_\star$ is 0·00035. Therefore, the maximum saturated depth will not exceed 30 cm provided that the inflow rate does not exceed 11 cm/year. If the actual inflow rate is expected to exceed the limiting value, the design should be modified by increasing the barrier slope, decreasing the maximum drainage distance, or using a drainage-layer material with a larger hydraulic conductivity.

LEAKAGE FRACTION FOR INSTANTANEOUS INFLOW

In this section, a simple formula is developed for the leakage resulting from a single instantaneous inflow to a drainage layer over a compacted-clay barrier.

Development of Leakage Formula

This analysis assumes that the drainage layer is fully drained and the barrier is saturated when the inflow occurs, and that the initial saturated depth is much smaller than the barrier thickness. Under these conditions, the leakage volume, V_L (expressed as an equivalent uniform depth of water), depends on six quantities:

- the inflow volume, V_o (also expressed as a depth of water);
- the barrier slope, S;
- the maximum horizontal drainage distance, L;
- the effective (drainable) porosity of the drainage layer, ϕ_e;
- the hydraulic conductivity of the drainage layer, K_d;
- the hydraulic conductivity of the barrier, K_b.

Using dimensional analysis, one can show that the leakage fraction, V_L/V_o, is determined by the values of three dimensionless parameters: S, K_b/K_d, and $V_o/(\phi_e L)$.

The relationship between the leakage fraction and these three dimensionless parameters was investigated by solving the governing differential equation numerically. This equation was expressed in the dimensionless form

$$\frac{\partial h_\star}{\partial t_\star} = y_\star \frac{\partial^2 h_\star}{\partial x_\star^2} - \frac{\partial y_\star}{\partial x_\star} \frac{\partial h_\star}{\partial y_\star} - q_\star = 0 \qquad (6)$$

Here t is the time since the inflow occurred, h is the elevation of the phreatic surface above the datum defined in Fig. 1, and q_b is the local rate of leakage through the barrier, with the dimensionless forms of these variables defined as follows: $t_\star = (K_d t)/(\phi_e L)$, $h_\star = h/L$, and $q_\star = q_b/K_d$. The other variables are as defined previously. Under the stated assumptions, the rate of leakage through the barrier equals the hydraulic conductivity of the barrier where the saturated depth over the barrier is nonzero, and zero where the saturated depth over the barrier is zero.

This problem was solved using a nonlinear implicit finite-difference scheme. In this scheme, $h_\star$ is considered the dependent variable and the $y_\star$ terms are considered coefficients. An iterative solution procedure is used to determine the values of these coefficients. A hydraulic gradient of -1 is maintained at the downstream boundary at all times. A hydraulic gradient of zero is maintained at the upstream boundary until the saturated depth at this boundary equals zero. As drainage continues, the upper boundary of the finite-difference grid is moved down the sloping barrier, tracking the upper end of the saturated zone.

Figure 4 shows some numerical results for a typical case. In this example, the inflow volume is 10 cm, the effective porosity is 0·40, the barrier slope is 0·02, the maximum horizontal drainage distance is 30 m, the hydraulic conductivity of the drainage layer is 10^{-2} cm/s, and the hydraulic conductivity of the barrier is 10^{-7} cm/s. The computed leakage volume is 0·86 cm, for a leakage fraction of 8·6%. Drainage ceases 128 days after inflow begins.

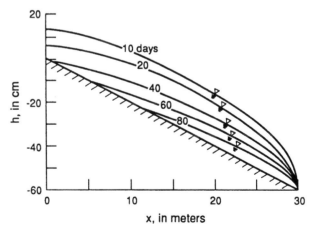

Figure 4. Recession of the saturated zone over a barrier following instantaneous inflow.

In a set of numerical experiments, leakage fractions were computed for all possible combinations of four barrier slopes (0·01, 0·02, 0·05 and 0·10), four barrier values of K_b/K_d (10^{-7}, 10^{-6}, 10^{-5} and 10^{-4}), and five values of $V_o/(\phi_e L)$ (0·0001, 0·0003, 0·001, 0·003 and 0·01). Multiple linear regression in log space produced an equation for leakage fraction as a power function of these three parameters. However, a simpler and equally satisfactory relationship was obtained by combining these three dimensionless parameters into a single parameter. This relationship is

$$\frac{V_L}{V_b} = \begin{cases} 0.64M^{0.91}, & M \leq 0.8 \\ 1 - \dfrac{1}{2.6M}, & M \geq 0.8 \end{cases} \tag{7}$$

where

$$M = \frac{K_b L \phi_e}{K_d S V_o} \tag{8}$$

Figure 5 shows a graph of eqn (7) and the numerical results upon which it is based. The equation has a standard error of estimate of about 26% (0·11 log cycles) for $M \leq 0.8$. Because it is a good

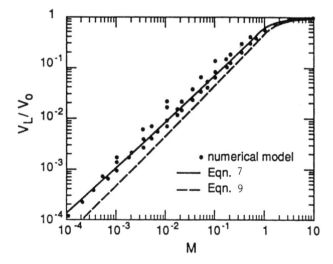

Figure 5. Leakage fraction for an instantaneous inflow.

indicator of the leakage fraction, the dimensionless number M is termed the leakage number.

Comparison with Wong's Approximate Analytical Solution

An approximate analytical solution for leakage following a single instantaneous inflow was first presented by Wong (1977) and later corrected by Kmet *et al.* (1981). Wong assumed that the depth in the saturated zone remains spatially uniform while the upper end of the saturated zone moves downslope at a constant speed, driven by a hydraulic gradient equal to the barrier slope. Where the initial saturated depth is much smaller than the barrier thickness, one can show that Wong's assumptions lead to the following simple formula for the leakage fraction:

$$\frac{V_L}{V_o} = \begin{cases} 0 \cdot 5M, & M \le 1 \\ 1 - \dfrac{1}{2M}, & M \ge 1 \end{cases} \tag{9}$$

It is interesting to note that this formula is quite similar in form to eqn (7). Figure 5 compares the two formulas with numerical solutions of the governing differential equation over the range of practical interest. This comparison shows that eqn (9) underestimates leakage. This is primarily because Wong's model overestimates the speed at which the trailing edge of the saturated zone moves downslope; a consequence of assuming the hydraulic gradient equal to the barrier slope.

Example

Problem. A proposed design for a landfill cover contains a compacted clay barrier $0 \cdot 9$ m thick with a hydraulic conductivity of 10^{-7} cm/s, and a drainage layer of coarse sand with a hydraulic conductivity of 10^{-2} cm/s and effective porosity of $0 \cdot 4$. The barrier slope is 2% and the maximum horizontal drainage distance is 100 m. Because the landfill is located in a semi-arid region, significant inflows to the cover drain are expected only infrequently. Estimate the leakage resulting from an instantaneous 4-cm inflow.

Solution. Applying eqns (7) and (8), one finds that the leakage number, M, has a value of $0·8$, and the corresponding leakage fraction is $0·32$. Therefore the estimated leakage is $1·6$ cm. Because the leakage fraction is so large, the cover should be redesigned. Leakage can be reduced significantly by making one or more of the following design changes: (1) increasing the hydraulic conductivity of the drainage layer, (2) reducing the hydraulic conductivity of the barrier layer, (3) increasing the barrier slope, or (4) reducing the horizontal drainage distance. The first option, which can be accomplished by using a coarser drainage material, is likely to be the most feasible. The hydraulic conductivity of a clay can be reduced only so far through compaction, and a large increase in slope or reduction in drainage distance is usually impractical or undesirable.

CONCLUSIONS

In designing a drainage system for a bottom liner or final cover, two primary concerns are the maximum saturated depth over the barrier and the amount of leakage through the barrier. The analytical solution presented here defines the entire saturated-depth profile over an impervious sloping barrier under steady-state conditions. The shape of the profile depends upon the barrier slope and the ratio of the inflow rate to the hydraulic conductivity of the drainage layer. The maximum saturated depth can be estimated directly using Fig. 3. Equation (7) provides a quick estimate of the leakage resulting from a single instantaneous inflow to a drainage system over a compacted-clay barrier. This estimate is based on the value of a single dimensionless parameter termed the leakage number. This parameter accounts for the hydraulic conductivities of the drainage and barrier layers, the lateral drainage distance, the barrier slope, and the inflow volume.

REFERENCES

Demetracopoulos, A. C. & Korfiatis, G. P. (1984). Design considerations for landfill bottom collection systems. *Civil Engineering for Practicing and Design Engineers,* **3**(10), 967–84.

Kmet, P., Quinn, K. J. & Slavik, C. (1981). Analysis of design parameters affecting the collection efficiency of clay-lined landfills. In *Proceedings of the Fourth Annual Madison Waste Conference of Applied Research and Practice on Municipal and Industrial Waste*. University of Wisconsin Extension, Madison, pp. 250–65.

Wong, J. (1977). The design of a system for collecting leachate from a lined landfill site. *Water Resources Research*, **13**(2), 404–10.

6.3 Design of Geosynthetic Drainage System in Landfills

ANDREA CANCELLI[a] & PIETRO RIMOLDI[b]

[a] Department of Earth Sciences, University of Milan, Via Mangiagalli 34, 20133 Milano, Italy
[b] Geosynthetics Division, TENAX S.p.A., Via Industria 3, 22060 Viganò (CO), Italy

INTRODUCTION

Drainage of leachate in landfills is mandatory in order to ensure the geotechnical stability of the waste and to prevent release of leachate into surface and ground water. The significance of these issues demands that drainage systems be properly designed and carefully installed.

Geosynthetics (geomembranes, geotextiles, geogrids, geonets and geocomposites) have during recent decades found increasing uses in landfill construction. Compared to traditional liners and drains, geosynthetic materials present some technical advantages:

- reduced thickness allowing additional volume for waste;
- constancy of physical, hydraulic and mechanical properties and related quality control;
- ease and reduced time of installation.

An example of a landfill equipped with geosynthetics is schematically shown in Fig. 1.

Designing geosynthetic drainage systems can present peculiar difficulties and uncertainties when these hydraulic structures come into contact with wastes instead of natural soils. For instance, in many

543

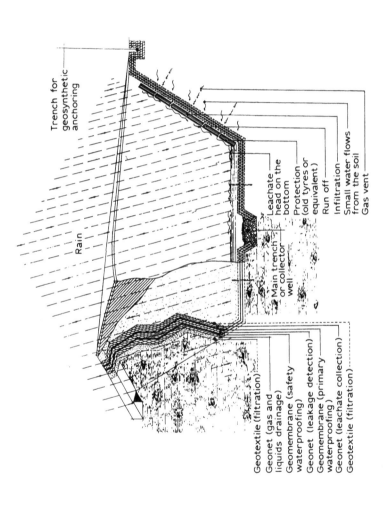

Figure 1. General scheme of a landfill using geosynthetics.

applications geotextiles should not come into contact with leachate, and interactions can occasionally originate from defective installation, or from long-term deterioration of the geomembrane; in other cases, a geotextile is used as a filter and drainage for leachate collection systems, and comes directly into contact with leachate from the start of the landfilling operation. Therefore it should be properly selected and designed for such working conditions (Cancelli & Cazzuffi, 1987). Supplementary problems, due to previously unforeseen leachates, are experienced in landfills in which chemical wastes are mixed with domestic garbage.

Some general criteria for designing with geosynthetics are well known in international literature (see, e.g. Giroud, 1984; Van Zanten, 1986). However, the increasing use of geosynthetics (including geonets and geonet-based geocomposites) requires theoretical knowledge of the behaviour of geosynthetic drainage systems in waste landfills and the development of design criteria for practical applications.

THEORY OF LEACHATE COLLECTION AND DRAINAGE

Evaluation of Input Flow

The quantity of liquid reaching the primary synthetic drainage system depends mainly on local conditions such as rainfall, type of soil cover, and actual state of the landfill (in use or after capping). All the possible situations can be reduced to the scheme presented in Fig. 2.

The rainfall per unit area (q_r) is given by:

$$q_r = P/A \tag{1}$$

where

P = rainfall (m^3/s),
A = horizontal area (m^2), and
q_r = rainfall per unit horizontal area (m^3/s/m^2).

Since the actual surface is sloping, the effective area is:

$$q_1 = P/A_1 = P/(A/\cos \alpha) = P \times \cos \alpha /A = q_r \times \cos \alpha \tag{2}$$

where q_1 = rainfall per unit area on slope (m^3/s/m^2).

Not all the rainfall reaches the drainage system, since a part of it becomes runoff; therefore the input rainfall is given by:

$$q_d = q_1 \times f = q_r \times \cos \alpha \times f \tag{3}$$

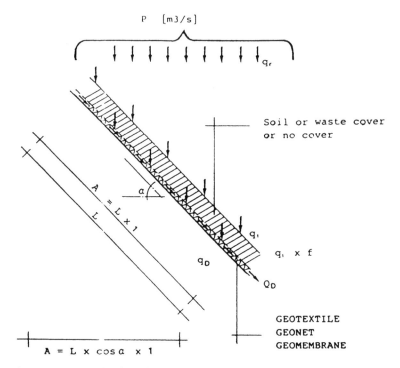

Figure 2. Scheme for the calculation of input flow.

where

f = coefficient of infiltration of the cover material, and

q_d = rainfall, per unit area of slope, entering the drainage system $(m^3/s/m^2)$.

The input rainfall collects along the slope to give the total input flow per unit width across the slope:

$$Q_d = q_d \times L \times 1 = q_d \times L \tag{4}$$

where

L = length of the slope (m), and

Q_d = input flow (m^2/s).

For further evaluation of the input flow, the state of the landfill, whether in use or finally capped, must be considered.

A landfill in use. For a landfill in use the most severe conditions for the drainage system occur during intense rainfalls: therefore q_r corresponds now to the rainfall intensity j (m/s). According to the principles of hydrology, it is possible to compute j from rainfall statistics:

$$h_r = a \times t^n \tag{5}$$

therefore:

$$j = h_r/t = a \times t^{n-1} \tag{6}$$

where

h_r = height of rainfall (mm),
t = duration of the rainfall (h), and
j = rainfall intensity (mm/h).

Finally to pass from j to q_r, expressed as in m³/s/m², (equivalent to m/s), the following equation applies:

$$q_r = j \times 2 \cdot 777 \times 10^{-7} \, \text{m}^3/\text{s}/\text{m}^2 \tag{7}$$

Equation (4) now becomes:

$$Q_d = 2 \cdot 777 \times 10^{-7} \times a \times t^{n-1} \times \cos \alpha \times f \tag{8}$$

For the coefficient of infiltration f, the following values can be applied:

$f = 0 \cdot 2 - 0 \cdot 35$ for drainage system covered with soil or wastes
$f = 1 \cdot 0$ for no coverage of the drainage system

The value of f in the case of a soil/waste cover is extrapolated from measurements taken at different waste disposal sites, as shown in Fig. 3 (Wiemer, 1987). The coefficients a and n of the rainfall possibility curve can change considerably from zone to zone. As an example, for the area of Milan, Italy, the values, relative to a 'return time' of 10 years, are:

$a = 53 \cdot 46$ $n = 0 \cdot 407$ if $t = 0 \cdot 7$ h
$a = 50 \cdot 06$, $n = 0 \cdot 222$ if $t > 0 \cdot 7$ h

Most of the codes that are actually available for landfilling (e.g. in USA and Italy) require that the drainage system must be designed for a rainfall having a 'return time' of 10 years. It appears reasonable to use in eqn (8) a duration of rainfall $t = 0 \cdot 5 - 1 \cdot 0$ h.

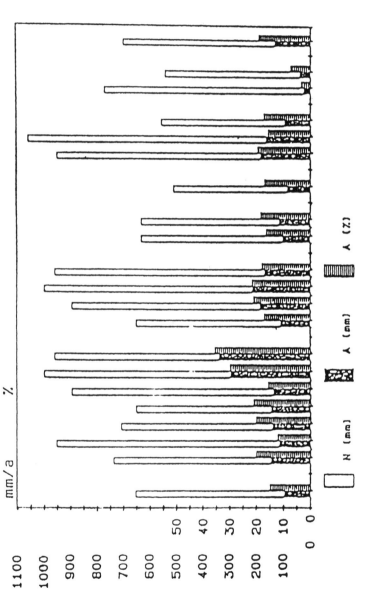

Figure 3. Rainfall (N) and leachate production (A) in different landfills (Wiemer, 1987).

A capped landfill. In a capped landfill the leachate originates from the water contained in the waste materials and released during the mineralization process. It is possible to estimate the following figures (Andretta *et al.*, 1988), valid for a 30-m deep landfill:

Q_p = total quantity of released leachate = 6 m³/m² = 6 m

t_p = time to release all the leachate = 10 years = $3 \cdot 1 \times 10^8$ s

Then, in this case:

$$q_r = Q_p/t_p = 1 \cdot 9 \times 10^{-8} \ (\text{m}^3/\text{s/m}^2) \tag{9}$$

This value of q_r appears to be far less than the value obtainable from eqn (7); therefore the situation of waste disposal in use is more critical than the situation after capping for designing the drainage system. The design of the geosynthetic drainage system on slopes and the bottom of the landfill should therefore be done for the landfill in use.

Critical Input Flow

The critical input flow must also be evaluated both for the landfill in use and for the completed landfill.

A landfill in use. For a landfill in use (Fig. 4) the input flows Q_b and

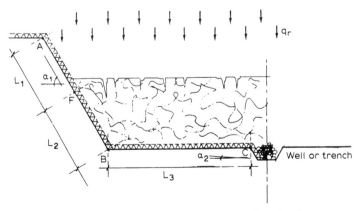

Figure 4. Scheme for the calculation of critical input flow for the situation of waste disposal in use.

Q_c at points B and C are critical. Therefore the input flow to design the geosynthetics drainage system on the slopes is given by:

$$Q_b = q_r \times f_1 \times \cos \alpha_1 \times L_1 + q_r \times f_2 \cos \alpha_2 \times L_2 \tag{10}$$

where

α_1 = angle of side slopes (degrees),
α_2 = angle of the landfill bottom (degrees),
L_1 = length of slope over the waste mass (m),
L_2 = length of slope under the waste mass (m),
f_1 = coefficient of infiltration on AF (see Fig. 4) = 1·0, and
f_2 = coefficient of infiltration on FB (see Fig. 4) $\approx$ 0·2–0·3.

The input flow to design the drainage system on the bottom is:

$$Q_c = Q_b \times q_r \times f_3 \times \cos \alpha_2 = q_r \times (f_1 \times L_1 \times \cos \alpha_1 + f_2$$
$$\times L_2 \times \cos \alpha_1 + f_3 \times L_3 \times \cos \alpha_2) \tag{11a}$$

where

L_3 = distance between the top of the slope and the main trench or well (m), and
f_3 = coefficient of infiltration on BC (see Fig. 3) $\approx$ 0·2–00·3.

In general, to take into account a situation with different lengths L_i, different inclinations α_i and a different coefficient of infiltration f_i, the following equation can be used:

$$Q_c = q_r \times \sum_{i,j} f_i \times L_i \times \cos \alpha_j \tag{11b}$$

A capped landfill.   For a capped landfill (Fig. 5), the input flows Q_b and Q_a are critical to the design of the drainage system between the capping waterproof layer and the soil cover. They are given by:

$$Q_e = q_r \times f_4 \times L_4 \times \cos \alpha_3 \tag{12}$$
$$Q_a = Q_e + q_r \times f_4 \times L_5 \times \cos \alpha_4$$
$$= q_r \times f_4 (L_4 \times \cos \alpha_3 + L_5 \times \cos \alpha_4) \tag{13}$$

where

α_3 = inclination in ED (see Fig. 5) (degrees),
α_4 = inclination in EA (see Fig. 5) (degrees),

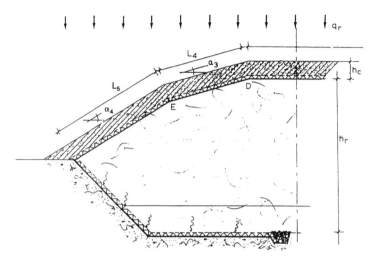

Figure 5. Scheme for the evaluation of the input flow for the situation of waste disposal after capping.

L_4 = distance between points E and D (see Fig. 5),
L_5 = distance between points E and A (see Fig. 5), and
f_4 = coefficient of infiltration of the soil cover.

For the coefficient f_4, the following empirical values are generally assumed in common practice:

$f_4 = 0.8$ for uniformly grassed soil cover
$f_4 = 0.2$ for bare soil cover

Therefore, since in general the soil cover will not be permanently covered with grass, an average value can be constantly assumed for f:

$f_4 = 0.5$

Drainage of Leakage

Leaks of leachate from the primary waterproof liner can occur for two reasons: defects in the seams of the geomembrane or damage to the liner during placement and compaction of waste materials. Both cases are extremely difficult to predict and therefore some simplifying hypotheses are necessary to evaluate the leakage flow rate.

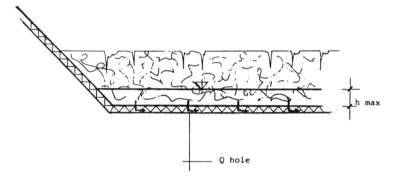

Figure 6. Drainage of leakage.

With reference to Fig. 6, the first hypothesis is that the leachate drainage system is designed in such a way as to maintain the hydraulic head on the bottom of the waste disposal under a maximum limit value; according to USA regulations (US EPA, 1985), the maximum hydraulic head allowed on the bottom is;

$$h_{max} = 0{\cdot}30 \text{ m} \tag{14}$$

The second hypothesis is that the holes produced in the primary liner have an average size, as measured in different practical situations in many landfills (Bonaparte *et al.*, 1987), given by:

$$A_{hole} = 1 \text{ cm}^2 = 0{\cdot}0001 \text{ m}^2 \tag{15}$$

The third hypothesis is that the liquid flows through the holes according to Bernoulli's law for free flow through orifices; the average speed of the flow through the holes is then:

$$v = C\sqrt{2g \times h_{max}} \tag{16}$$

where C = coefficient of flow through orifices = $0{\cdot}60$ (sharp-edged holes).

The average flow rate through a single hole Q_{hole} (m^3/s) is then:

$$Q_{hole} = v \times A_{hole} = C \times A_{hole}\sqrt{2g \times h_{max}} \tag{17}$$

Q_{hole} from eqn (17) is an upper boundary value, since if there is liquid pressure under the hole, the flow rate through the hole decreases: therefore Q_{hole} is the flow rate to use in designing the leakage drainage system.

In order to design the well to collect all the leakage from all the holes in the primary liner, the following hypothesis is needed: the average frequency is of 1 hole every 4000 m² of primary liner, as measured in different landfill (Faure, 1984; Kastman, 1984; Giroud & Fluet, 1986; Bonaparte *et al.*, 1987). The total flow-rate arriving to the collector well is then:

$$Q_1 = v \times A_{\text{hole}} \times S \times F_{\text{hole}} \tag{18}$$

where

Q_1 = total leakage flow rate (m³/s),
S = total surface of primary liner (side slopes and bottom 3 (m²)), and
F_{hole} = average frequency of holes = 1/4000 (m⁻²).

Drainage between Soil and Bottom Liners

With reference to Fig. 7, it is not possible in this case to define a

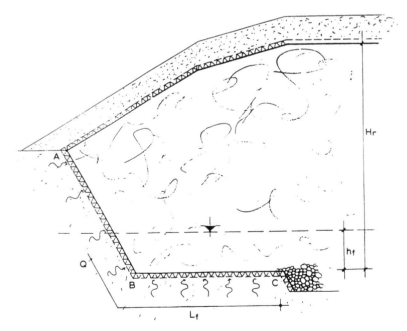

Figure 7. Drainage between soil and bottom liner.

general method to evaluate the flow Q_t, reaching the drainage system between the soil and the bottom liner. In general, if the water table is some metres under the bottom of the waste disposal, then Q_f is due only to humidity and capillarity and it can be assumed equal to zero. In this case, only a layer to protect the bottom liner is needed between the soil and the bottom liner: therefore the protective geosynthetic (geonets or nonwoven geotextiles) should be selected based on mechanical properties like compression and puncture resistance.

On the contrary, if there is a small water flow outcoming on the side slopes or at the bottom, it is needed to evaluate the specific flow rate Q_f (m³/s/m) case by case.

Flow-Rate and Transmissivity of the Geonets

In order to design a geosynthetic drainage system, the discharge which every type of geonet can carry under specified conditions of applied pressure and hydraulic gradient needs to be known.

The hydraulic gradient i (dimensionless) is defined as:

$$i = \delta h / L \tag{19}$$

where

$\delta h = $ loss of hydraulic head of the fluid flowing in the geonet (m), and

$L = $ distance between two points along the average direction of flow in the geonet (m).

The discharge capacity can be given in terms of transmissivity, defined as the discharge per unit width of the geonet and per unit of hydraulic gradient; or better in terms of specific flow-rate, defined as the discharge per unit width in the geonet, under a specified hydraulic gradient.

Under the hypothesis that the fluid flow is laminar and on the basis of the Darcy's law, transmissivity and specific flow-rate are given by:

$$\theta = (q/B)/i \tag{20}$$
$$Q = (q/B) \tag{21}$$

where

$\theta = $ transmissivity (m³/s/m = m²/s),
$Q = $ specific flow-rate (m³/s/m = m²/s),

q = discharge (m³/s), and
B = width of the geonet (m).

In the authors' opinion, the specific flow-rate is more suitable for practical use. The specific flow-rate of geonets can be measured with an apparatus similar to the one originally used by Darcy to measure the permeability of sands. As shown in Fig. 8, the apparatus consists of two water reservoirs, which allow it to maintain a constant head, a central housing for the geosynthetic specimen, upon which a constant predefined pressure is applied by means of a rigid plate, and of a system for measuring the steady state flow (Cancelli *et al.*, 1987). The results of these tests are usually presented in the form of diagrams (see Fig. 9), giving the specific flow-rate Q (m²/s) versus the applied pressure σ (kPa), with different curves having the hydraulic gradient i as parameter. The tests are usually performed in a laboratory using water at a temperature of 20°C.

It is possible to calculate the specific flow-rate for another temperature with the equation:

$$Q_{20} = Q_T \times (\eta_{20}/\eta_T)$$

where

Q_{20}, Q_T = specific flow-rate at 20°C and T°C, and
η_{20}, η_T = viscosity of water at 20°C and T°C.

Selection of the Geonet

The required specific flow-rate for the geonet is:

$$Q_{req} = Q_{in} \times FS \tag{23}$$

where

Q_{req} = required specific flow-rate (m²/s),
Q_{in} = input flow-rate (m²/s), and
FS = Factor of Safety.

In order to select the geonet for a specific project, evaluation of the pressure applied to the geonet and the hydraulic gradient i under which the fluid flows in the geonet is required. At this point, it is possible to enter the diagrams of different geonets: as shown in Fig. 10, if the condition is (i_1, σ_1) the geonet is unacceptable, because

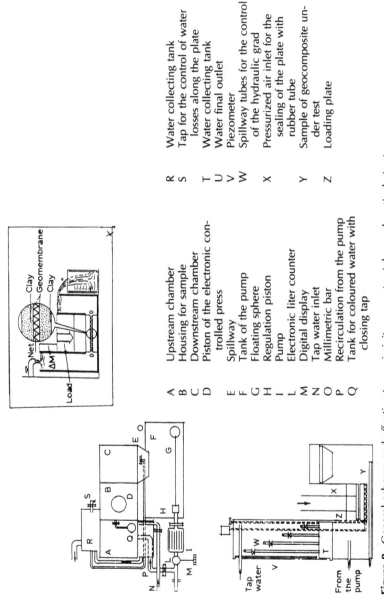

Figure 8. General scheme and effective transmissivity apparatus (plan and vertical view).

A Upstream chamber
B Housing for sample
C Downstream chamber
D Piston of the electronic controlled press
E Spillway
F Tank of the pump
G Floating sphere
H Regulation piston
I Pump
L Electronic liter counter
M Digital display
N Tap water inlet
O Millimetric bar
P Recirculation from the pump
Q Tank for coloured water with closing tap

R Water collecting tank
S Tap for the control of water losses along the plate
T Water collecting tank
U Water final outlet
V Piezometer
W Spillway tubes for the control of the hydraulic grad
X Pressurized air inlet for the sealing of the plate with rubber tube
Y Sample of geocomposite under test
Z Loading plate

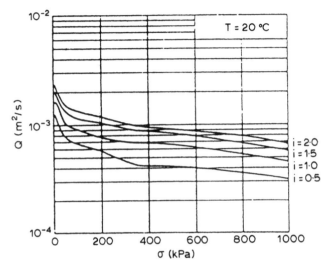

Figure 9. Diagram showing the specific flow-rate Q versus the applied pressure and the hydraulic gradient i for a geonet.

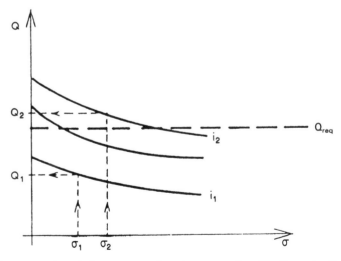

Figure 10. Use of diagrams to select the geonet (specific flow-rate Q versus applied pressure σ for various hydraulic gradients i).

$Q_1 < Q_{req}$; if the condition is (i_2, σ_2) the geonet is acceptable, since $Q_2 > Q_{req}$. If the value of i is different from the values of the diagram, it is possible to evaluate the specific flow-rate for the actual hydraulic, gradient i by the experimental formula:

$$Q_{i2} = Q_{i1} \sqrt{(i_2/i_1)} \tag{24}$$

where

Q_{i1} = specific flow-rate from the diagram (m^2/s),
Q_{i2} = specific flow-rate for the i_2 gradient (m^2/s),
i_1 = the hydraulic gradient on the diagram of the geonet, immediately above the actual hydraulic gradient, and
i_2 = the actual hydraulic gradient.

As an example, if the bottom has an inclination $\alpha_2 = 1\cdot5°$ and the minimum value on the diagram of the geonet is $i_1 = 0\cdot25$, then:

$$i_2 = \sin \alpha_2 = 0\cdot020$$

and

$$Q(i = 0\cdot020) = Q(i = 0\cdot25) \times \sqrt{(0\cdot020/0\cdot25)} = 0\cdot28 \times Q(i = 0\cdot25)$$

PRACTICAL SITUATIONS

Primary Drainage System for Leachate Collection

With reference to Fig. 4, the input flow must be evaluated with eqns (10), (11a) or (11b), in the situation of a landfill in use, just before capping, i.e. in the situation of the maximum height of waste overloading the drainage system. Then the required flow-rate Q_{req} must be evaluated by eqn (23), where the Factor of Safety FS_1 to be used depends on the applied pressure and the type of draining geosynthetic, according to Table 1.

For the selection of the geonet, the input values for the (Q, σ, i) diagrams and, when needed, for eqn (24) will be as follows.

TABLE 1. Factors of Safety (FS) for the Design of the Geosynthetic Drainage System

Applied pressure	Type of geosynthetic	FS_1	FS_2	FS_3	Notes
$\sigma \leq 400$ kPa	HDPE geonet	3–5	2–3	3–5	FS takes into account the elastic compression and the occlusion of the geonet by the geotextile.
$\sigma > 400$ kPa	HDPE geonet	5–10	3–6	—	FS takes into account also the compression creep of the geonet.
$\sigma \leq 400$ kPa	LDPE of foamed geonet; nylon or PP mat	5–10	3–6	5–8	FS takes into account also the compression creep of these geodrains occurring at low pressure.
$\sigma > 400$ kPa	LDPE or foamed geonet; nylon or PP mat	10–15	6–10	—	FS accounts also as for the high compliance of the structure of those geodrains.

Side slopes. On side slopes, in point B:

$$i = i_s = H/L_s = \sin \alpha \tag{25}$$

$$\sigma = \sigma_b = \gamma_r \cdot H_{rb} - tg^2(45° - \Phi_r/2) \tag{26}$$

$$\Phi_r = 20°-27° \tag{27}$$

where

γ_r = unit weight of compacted wastes (kN/m³),
H_{rb} = maximum height of waste in point B (see Fig. 4) before capping (m), and
Φ_r = internal friction angle of waste material (degrees).

Designing according to eqn (25) guarantees that the liquid flow will always be at atmospheric pressure, also with full channels flow in the geonet.

Equation (26) assumes a Rankine-type active pressure of the

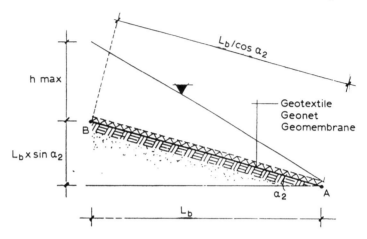

Figure 11. Scheme to evaluate the hydraulic gradient on the bottom of the waste disposal.

compacted wastes against the drainage system, the angle of internal friction being given by eqn (27) (Cancelli, 1987).

Bottom. On the bottom, with reference to Fig. 11, the hydraulic gradient is given by:

$$i_f = \delta h/L = \frac{h_{max} + L_b \times \sin \alpha}{L_b/\cos \alpha_2} \tag{28}$$

Therefore, at point C:

$$i = i_f = \frac{h_{max} + L_b \times \sin \alpha_2}{L_b} \cos \alpha_2 \tag{29}$$

where h_{max} given by eqn (14) is the maximum hydraulic head allowable on the geomembrane, according to the American regulation (US EPA 1985).

Using eqns (29) and (14) for the design of geonets guarantees that the hydraulic head on the geomembrane will always be modest. The pressure applied to the drainage system is:

$$\sigma = \sigma_b = \gamma_r \times H_{rc} \tag{30}$$

where H_{rc} = maximum height of waste at point C (see Fig. 4) before capping (m).

Drainage of Leaks from the Primary Geomembrane

With reference to Fig. 12, the liquid leaking from a hole in the primary geomembrane spreads laterally until it flows with a fixed width B_f. In this condition and allowing a certain excess flow near the hole, the required specific flow-rate is:

$$Q_{req} = FS_2 \times Q_{hole}/B_f \tag{31}$$

where Q_{hole} is given by eqn (17).

The final flow width B_f depends on the flow-rate Q_{hole}, the

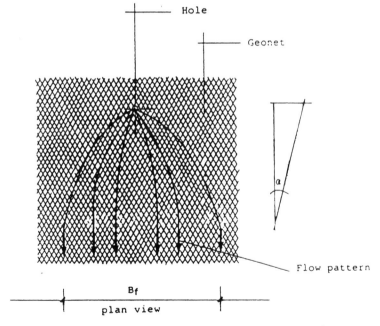

Figure 12. Flow patterns of liquids leaking from a hole in the primary geomembrane.

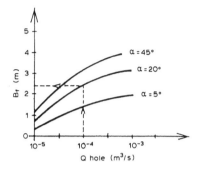

Figure 13. Diagram for a typical geonet, showing the constant width of flow B_f versus the flow-rate from a hole and the slope angle α.

inclination α and the type of geodrain: for a typical HDPE geonet, B_f can be evaluated from Fig. 13.

The Factor of Safety FS_2 to be used must be obtained from Table 1: FS_2 values are lower than FS_1 values because the hazard of overflow in this case is lower.

For the selection of the geonet, the applied pressure must be obtained by eqn (30), while the hydraulic gradient must be evaluated in the most critical condition, that is when the hole in the geomembrane occurs at point B. In this case the hole is under the maximum hydraulic head and is placed in the farthest point from the collector well. Therefore i must be calculated using eqn (29).

Drainage between Geomembranes and Foundation Soil

In this situation, with reference to Fig. 7, the required specific flow-rate is:

$$Q_{req} = Q_f \times FS_2 \tag{32}$$

where Q_f = specific flow-rate reaching the drainage system between the soil and the bottom liner.

The Factor of Safety FS_2 is reported in Table 1. To select the geonet, with reference to Fig. 7, the hydraulic gradient is:

$$i = i_f = \frac{p_f + h_f}{L_f} \tag{33}$$

where

p_f = piezometric head of the water flowing in the soil (m),

h_f = vertical distance between the water inlet and the entrance in the collector well (m), and

L_f = total distance along the geonet between the water inlet and the collector well (m).

If p_f is unknown, for the sake of safety it is better to assume:

$$i = h_f/L_f \tag{34}$$

The most severe condition for the applied pressure occurs after capping, therefore the pressure to be used to select the geonet is:

$$\sigma = \sigma_f = \gamma_r \times H_r + \gamma_c \times H_c \tag{35}$$

where

γ_r = unit weight of compacted wastes (kN/m^3),

γ_c = unit weight of cover soil (kN/m^3),

H_r = maximum height of waste (m), and

H_c = thickness of soil cover (m).

Drainage of the Capping

With reference to Fig. 14, Q_e and Q_a, at points E and A, can be obtained with eqns (12) and (13). The required specific flow-rates at

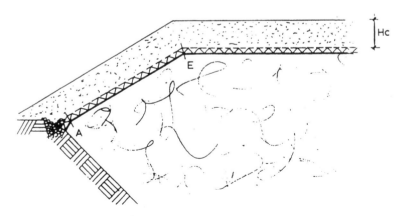

Figure 14. Drainage of the capping.

points E and A can be obtained with eqn (23), where for the Factor of Safety the value FS_3 from Table 1 must be used.

The pressure applied to the capping drainage system is low, but there is a hazard of the cover soil sliding if the water goes under pressure: therefore, according to Table 1, FS_3 must be rather high for compressible, and subject to high compression creep, drainage products.

To select the geonet the following values are to be used:

$$\sigma_e = \sigma_a = \gamma_c \times H_c \tag{36}$$

$$i_e = \sin \alpha_3 \tag{37}$$

$$i_a = \sin \alpha_4 \tag{38}$$

with Y_c, H_c, α_c, α_4 as defined above.

Using eqns (37) and (38) for the hydraulic gradient, at points E and A, guarantees that the liquid flow is always at atmospheric pressure.

CONSTRUCTION DETAILS

After having completed all the calculations, it is necessary to prepare the design drawings with all the details needed to construct the landfill

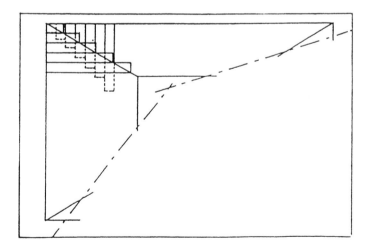

Figure 15. Example layout for adjacent rolls of geonet.

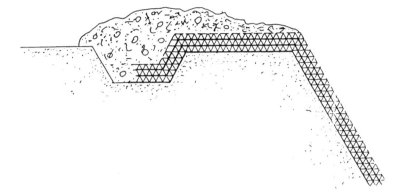

Figure 16. Anchor trench.

in the correct way. The design details are mostly common practice, therefore only a few simple design details relating to the specific use of geosynthetics need to be shown here. Figure 15 shows the correct layout to overlap the geonets and the geotextiles at the edges along the slopes. In order to avoid downward movement of the different layers along the side slopes, all the geosynthetics must be anchored in the top trench, as shown in Fig. 16. The main trench at the bottom of the landfill is very important as it collects all the liquids toward the collector well, where they are pumped to the ground level and then

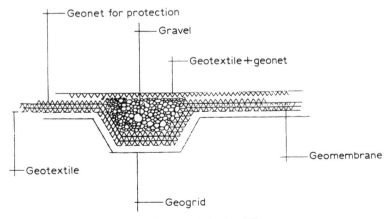

Figure 17. Main trench on the bottom of the landfill.

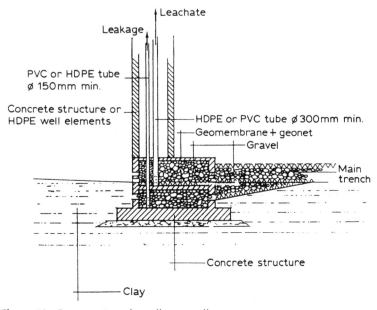

Figure 18. Cross-section of a collector well.

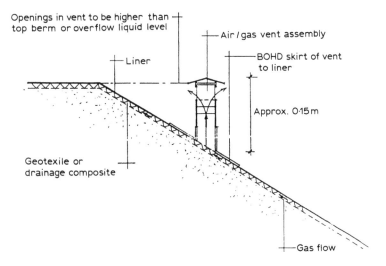

Figure 19. Gas vent detail.

treated or recirculated. Figures 17 and 18 show how the main trench and the collector well can be constructed. Finally, Fig. 19 shows the gas venting system, which allows a free flow toward the atmosphere for gases coming from the soil beneath the bottom liners.

REFERENCES

Andretta, F., Cómolli, P. & Guglielmetti, M. (1988). Aggiornamenti sulle problematiche di trattamento e smaltimento del percolato di scarico controllato dei rifiuti urbani. *Acqua-Aria*, **3-88**, 355–61.

Bonaparte, R., Beech, J. F. & Giroud, J. P. (1987). Background document on bottom liner performance in double-lined landfills and surface impoundments. US EPA, Report EPA/530-SW-87-013, Washington, DC.

Cancelli, A. (1987). Soil and refuse stability in sanitary landfill. In *Proceedings International Sanitary Landfill Symposium*, Cagliari, Italy.

Cancelli, A. & Cazzuffi, D. (1987). Permittivity of geotextiles in presence of water and pollutant fluids. In *Proc., Geosynthetics '87 Conference*, New Orleans.

Cancelli, A., Cazzuffi, D. & Rimoldi, P. (1987). Geocomposite drainage systems: mechanical properties and discharge capacity evaluation. In *Proc. Geosynthetic '87 Conference*, New Orleans.

Faure, Y. H. (1984). Design of drain beneath geomembranes: discharge estimation and flow patterns in the case of leak. In *Proc. Int. Conf. on Geomembranes*, vol. 2. Denver, CO.

Giroud, J. P. (ed.) (1984). *Geotextiles and Geomembranes—Definitions, Properties and Design*. IFAI, St Paul, MN.

Giroud, J. P. & Fluet, J. E., Jr. (1986). Quality assurance of geosynthetic lining systems. *Journal of Geotextiles and geomembranes*, 3(4), 249–87.

Kastman, K. H. (1984). Hazardous waste landfill geomembrane: design, installation, and monitoring. In *Proc. Int. Conf. on Geomembranes*, Denver, CO.

Koerner, R. M. (ed.) (1986). *Designing with Geosynthetics*. Prentice-Hall, Englewood Cliffs, NJ.

US EPA (1985). Minimum technology guidance on double liner systems for landfills and surface impoundments—design, construction and operation. EPA 530-SW-85-012, Cincinnati, OH.

Van Zanten, R. V. (ed.) (1986). *Geotextiles and Geomembranes in Civil Engineering*. A. A. Balkema, Rotterdam.

Wiemer, K. (1987). Technical and operational possibilities to minimize leachate quantity. In *Proceedings International Landfill Symposium*, Cagliari, Italy.

6.4 Incrustation Problems in Landfill Drainage Systems

MATHIAS BRUNE, HANS GÜNTER RAMKE,
HANS JÜRGEN COLLINS & HANS H. HANERT

*Institut für Mikrobiologie und Leichtweiss-Institut für Wasserbau,
Technische Universität Braunschweig, Postfach 3329, D-3300
Braunschweig, Germany*

INTRODUCTION

A long-term functioning drainage system is just as important for protecting surface water and groundwater as the lining system of a landfill and additionally necessary for ensuring landfill stability. In the mid 1980s an increasing number of failures of drainage systems of sanitary landfills became known. The formation of hard, insoluble deposits in drain pipes and in the drainage layer often led to a loss of efficiency of the drainage system. The cause behind the sometimes very extensive incrustation process was largely unknown (see Ramke, 1986, 1989).

The investigations reported here thus had two aims:

* elucidation of the causes of incrustation;
* deduction of recommendations for the construction and operation of sanitary landfills.

In order to derive an understanding of the incrustation processes in landfill drainage systems both field investigations and laboratory experiments (with columns) were carried out.

SURVEY OF LANDFILL DRAINAGE EXPERIENCES

In order to get an idea of the degree of damage to landfill drainage systems, 38 landfill operators in Germany were contacted and

questioned. The data provided by 28 operators on 31 landfills, two of
them without a drainage system, were evaluated and incorporated into
this study.

Data were requested on principal damage, incrustation in and
mechanical damage of the pipes. Furthermore the operators were
asked about subsequent damage, such as ponding of water directly on
the liner or in the body of the landfill or an untapped intrusion of
water at the base or the slopes of the landfill. The following remarks
are based on data from the 29 landfills with a drainage system.

The clogging of the pipes was ascertained either by flushing or
inspection. The extent of deposition of clogging material ranged from
just a layer of sludge on the floor of the pipe to hard incrustations
which drastically reduced or blocked the pipes' cross section. Deposits
of incrustation material were shown to be present in more than half of
the cases investigated, while in only three cases was deposition
definitely not observed.

Mechanical damage to the drainage pipes was reported for one-third
of all the sites. However, the actual degree of damage varied greatly,
from just a single break in a pipe or sleeve over the whole landfill to a
large number of seriously damaged pipes which were ruptured along
most of their length. It did become clear in the course of the
evaluation that stoneware pipes are particularly susceptible to mechanical damage.

Water discharge from the slope of the landfill was reported for about
40% of the sites. Again, the degree of damage varies and can range
from local damp spots to slope failures and greater discharges of
leachate. An accumulation of water on the bottom of the landfill was
reported in 8 cases.

More frequent than the accumulation of water on the bottom is the
accumulation of leachate within the body of the landfill, the causes of
which are still not exactly known. Possible causes are impermeable
layers within the landfill and the impairment of gas exchange.
Assuming that every discharge of leachate from the slope results from
water accumulating in the waste, 60% of the evaluated sites were thus
affected. Thus, besides the long-term efficiency of the drainage system
at the base of the dump, in future more consideration will need to be
given to drainage within the landfill body itself.

Following the survey by questioning landfill operators, about half of
the sites were visited to verify the data provided and to obtain more
detailed information. At seven of the sites the operators had performed

excavations. It was revealed that layers of waste deposits above the drainage system could become consolidated and impermeable to water. Incrustation processes apparently occur not only in the drainage system but also in the refuse material above it. According to those who conducted the excavations, a washing out of fine particles from the refuse into the drainage system was not observed, even when coarse gravel was employed as drainage material.

The damage to drain pipes—clogging and mechanical damage— reported in the survey could in most cases not have led to actual failure of the drainage system with subsequent accumulation of leachate. However, the scope of the investigation was not sufficient to establish any clear links. The data provided by the operators and the on-site investigations clearly showed that incrustation processes occurred at most of the landfill sites and that, together with insufficient drainage of the landfill body, they are the biggest problem for the long-term efficiency of the drainage system.

INVESTIGATION OF INCRUSTATION PROCESSES IN SANITARY LANDFILLS

Investigations Performed

Within this research programme, investigations were conducted at a total of 10 sanitary landfills. The investigations, which for various reasons could only be completely implemented in some cases, consisted of the following steps:

- chemical and microbiological analyses of the leachate
- determination of the composition of gases in the drain pipes
- temperature measurements in the drain pipes
- camera inspection and sampling of material flushed out of the pipes
- experiments on microbial biofilm formation in drain pipes
- excavation of the drainage system
- chemical and microbial analyses of incrustation materials

The aim of the study was to obtain information about the physicochemical and microbiological milieu conditions in the drainage system. Further emphasis was placed on the investigation of the

TABLE 1. Overview of Investigations Carried out at 10 German Landfills

Landfill site	*Leachate analysis*	*Gas analysis*	*Temp- erature measure- ment*	*Camera inspect- ion*	*Analysis of flushing material*	*Excav- ation*	*On- growth experi- ments*	*Remarks*
Altwarmbüchen	X	X	X	X	X		X	3 m'ments
Venneberg	X	X	X	X	X	X	X	2 m'ments
Geldern-Pont						X		2 ex.
Hessental	X	X	X	X	X	X		
W1	X	X		X	X	X		
D1	X	X	X		(X)			
H2	X	X	X	X	X			
S1	X				X	X		
W2	X				X			
S3	X	X	X	X	X			

incrustation material. Table 1 gives an overview of investigations carried out at the different sites. The methods are described in detail by Ramke and Brune (1990).

Milieu Conditions in Landfill Drainage Systems

Overview. The investigation of milieu conditions in landfill drainage systems will be explained with examples from two different sites, each being typical of particular methods of landfill operation and states of biological decomposition:

- Altwarmbüchen landfill: very rapid filling, ~10–20 m/annum; waste depth 10–40 m; intensive and long-lasting phase of acidic fermentation.
- Venneberg landfill: slow filling, ~2 m/annum; waste depth ~10 m; filling is at a standstill, stable methane phase.

Altwarmbüchen landfill. The investigation of this site was commissioned by the city of Hannover, and began in 1986 (see Steinkamp, 1987). Of all the sites investigated, it was the most rapidly filled. This resulted in a strong, continuous pollution load of leachate. As an example of a drain with a highly polluted leachate the data obtained from measurements in drain 30 in the summer of 1989 are listed in Fig. 1.

An oxygen concentration of about 0·2 mg/litre (owing to contamination with air during sampling), a redox potential (E_h) of between −150 and +100 mV and the presence of sulphide in all the drains provide a clear indication of the anaerobic conditions prevailing in the drainage system, which was also confirmed by the gas analyses. While the leachate from comparatively new regions of the landfill (drains 29–35) remained acidic during the course of the investigation (pH 5·5–6·5), the water from relatively old regions of the dump · (drains 3–24) was neutral to slightly alkaline (pH 7–8).

The concentration of organic substances peaked in the acidic leachate from the relatively new regions of the dump (drains 29–35) in which 35 000 mg/litre butyric acid and 7000–8000 mg/litre caprionic acid were measured; this heavy organic load was also reflected in COD values of up to 80 000 mg O_2/litre and in BOD_5 values as high as 45 000 mg O_2/litre. Noteworthy is that the section served by the highly burdened drain 30 had been completely filled during the period 1983–1985, but still continued to produce leachate with such a high organic content. Leachate from older regions of the site (drains 3–16), where filling had been completed in 1983, were also highly burdened with organic substances: butyric acid up to 4000 mg/litre, caprionic acid up to 1350 mg/litre, COD values of about 10 000 mg O_2/litre and BOD_5 values well over 1000 mg O_2/litre. The high proportion of biodegradable substances in the leachate of this dump which was rapidly being filled with domestic waste was reflected in the high BOD_5/COD ratio and in the ratio of the concentration of volatile fatty acids to COD.

As indicated by the analysis data presented in Fig. 1, the Altwarmbüchen landfill represents an extreme among the landfills investigated, not only in the respect of the quantity of organic substances, but also in the respect of the amount of inorganic substances contained in its leachate. This is also true for leachate from older sections of the dump. The high values obtained for electrical conductivity and chloride (20–110 mS/cm and 2500–90 000 mg/litre respectively) result at least in part from salt slag having been dumped in these sections. Leachate from relatively new sections of the landfill, with the highest content on organic substances, also had the highest

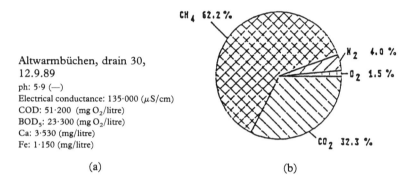

Altwarmbüchen, drain 30,
12.9.89

ph: 5·9 (—)
Electrical conductance: 135·000 (μS/cm)
COD: 51·200 (mg O_2/litre)
BOD_5: 23·300 (mg O_2/litre)
Ca: 3·530 (mg/litre)
Fe: 1·150 (mg/litre)

(a)   (b)

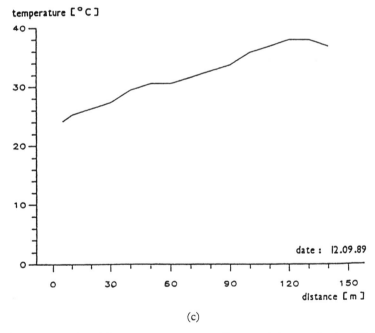

(c)

Figure 1. Results of the investigation into milieu conditions at the landfill in Altwarmbüchen. (a) Analysis of leachate; (b) composition of drain gas; (c) course of temperature in the drain pipe.

concentrations of inorganic substances, in particular, the iron and calcium concentrations were high (see drain 30, Fig. 1).

Analysis of the gas phase revealed a composition of 60–65% methane and 30–35% carbon dioxide in most of the drains at the Altwarmbüchen site—evidence that the methane phase had already been reached. Even drains with acidic leachate, such as drain 30, had a constant ratio of methane to carbon dioxide. The 5% proportion of air shown in Fig. 1 is accounted for by unavoidable contamination during sampling. When the ratio of oxygen to nitrogen in the dump gas corresponds to the natural ratio found in the atmosphere it may be assumed that both gases entered the sample as contaminants during sampling.

The temperature gradient measured in a drain pipe shown in the lower part of Fig. 1 may be considered typical for this site. Temperature measurements to a depth of 150 m inside a length of a drain pipe showed a maximum temperature of 40°C and a temperature gradient of 15°C over 100 m. However, since for technical reasons it was not possible to probe deeper into the pipe, the temperature curve is not complete and the actual temperature maximum may have been higher.

Microscopic examination of the samples of leachate revealed a high concentration of bacteria which was not much lower than that found in municipal sewage. Regardless of which section of the dump the sample were taken, as a rule, total direct counts of between 10^6 and 10^7 bacteria per millilitre were determined. Such pronounced concentrations of bacteria in the leachate led from the very beginning of the investigation to the expectation that microorganisms very probably played an important role in causing precipitation and incrustation in the drainage system.

The bulk of the bacteria population belonged to physiological groups that are involved in the degradation of the original organic waste to methane and carbon dioxide: methane bacteria with 10^3–10^5 cells/ml, sulphate-reducing bacteria with 10^4–10^6 cells/ml, anaerobic organo-trophic (fermentative) bacteria with 10^3–10^5 cells/ml, iron-reducing bacteria with 10^4–10^6 cells/ml and manganese-reducing bacteria, also with about 10^4–10^6 cells/ml. Despite the varying composition of the leachate no correlation could be found between the concentration of a particular group of bacteria and the corresponding section of the site.

From Fig. 2 it can be seen that the same groups of bacteria were also present in similar concentrations in the leachate of the other sites

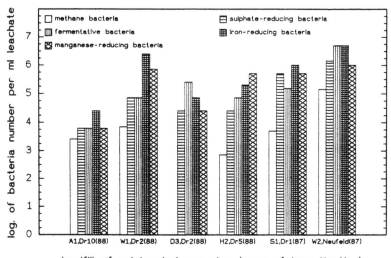

Figure 2. Bacterial flora in leachate from six landfill sites.

under investigation. Considering the short generation period of bacteria and the standard deviation, the variation in the concentration of one of these groups of bacteria by up to two orders of magnitude is not considered to be very significant.

Venneberg landfill. In contrast to the Altwarmbüchen landfill, the Venneberg landfill receives annually much less refuse and is being filled at a much slower rate. On one-half of the area of the section investigated the undermost refuse layers had been subjected to uncontrolled pre-composting before being highly compacted. In addition to the activities dealt with here, this landfill was also systematically investigated with respect to the permeability of its drain pipes and to pipe environment (see Ramke *et al.*, 1991).

While at the beginning of the landfill operation the load of leachate in the two areas had been markedly different (lower pollution in the area where pre-composting was carried out), at the time of the investigation, about seven years later, the load of leachate in both areas was much the same. Examination of the leachate showed a maximum oxygen content of $0.3 \, \text{mg} \, O_2$/litre and a redox potential (E_h) ranging between 0 and $+100 \, \text{mV}$ and neutral to slightly alkaline

Venneberg, drain 15,
4.7.88

pH: 7·0 (—)
Electrical conductance: 10·900 (μS/cm)
COD: 1·000 (mg O$_2$/litre)
BOD$_5$: 40 (mg O$_2$/litre)
Ca: 132 (mg/litre)
Fe: 28 (mg/litre)

(a)

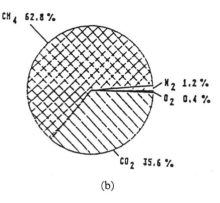

CH$_4$ 62.8 %

N$_2$ 1.2 %
O$_2$ 0.4 %

CO$_2$ 35.6 %

(b)

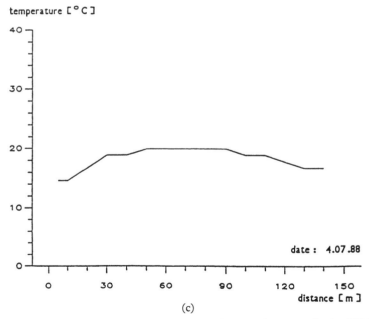

temperature [°C]

date : 4.07.88

distance [m]

(c)

Figure 3. Results of the investigation into milieu conditions at the landfill in Venneberg. (a) Analysis of leachate; (b) composition of drain gas; (c) course of temperature in the drain pipe.

pH values of between 7 and 8. The composition of leachate from drain 15 at the time of the investigation is shown in Fig. 3.

The COD values were on average an order of magnitude lower than those found for leachate from the landfill at Altwarmbüchen. The proportion of biodegradable substances, characterised by a BOD_5/COD ratio of 0·1 and under, was also much lower. In addition, the concentration of inorganic substances was lower than in leachate from the Altwarmbüchen landfill, as the analysis results for drain 15 in Fig. 3 show.

The above-mentioned landfill section was completed four years before the time of the investigation and was in the stable methane phase, this being confirmed by analysis of the gases from the investigated drains. All of them had a gas composition of ~65% methane and ~35% carbon dioxide. The portion of air, also shown in Fig. 3, is due to contamination during sampling, too.

The bottom part of Fig. 3 shows the temperature curve in drain 15 in its full length. Although the temperature rises with increasing height of the dump, the maximum temperature and the temperature gradient are lower than at the Altwarmbüchen site. The maximum temperature was 20°C, while the temperature gradient over a length of 60 m was 6°C. The reason for the lower temperature lies in the lower intensity of exothermal processes, which in turn results from the practice of pre-composting and from the smaller quantity of refuse deposited.

The total direct counts of bacteria, at 10^5–10^6 cells per ml leachate, were found to be on average an order of magnitude smaller than those found in leachate from Altwarmbüchen. However, considering the range of deviation inherent in determining bacteria numbers, as well as the small number of counts performed, the difference is of no great significance, so that no correlation can be drawn between the bacterial counts and the different states of decomposition or the different operational conditions at the two sites.

Extent of the Incrustation Processes

Incrustations were found in varying degrees at virtually all the landfill sites investigated. The large variation in the extent of incrustation is illustrated in the following examples:

- In the drainage pipes of the Altwarmbüchen dump, with its highly loaded leachate, incrustation was particularly intense. Although

the pipes were flushed at least once a year, it was usually necessary to repeat the process a number of times and occasionally the incrustations could only be removed by milling.
- During the investigation the slowly-filled Venneberg site, which was already in the stable methane phase and had only a lightly loaded leachate, showed practically no more deposition of clogging material in the drain pipes between annual flushings.
- A landfill with a aerobic pre-treatment of the refuse, which was also investigated, showed a very low degree of incrustation.

Impairment of the drainage systems ranged from moderate deposits on the pipe bottom to extensive incrustation of the drainage layer in the vicinity of the pipe and of the whole drainage layer. Although in all cases the drainage system still functioned, the danger to the sites through its possible failure was apparent.

Typical for the formation of incrustations were the observations made at the site in Altwarmbüchen. Here investigations per camera revealed three forms of deposits:

- incrustations on the bottom of the pipes,
- incrustations on either side of the pipe walls, and
- flat incrustations, located horizontally in the pipe and reaching from one wall to the other.

Impairment of the drainage systems' efficiency could range from light deposits on the bottom of the pipe to drastic reduction of the pipe's cross-sectional area. The incrustations on the pipe wall could in extreme cases totally reduce its permeability.

The extent of impairment of the drainage layer by incrustations was clearly demonstrated by excavations carried out in an experimental area of the landfill Geldern Pont (see Düllmann & Eisele, 1989). This extensive excavation of the bottom liner system revealed that not only the pipe casing had become consolidated and impermeable through heavy and extensive incrustation, but that the whole drainage layer was affected. Large areas of the drainage layer, which consisted of gravel and sand (20% at 0–2 mm and 80% at 2–9 mm, respectively), were incrusted and consolidated to between 1/3 and 2/3 of its thickness. The area affected amounted to between 30 and 80% of the area excavated. Permeability was reduced by several orders of magnitude down to values as low as 10^{-8} m/s.

Analysis of the Incrustation Material

The structure of the surface of the incrustation material from the drain pipes varied:

- deposits of spherical particles with a maximum diameter of 15 mm, but usually with a diameter under 5 mm;
- flat material with a thickness of some millimetres, attached by one surface to the pipe wall;
- lumpy deposits, either amorphous or of fine-grained structure;
- cuboid deposits with coral-like structures on their surface (vertical tubes, 2–3 mm diameter, some centimetres long).

The material that was flushed out of the pipes was generally hard, black, pulverisable and homogeneous. Testing with acid showed sulphide and carbonate as chemical components.

The structure of the incrusted drainage material ranged from grains of gravel, covered with a thin layer of fine material, to complete filling of the pores between the gravel grains forming a structure like that of concrete. Macroscopically, the fine material corresponded to that of the pipe deposits. Although drainage gravel covered with just a layer of sludge was found, usually it was consolidated to a hard mass which could extend over an area of several metres diameter.

The main constituents of the incrustation material are calcium and iron combined with carbonate and sulphur (mainly in the form of sulphide). The composition of some typical samples of incrustation material is presented in Fig. 4. Typical for the material flushed out of the drains is the small amount of insoluble residue after dissolving in aqua regia. The insoluble residue left from the incrustated drainage gravel is considerably greater, no doubt on account of the incomplete separation of incrustation material and drainage material. The composition of the incrustation materials ranges from those consisting primarily of calcium carbonate to others with a high iron and sulphur content. Generally a significant amount of organic material is found in all the deposits (loss on ignition or organic carbon).

The fine structure of a typical piece of incrustation material is shown at increasing magnification in Fig. 5(a)–(c) (Ramke & Brune, 1990). The example shown, from drain pipe incrustations in a sewage sludge deposit at the landfill Altwarmbüchen, Hannover, was photographed using a scanning electron microscope. The particles of material, which had been flushed out of the drain, are a three-dimensional network of tightly-packed aggregates of bacteria. The

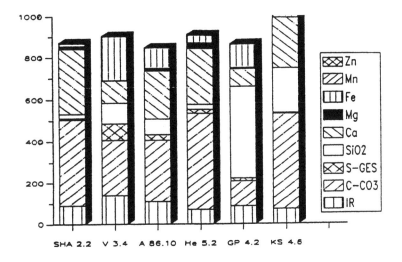

	Units[a]	Samples					
		SHA 2.2 (Flushing material)	*V 3.4* (Flushing material)	*A 86.10* (Flushing material)	*He 5.2* (Flushing material)	*GP 4.2* (Incrust-ation)	*KS 4.6* (Incrust-ation)
Ignition residue	g/kg TS	88·5	140·6	110·0	69·2	85·0	72·3
Carbon							
Total	g/kg TS	133·0	72·1	110·0	117·5	53·5	97·7
Organic	g/kg TS	50·0	19·0	51·0	24·8	28·9	6·4
Carbonate (CO_3)	g/kg TS	415·0	265·5	295·0	463·5	123·0	456·5
Total sulphur	g/kg TS	6·3	78·3	26·0	20·6	13·8	5·4
SiO_2	g/kg TS	20·8	100·2		22·4	437·4	212·0
Insoluble residue	g/kg TS			74·0	5·0		236·8
Ca	g/kg TS	308·0	102·0	227·0	264·0	84·0	245·4
Mg	g/kg TS	18·2	2·1	16·0	32·2	7·2	1·6
Fe	g/kg TS	10·9	212·0	97·0	34·0	112·0	14·8
Mn	g/kg TS	3·1	2·9	3·0	4·9	2·0	
Zn	g/kg TS	0·5	0·5		0·4	2·0	0·6

[a] TS: Dry solid matter.

Figure 4. Composition of selected incrustation material from six German landfills.

(a)

(b)

Figure 5. Typical structure of incrustation material in landfill drainage systems—progressive magnification in scanning electron microscope. (a) Low magnification; (b) medium magnification; (c) high magnification.

(c)

Figure 5. (Contd.).

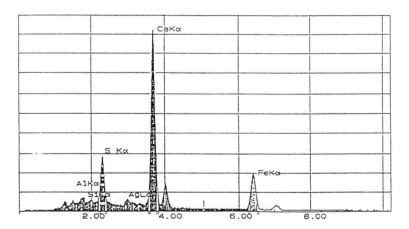

Figure 6. EDX analysis of incrustation material (material flushed out of drain A1/3, landfill Altwarmbüchen, August 1988).

aggregates, with a diameter of about 10 μm, are covered to varying degrees with inorganic precipitates. The network of bacterial aggregates encloses at least an equal volume of pore space, the individual pores of which have a similar diameter to those of the aggregates.

The deposits formed on the bacteria have a diameter well under 1 μm and appear as pimples on the cell surface. At a more advanced stage of incrustation the precipitates enclose the whole aggregate of bacteria. On a single piece of flushing material incrustations at various stages of development can be seen. In general these structures could be observed on flushing material from all the investigated sites, as well as on the incrustated drainage material.

According to the EDX analyses (energy dispersive X-ray analysis) which were carried out in conjunction with the SEM (scanning electron microscope) examination, the principle elements contained in the deposits precipitated by the bacteria were sulphur, calcium and iron. Although their proportions varied, they always formed the bulk of the elements present both in the pipe deposits and in the incrustation material from the drainage gravel. Figure 6 shows a typical EDX analysis of flushing-material (drain A1/3 of the Altwarmbüchen site, August 1988) at 81-magnifications in the SEM. The peaks for aluminium and silver are due to these elements being present in the sample holder and the silver paint used to fix the sample to the sample holder. The peak for silicon inicates a small amount of fine sand grains being mixed in with the material.

Microbiological On-Growth Experiments on Biofilm Formation

It was possible to observe the initial stages of the incrustation process in landfill drainage systems in on-growth experiments by placing sterilised glass microscope slides and gravel for varying periods of time in the drain pipes of two sanitary landfills. In seven such experiments (one at the Venneberg site and six at the Altwarmbüchen site) the progressive stages of incrustation were followed, as well as the influence of leachate quality and the period of exposure on the intensity of its formation. Following exposure in the drain pipes, the glass slides were subjected to fluorescence and phase contrast microscopy and, if called for, to a determination of iron content, while the particles of gravel were subjected to SEM and EDX analyses. As regards the development of incrustations, all the experiments provided similar results.

Examination with the light microscope showed that the microorganisms present in the leachate colonised the glass slides predominantly in aggregates. Precipitation of inorganic material only occurred on the microorganisms and their excretion products (slime fibrils)—never on the uncolonised spaces in-between. Examination in the scanning electron microscope provided more detailed information. The newly-forming deposits on the bacteria were well under 1 μm in diameter, without any discernible crystalline structures (see Fig. 7(a)). Within a very short space of time (see below) these microscopically tiny 'seeds', consisting of bacteria and pimples of precipitates beginning to form on their surfaces, can develop into macroscopically-recognisable incrustations. Figure 7(b), a SEM-micrograph of incrusted biofilm on gravel, from the on-growth experiments at the landfill Venneberg, shows that:

• the primary inorganic deposits, which initially form on the bacteria and their slime fibrils, subsequently increase in size (secondary growth) and fuse together;
• the large surface area thus created provides very favourable conditions for renewed colonisation. The continual renewal of this active layer of microorganisms is necessary for the process of continued precipitation.

In all cases SEM-examination showed that the three-dimensional structure of the deposits on the experimentally-exposed gravel was the same as that of the drain pipe deposits—a clear indication that large-scale incrustation results from the experimentally-demonstrated colonisation and precipitation processes. Another significant correlation between large-scale and experimental incrustations is the similar chemical composition of their precipitation material: EDX-spot analyses of the precipitates on the microorganisms revealed them to contain sulphur, calcium and iron as their principal elements.

The speed at which the experimental surfaces were colonised and precipitates formed varied depending on the quality of the leachate. These processes were most intensive in those drain pipes where the leachate was highly loaded with organic and inorganic substances and where it had already reached a pH of about 7. In these cases the freshly exposed glass slides became colonised over their entire surface within one day, combined with a precipitation rate for iron of 96 μg/cm^2 day. This is comparable to the precipitation rate in rapid filters employed for the biological removal of iron from drinking water. Colonisation and precipitation were far less rapid when the leachate

(a)

(b)

Figure 7. Fine structure of forming incrustation (experiments on film formation)—scanning electron micrographs. (a) Precipitate formation on bacteria (gravel from column experiment); (b) bacterial film with precipitates (gravel, landfill Venneberg).

originated from a landfill in the stable methane phase with a low organic and inorganic load and neutral pH. In leachate that was highly burdened with organic and inorganic substances, but which had an acidic pH of between 5·5 and 6·5, only an insignificant level of colonisation and precipitation was observed.

Results of the Field Investigators

The results of the field investigations can be summarised as follows:

- It was found that varying degrees of incrustation took place in the drainage systems of nearly all the landfills investigated.
- In the drain pipes the degree of incrustation ranged from thin layers on the pipe wall to a significant reduction in pipe cross section. Generally, incrustations in the pipes could be removed by flushing or, in some cases, by milling.
- Incrustation of the drainage layer, particularly in the casing of the pipe, can in extreme cases lead to a complete loss of permeability, although a certain residue permeability usually remains.
- Silting of the drainage layer by fine particles being washed out of the overlying refuse was seldom observed in the course of the excavations. The transition from the lowest layer of refuse to the drainage layer was usually sharply defined. Only in cases of intensive water flow might a degree of silting be assumed.
- Physicochemical analyses of the leachate and gases in the drain pipes, as well as the colour and chemical composition of the deposits, show that milieu conditions in the drainage system are determined by the anaerobic microbial processes in the landfill.
- Microscopic examination of the incrustation material, of the microflora in the drainage system and of the experimentally-exposed pieces of glass and gravel provided evidence for the metabolic activity of anaerobic bacteria being the cause of the incrustations (see below).
- Purely physicochemical causes for the incrustations, such as a reduction in temperature or an altered partial pressure of carbon dioxide, cannot be ruled out, but considering the proven role of microbiological processes, the physicochemical causes can only be of minor significance.
- The main components of the incrustation material are the cations

of calcium and iron, combined with carbonate and sulphur (mainly in the form of sulphide).

- Of all the landfill sites investigated, the one being most rapidly filled was also the one with the highest concentrations of organic and inorganic substances in its leachate and with the most rapid formation of incrustations. It was here that the greatest annual amounts of drain deposits were repeatedly formed.
- Once a landfill has reached the stable methane phase with its lightly-loaded leachate, incrustation hardly occurs.
- The landfill with the aerobic pre-treatment of the refuse was found to have a very low intensity of incrustation formation.
- Excavation of a sewage sludge deposit revealed that over large areas the drainage system had become more or less impermeable due to massive incrustations.
- Limestone gravel is absolutely unsuitable as a drainage layer material. It decomposes under the milieu conditions prevalent on the bottom of a sanitary landfill.

SIMULATION OF INCRUSTATION PROCESSES IN THE LABORATORY

Purpose and Construction of the Experimental Set-Up

Parallel to the investigation of the landfill sites, laboratory experiments were conducted with the aim of simulating the incrustation process. It was intended to isolate the influences of individual factors on the incrustation process and to analyse the course of its development in time. The experiments are described in detail in Ramke & Brune (1990).

Two series of experiments with different objects were conducted:

Preliminary series: —the development of a suitable method for simulating the incrustation process in a landfill.

Main series: —the influence of leachate quality on the incrustation process,
—the influence of grain size distribution of drain material on the longevity of the drainage layer.

Apart from the geometrical similarity, the laboratory simulation of the incrustation process also needed to reproduce the physicochemical and microbiological conditions on the bottom of a sanitary landfill. Furthermore, the leachate employed could not be synthesised satisfactorily because of its complex composition and had to be obtained for this reason from landfill sites in operation.

The experimental apparatus is illustrated in Fig. 8. The experiments

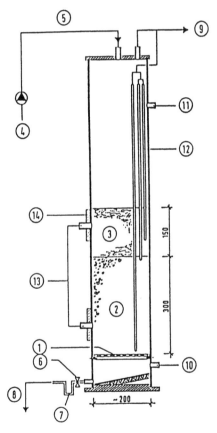

Figure 8. Experimental apparatus for simulation of incrustation processes in the laboratory. 1, Perforated plate; 2, gravel; 3, refuse compost; 4, from leachate storage container; 5, leachate inflow; 6, tap; 7, syphon; 8, outflow; 9, to gas catchment bag; 10, flush gas inlet; 11, flush gas outlet; 12, piezometer; 13, sampling ports; 14, glove box connector.

were conducted in closed plastic columns with a diameter of up to 200 mm and a height of ~1 m. The drainage material was placed on a perforated plate in the column to a height of 30 cm, on top of which was placed a further 15-cm layer of refuse compost. Leachate, supplied by a peristaltic pump, was fed into the column from above and permeated the unsaturated layers of compost and drainage material before passing through the perforated plate.

The leachate was collected once a week from a landfill site and stored in a sealed container to prevent it taking up oxygen from the air. It only passed through the column once before being collected separately. The generated biogas was collected in an airtight bag. Samples of drainage material could be taken from the column through sampling ports enclosed for the procedure in a glove box containing an oxygen-free atmosphere. In this way the chemical and microbiological analysis of all phases of the column was possible. Water retention in the column was measured by piezometers, which were fitted to the columns at various heights.

Experimental Procedure

Simulation of milieu conditions. During the first 2 weeks only lightly-loaded leachate was used so as not to impair the development of an adapted microflora by immediately applying a highly-loaded organic leachate with a low pH. The experiments were conducted at a temperature of 30°C in order to provide the most favourable conditions for the landfill bacteria (with an optimum growth temperature of about 35°C). The columns were flushed at least once a day with a gas mixture of 50% nitrogen plus 50% carbon dioxide, until the endogenic production of methane and carbon dioxide ensued. The volume of gas exchanged at each flushing amounted to at least twice the column volume.

Once gas production had commenced, the composition of the gas in the columns and the oxygen content of the leachate were continually monitored, in order to check on any contamination from the atmospheric oxygen. It was not possible to prevent oxygen from entering the columns entirely, but the amount measured in the leachate (~0·1 mg/litre) and the black colour of the biofilm covering the drainage gravel showed that the effects of oxygen on the milieu conditions in the columns were negligible.

TABLE 2. Overview of Main Series of Laboratory Column Investigations

Column	Drainage material —grain size (mm)	Geotextile	Leachate pollution	Duration of experiment (days)	Flowrate (litres/day)	Remarks
1	8–16	—	high	462	7·4	standard
2	1–32	—	high	394	6·8	
3	8–16	—	low	462	7·6	
4	8–16	—	high	462	8·1	standard
5	16–32	coarse pores	high	421	7·8	
6	16–32	coarse pores	low	462	9·2	
7	16–32	—	high	462	9·4	
8	16–32	—	low	462	9·8	
9	8–16	—	high	433	15·2	double flowrate
10	2–4	—	high	406	9·5	
11	2–8	fine pores	high	289	9·1	
12	16–32	fine pores	high	289	8·0	

When the water in a column ponded, the inflow was interrupted for a few days. The inflow of leachate was finally turned off, when the height of the water level did not change during this period.

Overview of Laboratory Experiments. In the following overview, only the procedure and results of the main series will be shown. The specifications in the main series of experiments are shown in Table 2. In order to investigate the influence of leachate quality on the incrustation process, two leachates with different pollution loads were used. One leachate was highly loaded with organic and inorganic substances from sections of a landfill which were still in the phase of acidic fermentation. The other leachate was relatively lightly loaded and typical for sections of landfill in the stable methane phase of decomposition. The influence of grain size on the effective life of columns was ascertained by employing five different drainage materials. In four columns the long-term behaviour of two different non-woven HDPE geotextiles, placed between compost and drainage layer, in relation to leachate quality was additionally studied. The variation

- highly loaded leachate

- drain material with grain size 8–16 mm
- without geotextile

was used as the standard against which the other columns were compared.

Decomposition Processes in the Columns

The concentration range of leachate constituents is listed in Table 3. The sources of the two kinds of leachate had to be changed after 5 and 10 months respectively, since the pollution load of the original leachates decreased too much.

The course of decomposition processes in the columns are decisive for explaining the process of incrustation and shall be shown in some detail. In the lightly-loaded columns the degradation of the organic substances only began after changing the source of leachate to one with more organic constituents. Only with this higher organic load did the microbiological decomposition processes start to the extent that they were clearly indicated by the on set of gas production. After 150–200 days the biodegradation processes stopped, which was to be seen by the same concentration of the organic load (COD) in the

TABLE 3. Overview of Leachate Composition in Column Experiments

		Slightly polluted leachate		Highly polluted leachate	
		Phase I	*Phase II*	*Phase 1*	*Phase 2*
pH		7·4–7·9	7·2–7·9	7·2–7·8	5·7–8·1
Electrical conductance	(μS/cm)	13 450–20 300	8 950–21 900	8 950–21 900	14 000–65 600
BOD_5	(mg/litre)	57–930	145–2 700	400–4 365	3 840–45 900
COD	(mg/litre)	1 450–2 070	1 610–6 340	1 630–8 890	3 350–63 700
$NH_4^- N$	(mg/litre)	740–1 440	620–2 080	620–1 690	1 660–3 500
Mg	(mg/litre)	170–270	100–270	100–360	200–840
Ca	(mg/litre)	70–120	120–290	130–380	150–4 000
Fe	(mg/litre)	10–28	8–79	8–85	9–870
Mn	(mg/litre)	0·2–0·6	0·4–4·0	0·6–4·1	0·3–28
Cl	(mg/litre)	2 590–3 830	1 490–3 550	1 490–3 540	3 545–21 700
SO_4	(mg/litre)	29–115	1–68	1–121	2·6–118

inflow and outflow. The cause was a continual reduction in available organic substrate in the leachate of the second source, too.

A reduction in the iron and calcium concentrations in the outflow was only observed once the microbial degradation processes had started. Before this the concentrations in the inflow and outflow were identical. The concentration of magnesium was hardly altered by the microbial activity. When the intensity of the decomposition processes later decreased, the concentrations of iron and calcium in the outflow increased again. Only in the case of calcium was the concentration in the final phase somewhat lower in the outflow than in the inflow. Gas production in the lightly-loaded columns, at about 2 litres/day, was significantly lower than in the highly-loaded columns. The composition of generated gas, which was ~70% methane, remained fairly stable.

In Fig. 9, for the highly-loaded columns 2, 5, 7 and 12, the concentration curves for COD, calcium and iron in the inflows are compared. For all the columns a significant elimination of the organic and incrustation-relevant inorganic substances is seen. At the beginning, the organic load of leachate (the inflow) dropped slowly. The use of leachate from another landfill caused a slight increase in COD, which then rose steeply, to values of up to 60 000 mg/litre, when leachate from a newly-filled section of the landfill was applied. This sudden surge of highly-loaded leachate killed the biocoenosis in most of the columns, while in other columns the experiment had already been terminated because of heavy incrustations and resulting impermeability.

The COD curves illustrate the partial microbial decomposition of degradable organic substances contained in the applied leachate. The COD values of the outflows were always lower than those of the inflow. The rate of COD elimination, up until the surge of acidic and very highly loaded leachate, had nearly the same values in all the columns. After applying the strongly acidic leachate to the columns, which were still permeable, the rate of degradation dropped drastically or stopped completely. The calcium and iron curves followed the COD curve in the outflow. Parallel to an increase in the organic burden of the inflow, an increase in the iron and calcium concentration occurred. This is accounted for by the very high content of volatile fatty acids and the associated drop in pH. Until the highly-loaded leachate was fed into the columns, they all showed more or less the same concentrations of calcium and iron in the outflow, which were significantly lower than

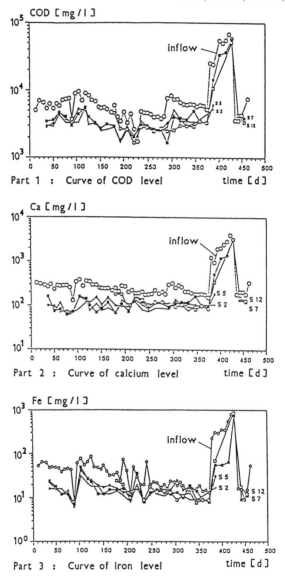

Figure 9. Main series of laboratory experiments for simulating incrustation processes—curves of inflow and outflow levels of columns with highly-loaded leachate. Curves of (a) COD level, (b) calcium level and (c) iron level.

those in the inflow. The reaction to the surge of highly-loaded leachate corresponded to its high load of organic substance.

A comparison of the inflow and outflow curves for the iron and calcium concentrations shows the degree of elimination of these incrustation-relevant cations during their passage through the columns. The similarity of the iron and calcium curves with those for COD provides a clear indication of microbial cause of the incrustation process.

As with the on-site samples of landfill leachate, inflow and outflow of the experimental columns were subject to total direct counts of bacteria and, at certain times, to differential cell counts. The total direct counts revealed bacteria concentrations in the inflow and outflow similar to those found in on-site landfill leachate (7×10^6–1×10^7 cells/ml). Fluorescence microscopy showed repeatedly that 20–40% of the cells were methane bacteria.

The differential cell count also showed the bulk of cells to belong to one of the same five groups of bacteria found in the on-site leachate (methane bacteria, sulphate-reducing bacteria, iron- and manganese-reducing bacteria and fermentative bacteria). They occurred in the inflow and outflow in similar concentrations to those found in the on-site leachate. Although no significant difference in concentration of any of these groups of bacteria between the inflow and outflow was found, the results of the experiment allow us to conclude that these bacteria were active in the columns. The proportions of the different physiological groups of bacteria in respect to the total number of bacteria was also typical of all the samples of landfill leachate investigated.

Development of Permeability and Pore Volume

The development of permeability in the columns was followed visually in the transparent columns (biofilm formation, water retention), by reading the piezometers and by determining pore volume.

Columns 3, 6 and 8, which were fed with lightly-loaded leachate, showed no significant reduction in their permeability during the whole course of the experiment. In none of these columns did water ponding occur.

Ponding of water only occurred in a number of the columns fed with highly-loaded leachate. In colunn 12 water retention above the geo-textile was occasionally observed 10 months after the beginning of the run; after 12 months it occurred continually. The runs of columns 2, 5 and 10 had to be terminated prematurely after 390–430 days, owing to their reduction in permeability prohibiting any further application of leachate. Reduced permeability was indicated by an increasing ponding of water in the column. Column 2 contained a well-graded gravel filter with a grain size 1–32 mm; column 10 contained a fine gravel of a grain size 2–4 mm and column 5 had a geotextile over coarse gravel with a grain size 16–32 mm.

To check that the ponding of water was not caused by incrustation and consolidation of the compost layer, this was removed from columns 2, 5 and 12 after ~410 days; from column 10 the compost layer was removed after about 455 days, nearly one week before the end of the run. In all the columns where the compost layer was removed before the end of the run, the compost was found to contain large areas of local consolidation, but only in column 12 was permeability re-established. In the other columns the upper layer of gravel or geotextile had become more or less impermeable. Directly after removal of the compost layer and the resumed application of leachate, a renewed ponding of water occurred, so that after 1–2 weeks the runs had to be terminated. Thus, the impermeability of the columns was accounted for by blockage of the drainage layer.

In the other columns at the end of the experiments it was observed that a thick biofilm had formed on the surface of the drainage material and that locally the pore volume seemed to be greatly reduced, but no ponding of water was to be seen. If we disregard column 11, all the columns, which remained permeable and were fed with strongly-loaded leachate, contained the coarse drainage materials with the grain size 8–16 mm or 16–32 mm without geotextile between drainage material and compost.

The pore volume of the columns was determined by carefully filling the columns with leachate and then weighing the amount which drained out from measured sections. The reduction of pore volume in the drainage layer (Fig. 10) corresponded to an increasing degree of incrustation. While the lightly-loaded columns showed virtually no reduction in pore volume, the pore volume of some of the highly-loaded columns was reduced considerably. In the columns with the coarsest drainage material the pore volume was reduced by between 5 and 10%, while in those with the finer 8–16 mm gravel the initial pore volume was lower and the reduction in pore volume was a little more. Far more significant was the difference between the columns filled

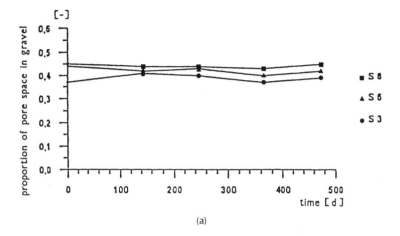

(a)

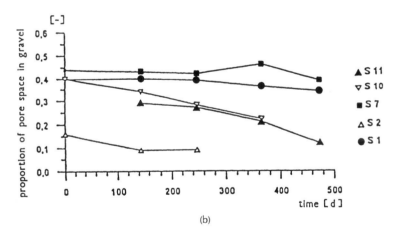

(b)

Figure 10. Development of the proportion of gravel pore space in the columns: (a) lightly-loaded columns; (b) highly-loaded columns.

with coarse drainage materials and those filled with fine drainage material. The measured reduction in pore volume for columns 10 and 11 was about 20% and thus two to three times greater than for the columns with 8–16 mm gravel, and four times greater than for the columns with 16–32 mm gravel. Column 2 (grain size 1–32 mm) had

an initial experimentally-determined pore volume of 16% (the calculated pore volume was considerably higher), which after running for
about 6 months was reduced to ~10%.

Observations Made during the Dismantling of the Columns

The observations made during the dismantling of the columns of the
main series of experiments are shown schematically in Table 4.

Dismantling the columns which had been subjected to lightly-loaded
leachate showed the compost to be locally consolidated and extensively interspersed with a fine tar-like material (incrustation material).
The compost in the highly-loaded columns only differed from that in

TABLE 4. Observations Made on the Columns when They were Dismantled

Column	Compost	Transition zone	Geotextile	Gravel
2	Locally consolidated	Upper layer of gravel very hard	—	Strongly consolidated at top, without pores, at bottom looser
4	Locally consolidated	Upper layer of gravel very hard	—	Strongly consolidated, still porous
5	Consolidated	'Cake' (3 cm), hard[a]	Strongly reduced permeability	Loose, with thin biofilm
6	Loose	'Cake', loose[a]	Little reduced permeability	Loose, hardly any biofilm
7	Consolidated, few pores	Compost–gravel: siltations	—	Loose, with biofilm, permeable
8	Loose	Loose, hardly any fine particles in gravel	—	Loose, hardly any biofilm, 'clean'
9	Locally consolidated	Upper layer of gravel very hard	—	Consolidated throughout, still permeable
10	Consolidated, few pores	Upper layer of gravel very hard	—	Consolidated throughout, few free pores at top
11	Consolidated	'Cake'[a]	Strongly reduced permeability	Consolidated
12	Locally consolidated	'Cake'[a]	Strongly reduced permeability	Loose, hardly any biofilm

[a] 'Cake': compost cemented to geotextile by fine incrustation material.

the lightly-loaded ones in the quantity of material deposited. The highly-loaded columns often showed a conically-formed area of strongly-incrusted compost, which was very difficult to remove. The larger pores were filled with incrustation material and the coarse structure of the compost was usually no longer discernable.

The drainage material of the highly-loaded columns could only be removed from the columns with a crowbar. The pore space between the grains, especially of the finer gravel, was distinctly reduced. The pores were clogged with the same black, tar-like incrustation material that had consolidated the compost. The drainage material of columns 2 (gravel 1–32 mm) and 10 (fine gravel) showed hardly any pores in the upper section. The drainage material of columns 4 and 9 (gravel 8–16 mm) were strongly covered with incrustation material and partly pasted up, but the coarser pores were distinctly perceptible.

The geotextile which had been inserted into column 5 could only be removed as a hard packet, which was ~3–4 cm thick. Viewed in profile from top to bottom it consisted of consolidated compost clogged with fine incrustation material, the geotextile itself which was likewise incrusted and, finally, pieces of gravel clinging to the underside. The top surface of the geotextile was glossy and more or less completely sealed by the tar-like incrustation material. The bottom side of the geotextile could still be seen to possess open pores. This 'geotextile-cake' extended over the whole cross section of the column and had a fairly constant thickness. The gravel lying below it was quite loose and only covered with a thin, black, slimy film.

The same observations were made in the other columns with a geotextile layer, although here, owing to the lower intensity of the biodegradation processes, the extent of caking was less. Thus, it can be said generally that a zone of reduced permeability is formed above the geotextile. The height of this zone, the proportion of precipitated incrustation material and the degree of hardening vary, depending on the experimental conditions, such as leachate quality, intensity of the biodegradation processes and duration of the run.

While the gravel in the columns with geotextile generally remained loose and covered with only a light film, the gravel in the columns without geotextile but coarse drainage material was covered with a thicker film and locally clumped together. However, the larger pores of the coarse gravel (16–32 mm) where not clogged and this coarse drainage material more or less maintained its permeability. In column 7, for example, at the transition between compost and coarse gravel

(16–32 mm), a zone had formed where compost material and black incrustation material had settled between the particles of gravel and caused some silting. The depth of this transition zone was only about 3 cm and beneath it no silting had taken place, despite the high flow rates during permeability measurement. The coarse drainage material with a grain size of 16–32 mm showed the least impairment of pore volume and permeability of all the drainage materials tested.

Composition of the Incrustation Material

Table 5 lists the composition of some selected samples of incrustation material from the columns. Besides a proportion of insoluble residue derived from the drainage material, the incrustation material from the pores in the gravel and from the geotextile consisted mainly of organic material, carbonate and calcium, whereby the calcium may be assumed to be mainly in the form of carbonate. However, significant amounts of iron and sulphur were also present. The addition of hydrochloric acid to the samples caused effervescence and the smell of hydrogen sulphide, which suggests the presence of iron sulphide, particularly. Thus, the incrustation material taken from the columns corresponds to that obtained from the drainage systems of the investigated landfill sites.

TABLE 5. Main Series of Laboratory Experiments for the Simulation of Incrustation Processes—Chemical Composition of the Incrustation Material

	Units[a]	Sample 2.1 (compost)	Sample 2.2 (gravel)	Sample 4.1 (gravel)	Sample 9.1 (gravel)
Ignition residue	g/kg TS	265·5	148·4	174·6	181·8
Carbon					
Total	g/kg TS	135·5	96·2	94·2	76·6
Organic	g/kg TS	77·6	31·3	38·3	44·8
Carbonate (CO_3)	g/kg TS	289·5	324·5	279·5	159·0
Total sulphur	g/kg TS	8·1	11·3	36·7	36·7
Insoluble residue	g/kg TS	261·9	309·9	204·9	266·0
Ca	g/kg TS	176·7	200·0	200·0	140·0
Mg	g/kg TS	13·4	15·2	11·0	8·2
Fe	g/kg TS	30·6	33·0	76·0	127·5

[a] TS: Dry solid matter.

Incrustation material from the surfaces and pore spaces of the drainage layer of various columns were examined by light microscopy and SEM and found to possess very much the same structure, which also corresponded to the structure of the incrustation material obtained from the drainage layer and drain pipes of the sanitary landfill sites. Also, the results of EDX analyses revealed a close correlation between the incrustation material from the landfill drainage systems and that from the experimental columns.

Results and Conclusions

Laboratory simulation of the incrustation processes which take place on the bottom of sanitary landfills led to the following results:

- Incrustations were not formed in significant amounts by the lightly-loaded leachate of the stable methane phase of decomposition.
- By the use of highly-loaded leachate, incrustation was extensive and corresponded to that observed at sanitary landfills.
- The incrustation material from the columns corresponded to that from the landfill sites not only in its chemical composition, but also in its microscopic structure.
- Independent of the degree of incrustation, the incrustation material always had a similar structure.
- At the end of the experiment the permeability of coarse drainage material, especially gravel of grain size 16–32 mm, was hardly affected, while that of finer drainage material, e.g. gravel of grain size 8–16 mm, showed a significant reduction in pore volume.
- Fine and well graded drainage material (2–4 mm and 1–32 mm, respectively) suffered almost a complete loss of permeability.
- By the use of highly-loaded leachate, the geotextiles placed between compost and the drainage layer largely lost their permeability.

From the above, the following conclusions can be drawn:

- In respect to incrustation formation, highly-loaded leachate is far more problematic than lightly-loaded leachate.
- Under similar conditions, coarse drainage materials retain their permeability far longer than finer drainage materials or the investigated geotextiles.

- Since sanitary landfills, at least initially, always produce highly-loaded leachate, drainage materials whose permeability can be seriously affected by incrustation processes are unsuitable.

CAUSES OF THE INCRUSTATION PROCESS

The overall conclusion of the microbiological investigations is that *anaerobic bacteria are responsible for the formation of incrustations in drain pipes and filter material.* This conclusion is based on the following observations:

- Anaerobic microorganisms are present in high concentrations in landfill leachate and readily colonise the surfaces of the drainage system.
- Analysis of the fine structure of the incrustation material shows that it consists of a network of aggregates of bacteria with deposits of precipitated inorganic material on their surfaces.
- On-growth experiments in which glass surfaces and gravel were exposed in the drain pipes of two sanitary landfills showed after microscopic examination that the incrustation process is only initiated on the bacterial cells and the slime fibrils which they excrete. Further precipitation centres around these seeds of incrustation until the whole biofilm becomes filled with inorganic precipitations. The aggregates of bacteria and the deposited inorganic material can accumulate to the degree already indicated.
- Chemical analysis of the incrustation material shows that its inorganic components consist of calcium, iron, magnesium and manganese combined with carbonate and sulphur.

Precipitation of the incrustation material is essentially caused by two processes (see Fig. 11):

- *Bacteriogenesis of sulphidic deposits.* Iron-reducing bacteria solubilise $Fe(III)$ by reducing it to $Fe(II)$, while sulphate-reducing bacteria reduce sulphate, for example gypsum, to sulphide. This bioreduction of sulphate causes the milieu in the vicinity of the sulphate-reducing bacteria to become more alkaline, which results in sulphur precipitating as its insoluble metal sulphide.
- *Bacteriogenesis of carbonates.* Before calcium carbonate can be

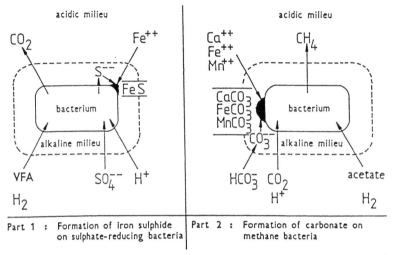

Figure 11. Schematic representation of processes leading to formation of incrustation material on bacteria.

precipitated, calcium must first be mobilised from the refuse material. This is achieved by fermentative organisms producing organic acids which lower the pH of the leachate and thus mobilise the calcium. Precipitation of calcium carbonate onto the surface of the methane and sulphate-reducing bacteria probably results from their metabolic consumption of hydrogen ions causing a local elevation of pH and consequently disturbing the balance between carbonate and hydrogen carbonate.

The formation of deposits in the drainage systems of sanitary landfills can thus be considered to take place in two stages:

- Fermentative bacteria together with iron- and manganese-reducing bacteria give rise to a process of mobilisation, whereby a part of the organic component of the refuse is converted into volatile fatty acids (VFA) being dissolved in the leachate. This leads to a lowering of the pH value, thus causing the increasing dissolution of parts of the inorganic components of the refuse.
- In the second stage, the precipitation process, predominantly methane- and sulphate-reducing bacteria in the drainage system, through their specific metabolism, cause the formation of insoluble sulphides and carbonates from metal ions dissolved in the

leachate. This is the essential process which leads to incrustations being formed.

Consequently, incrustations can only arise when the landfill leachate contains both easily-degradable organic substances (as substrate for the incrustation-forming bacteria) and inorganic ions (calcium, iron, sulphate and hydrogen carbonate, etc.) in solution.

RECOMMENDATIONS FOR THE CONSTRUCTION AND OPERATION OF SANITARY LANDFILLS

The main consequences to be derived from these investigations for site construction and operation are as follows:

- Incrustation of sanitary landfill drainage systems cannot be completely prevented, but can be greatly reduced by appropriate methods of landfill operation.
- The material of the drainage layer should be chosen as coarse as possible, in order to provide a sufficient proportion of pore space and pore diameter (nothing can be said, however, about the filter stability of materials with a grain size greater than 16–32 mm).
- Decisive for the effective life of the drainage system will be to limit the severity of incrustation through measures involving waste management and landfill operation, thus producing unfavourable conditions for the incrustation-forming bacteria.
- Intensity and duration of the acidic phase of biodegradation needs to be reduced. This can be achieved either by operational methods which result in a predominantly aerobic degradation of the organic refuse, such as

—pre-composting the refuse
—slower filling of the landfill

  or by methods of waste management, e.g.

—separate collection of organic waste and subsequent composting.

- The supply of inorganic incrustation-forming substances such as iron and calcium needs to be reduced. The separate dumping of building waste and rubble would probably be of good effect.

• According to the present state of knowledge, it is particularly critical when leachate with a high incrustation potential (heavily loaded with easily-degradable organic substances and a high iron and calcium concentration) coincides with a microflora in the stable methane phase. For this reason the establishment of a stable methane phase in the lower waste layers should be avoided when the site is going to be filled quickly and an intensive acidic fermentation has to be expected.

ACKNOWLEDGEMENTS

This work was supported by the German Federal Environment Agency (Umweltbundesamt) and the Ministry for Research and Technology (BMFT).

REFERENCES

Brune, M. (1991). Ursachen für die Bildung fester und schlammiger Sedimente in Entwässerungssystemen von Hausmülldeponien. Dissertation, Technische Universität Braunschweig.

Düllmann, H. & Eisele, B. (1989). Schadensanalyse von Deponiebasisab-dichtungssystemen aus Kunststoffdichtungsbahnen. Abschlußbericht. Umweltforschungsplan des Bundesministers für Umwelt, Naturschutz and Reaktorsicherheit, FE-Vorhaben 103 02 225.

Ramke, H.-G. (1986). Überlegungen zur Gestaltung and Unterhaltung von Entwässerungssystemen bei Hausmülldeponien. In *Fortschritte der Deponietechnik '86,* ed. K. P. Fehlau & K. Stief. Abfallwirtschaft in Forschung und Praxis, Band 16. Erich Schmidt Verlag, Berlin, pp. 251–91.

Ramke, H.-G. (1989). Leachate collection systems. In *Sanitary Landfilling: Process, Technology and Environmental Impact,* ed. Th. H. Christensen, R. Cossu & R. Stegmann. Academic Press, London, pp. 343–64.

Ramke, H.-G. & Brune, M. (1990). Untersuchungen zur Funktionsfähigkeit von Entwässerungsschichten in Deponiebasisabdichtungssystemen. Abschlußbericht. Bundesminister für Forschung und Technologie, FKZ BMFT 145 0457 3.

Ramke, H.-G., Kölsch, F. & Brune, M. (1991). Ermittlung geeignete Filtermaterialien für die Entwässerungssysteme von Abfalldeponien. Abschlußbericht zum Forschungsvorkaben 2091-BV4e 33185 im Auftrag des Niedersächsischen Ministers für Wissenshaft und Kunst.

Steinkamp, S. (1987). Untersuchungsergebnisse des Entwässerungssystems der Zentraldeponie Hannover. In *Möglichkeiten der Überwachung und Kontrolle von Deponien und Altablagerungen, Fachseminar,* ed. H.-J. Collins & P. Spillmann. Veröffentlichungen des Zentrums für Abfallforschung der Technischen Universität Braunschweig, Heft 2.

INDEX